Parametric Statistical Inference

Parametric Statistical Inference

J. K. Lindsey
University of Liege
and
Limburgs University
Diepenbeek
jlindsey@luc.ac.be

CLARENDON PRESS • OXFORD
1996

Oxford University Press, Walton Street, Oxford OX2 6DP
Oxford New York
Athens Auckland Bangkok Bombay
Calcutta Cape Town Dar es Salaam Delhi
Florence Hong Kong Istanbul Karachi
Kuala Lumpur Madras Madrid Melbourne
Mexico City Nairobi Paris Singapore
Taipei Tokyo Toronto
and associated companies in
Berlin Ibadan

Oxford is a trade mark of Oxford University Press

Published in the United States by
Oxford University Press Inc., New York

A catalogue record for this book is available from the British Library

Library of Congress Cataloging in Publication Data
(Data applied for)

ISBN 0 19 852359 9

Typeset by the author using LaTeX
Printed in Great Britain by
Bookcraft (Bath) Ltd
Midsomer Norton, Avon

Preface

Students of statistics who venture into applied work after years of study are often surprised to find out that they have been taught the wrong subject. (B. Efron, in Blyth, 1970)

For a long time I have thought I was a statistician, interested in inferences from the particular to the general. But, as I have watched mathematical statistics evolve, I have had cause to wonder and to doubt. (J.W. Tukey, 1962)

We can only conclude from the investigations here considered that the normal curve possesses no special fitness for describing errors or deviations such as arise either in observing practice or in nature. (K. Pearson, 1900)

Over the past years, two of the characteristics which have marked both applied and theoretical statistics have been the importance of the exponential family of probability distributions in model building and that of the likelihood function in statistical inference. Both were introduced by Fisher in the 1920s and 1930s, thus taking half a century to rise to prominence. However, when one surveys the textbooks available on statistical theory, one must be struck by the singular lack of systematic treatment of these topics. The present work attempts to fill this gap.

In recent years, the subject of statistical inference has seemed to become more and more divided, as many statisticians take a position on one side or the other of the debate between the frequentist and Bayesian schools. However, one area in which almost everyone seems to have come to be in agreement is on the central role of the likelihood function in inference procedures. The presentation in this text is based on this idea. The first six chapters are an attempt to present those aspects of statistical inference with which members of any school might be more or less in agreement (or in least disagreement!). The highlights, and some problems, of the two schools are presented in the two subsequent chapters, before closing with two simple unifying examples. Only *parametric* inference is covered; no attempt is made to introduce the methods of non-parametric approaches, which are, of course, primarily confined to the frequentist school.

The introduction of generalized linear models in the early 1970s brought to the attention of many statisticians the unifying position of the exponential family in much of statistical practice. Further study of this family has resulted in one of the more sophisticated areas of modern statistics. Details of some of these results are presented early in the text, so that the family can be used as an illustrative example throughout the rest of the exposition.

Thus, by the central place of the likelihood function and the exponential family in the text, I hope to bring to the student the underlying unity of current statistical *practice*, in spite of the apparent diversity and conflict. Hopefully, in this way, the present material will allow students to be presented with the viewpoints of the likelihood, frequentist, and Bayesian schools, indicating their respective strengths and weaknesses, within a single course, thus providing them with a wider foundation for subsequent work in statistics. At the same time, I mean to provide a firm basis for how much of applied statistics is actually practised. I develop this in Parts I and II, standing it on one simple, but fertile, postulate or principle concerning likelihood which is set forth in Section 3.1.3. Thus, this book attempts to look at how applied statisticians actually have come to make inferences, not how mathematical statisticians think they should.

A number of basic themes or theses run through this text. These include:

- Probability is appropriate for constructing models but much less so for making scientific inferences about them. Probability is fundamentally deductive: all possible outcomes, the sample space, must be delimited for it to be properly defined. But, by definition, all possible outcomes of a scientific inference procedure, from the particular to the general, cannot be foreseen, hence the limited applicability of probability in this context.
- Related to this is the fact that goodness of fit procedures generally cannot obey the likelihood or sufficiency principle because this principle assumes that the model function is 'correct', exactly what is being questioned.
- One of the most used tools of the statistician is hypothesis tests. Here, I propose that they are only appropriate for checking model adequacy (Fisherian tests) and for decision-making among clear alternatives (frequentist tests). Their wide use in model selection is unfortunate; such tests, and the related confidence intervals, do not hold the precision of individual parameters constant as the number of parameters simultaneously involved changes. Only proper model selection criteria, such as the AIC and the BIC, do this.

Many statisticians have the rather peculiar notion that scientists are primarily interested in some unknown parameter value, or some corresponding statistic, such as a mean or its estimate, in total abstraction from the circumstances which produce the data. (Occasionally, this may be true, if the scientist simply requires a precise measure of some directly observable quantity, in the presence of measurement error.) If these statisticians are correct, then perhaps this is because many scientists believe that statistics is capable of no more! For example, in a regression context, one often hears that some form of 'least squares' estimation is appropriate, because it makes few assumptions, and that the normal distribution is only required to make inferences. But, the prime role of such a distribution in a model is rather to attempt to describe how the data were stochastically produced. In my experience, scientists are primarily interested in the functioning of some phenomenon which can be regarded as a data generating mechanism. They represent this in some simplified way, as models, using their prior theoretical knowledge and their prior empirical information about the phenomenon, which, however, leaves some parameters in those models unspecified. These are the aspects of scientific inference to which statistical methods should be directed.

On the other hand, in engineering and other industrial settings, unknown parameter values, and tests, are often of principal interest. Perhaps, most prominently (for a statistician), this includes the pharmaceutical industry, where drugs must be tested before being allowed on the market. The primary goal here is not to understand the underlying data generating mechanism, but to aid in making a decision (whether a product should go to market or not), among a limited number of known possibilities, leading to actions, in the presence of uncertainty. This requires a model of human decision-making behaviour in the face of costs and benefits, which, in contrast to inference procedures, can be probability based. Thus, statistics is called upon to make contributions in both the scientific and industrial contexts. One may suspect that the techniques of most use may not be identical in the contexts of inference and of decision-making.

In scientific inference, *all models are wrong*, so that inference criteria depending on an assumption that a model is correct are generally irrelevant. On the other hand, in order to make a decision, one must usually assume that the model involved is correct for a decision to be possible.

Since antiquity, fundamental divisions have marked the ways in which human beings view the world, described by terms such as objective, subjective, empirical, rational, idealist, and so on. With the development of capitalism some centuries ago, and the dominance of market relations in most aspects of human activities, the primary world view of *all* human existence came to take various forms of monetary-dominated rationalism, although, as history and anthropology clearly show, it is an inappropriate description (model) in societies where such relations do not predominate. Thus, such models cannot have *universal* application, for example, to all human decision-making behaviour. These same philosophical cleavages which mark all science, from sociology to physics, are especially evident in statistics. This is not surprising because statistics is one discipline, with philosophy, which is primarily concerned with general ways to make sense of the world around us. But the disagreements have had the unfortunate consequence of leading to many futile and sectarian debates where the energy might better be spent on developing useful practical procedures for understanding the world. To resolve this problem, the approach that I take here is to look at the various schools of statistics and the conditions under which their respective points of view might be more appropriate.

All existing statistical methods can be criticized and found wanting on some criterion or another. Paradoxical examples can be constructed for all of them. A number of the most interesting of these problems, where they relate directly to common statistical applications, are used to help to illuminate a choice among possible approaches. Every effort is made to separate prior knowledge, information, and opinions, information obtained from the current observations, and knowledge of future effects resulting from decisions taken. In this way, the reader can more easily decide which types of knowledge and information should be combined to make inferences or decisions in any given circumstances.

This text is addressed to mathematically oriented statisticians and to advanced students in a mathematical statistics programme, who are interested in how statistics can actually be applied in practice, as well as to applied statisticians wishing to deepen their understanding of the inference process. It assumes of the reader a reasonable knowledge of probability theory and familiarity with the basic concepts of statistics, such as regression techniques.

The book is arranged as a research monograph following a coherent order of presentation. Part I covers the basic principles of modelling. Chapter 1 introduces, in a relatively simplistic manner, some of the main themes carried through the book. In contrast, Chapter 2 develops the theory of the exponential family in considerable detail. Part II is the core of the text, presenting likelihood-based inference, with Chapter 3 covering procedures for a set of available model functions and Chapter 4 model checking. Part III, the most difficult, introduces approximate techniques for complex likelihood functions, asymptotics in Chapter 5, and methods of factoring the likelihood to separate out parameters of interest in Chapter 6. Many experienced statisticians may consider that the book comes back down to earth in Part IV (although students may not!), where the two main approaches to statistical decision-making are presented, the frequentist in Chapter 7 and the Bayesian in Chapter 8. In such limited space, elaborate details cannot be discussed, but most of the material in the first six chapters is directly applicable to both approaches. Finally, in Part V, two simple examples illustrate, in a comparative way, many of the procedures.

Certain chapters are, intentionally, more difficult than others. In large part, this arises because some material is more recent and hence less familiar to many statisticians. In the interest of readability in an already long book, detailed formal proofs have not been given. Sketches of some of the more interesting and typical ones are, however, provided. Prior knowledge of probability theory must be assumed, but a summary of the important results is given in an appendix. Students, even those destined to work in applied statistics, should ideally have some exposure to measure theory and martingales if they are to feel at ease navigating in the modern world of survival analysis, stochastic processes, and Bayesian methods. Again, a short summary is annexed. Thus, generally, sections labelled as appendices contain summaries of technical material and leads into the relevant literature.

Too much material is presented for a single course, but one could be designed in a number of ways. For example, a basic, and fairly classical, course in statistical inference at the advanced undergraduate level would entail a detailed presentation of the material in Chapters 1, 3, 4, 7, and 8, with some selected topics from other chapters, especially the first two sections of Chapter 2, chosen according to the tastes and interests of the instructor. Much of Chapters 2, 5, and 6 is too difficult for undergraduates. A graduate inference course, which could assume previous knowledge of basic inference principles, might be built on Chapters 2, 5, and 6, picking up material in other chapters as necessary. On the other hand, an advanced course on direct likelihood inference would centre upon Chapters 2 to 6.

If a course based on the book is not to be too abstract, students should also be able to use some standard statistical package, such as GENSTAT, GLIM, Lisp-Stat, or S-Plus, which can, at least, fit generalized linear models, so that practical exercises can accompany the more theoretic work.

At this point, I cannot resist giving one anecdote. In 1969, I submitted my first paper to a major British statistical journal, applying direct likelihood inference in the exponential family. The referee's brief report, recommending rejection of the paper, succinctly pointed out that 'the examination of likelihood functions does not, by itself, constitute a method of inference'. (The dangers of the exponential family were not mentioned!) This was to be far from my last such encounter with statistics referees. Most recently, a referee for the same journal has informed me that I write for an audience of one. Hopefully, that will not be true of the present text!

In a work as complex as this, a certain number of errors will inevitably occur; reports of these will be greatly appreciated, as will any comments or suggestions for improvements.

Thanks go to Anthony Davison, Gentiane Haesbroeck, David Hinkley, Philippe Lambert, and Dave Sprott for discussion, comments, and suggestions which greatly helped to improve the text. Finally, I must extend my sincere thanks to Jim Kalbfleisch who first ignited my interest in statistical inference.

Diepenbeek and Liège J.K.L.
July, 1995

Contents

VI Appendices

Part I

Models

1
Model building

Parametric statistical inference is concerned with the construction and interpretation of *models* as applied to empirical data. It is *inference* because it involves making statements about particular observed phenomena which we hope will have more general import for a larger universe. It is *statistical* because the models used will have to describe random variability in the data as well as systematic effects. It is parametric because the models will have assumed functional forms, generally based on prior theory, but contain *parameters* whose values are unknown. (Scientists often have some idea of what these values may be, but seek independent empirical confirmation.) Thus, inference should provide the means for distinguishing the relative suitability of different model functions and, then, ways of providing a range of plausible values of those parameters involved, distinguishable from the ambient random variability. A model informs us about the universe, and hence about the functioning of the phenomenon under study, but also allows us to check if its functional form, itself, is adequate to the purpose for which it is employed.

1.1 Data generating mechanisms

Models are simplistic representations of some *data generating mechanism.* The latter is some complex process whereby empirical observations are actually produced. If everything could be known about the functioning of the universe, it could be taken to be deterministic. Not only would its complete history need to be known, but all measurements would have to be made to infinite precision! However, given the limited knowledge of humanity, in most situations it is more useful to consider phenomena to be partly systematic and partly random. Generally, we are not interested in all of the complexities of the process which produced what was observed. Instead, we want to isolate certain pertinent aspects; these will depend on the goal of the study, i.e. the problem to be solved, and should still be present if observations were made in other similar circumstances. This is an important simplifying role of models.

Example
The classical toss of a coin depends on enumerable factors, including how the coin happens to be held, the form of the surface upon which it falls, the direction of the wind, the air temperature, nicks in its surface, and so on. A model will make abstraction of much of this, assuming the coin to be a symmetric, infinitesimally thin mathematical disk with distinguishable sides, always thrown under identical conditions and not subject to wear. Then, the pattern of interest would involve how often each side comes up in a small number of trials. This is highlighted in the classical model describing independent identical Bernoulli trials,

$$f(y;\pi) = \pi^y(1-\pi)^{1-y} \qquad y = 0,1 \quad 0 \leq \pi \leq 1 \tag{1.1}$$

but where both the assumptions of independence and unchanging distribution are clearly violated by any empirical observations. □

The type of data generating mechanism which we consider to be operating will depend on the level of analysis of interest in the problem at hand. We can expect differing results when we construct the corresponding models to work at the levels of the atom, molecule, cell, organ, living being, or community.

Example
In classical mechanics, all laws are time reversible whereas, in statistical mechanics, there is irreversibility because entropy always increases. The former identifies the behaviour of individual molecules, while the latter looks at groups of indistinguishable molecules. □

In many situations, all of the factors not directly accounted for in the model cannot be controlled; they vary over time. Then, as we accumulate empirical observations, this uninteresting variability will also accumulate, and may drown that in which we are interested. (In probabilistic terms, collection of large quantities of data is almost invariably not stable even in the long run, what is called a non-ergodic process.) Thus, as we shall see, it is often preferable to collect a small, but adequate, number of observations for the problem at hand. But, at the same time, one must be assured that these aspects of the data generating mechanism of interest are acting stably throughout and that they are not distorted by the process of data collection. Evidently, the pattern studied should not have any fixed link with the amount of data collected, such as through the number of parameters.

Example
A clinical trial is used to compare the relative efficacy of two medications in treatment of an illness. If the trial is later repeated, both doctors and

patients will have available the accumulated knowledge about the drug and the illness, even though they did not participate in the first trial. Many other conditions will also have changed. □

Two quite distinct types of data generating mechanism commonly occur and should be clearly distinguished. In the natural sciences, such as physics and chemistry, and in engineering, deterministic scientific laws may be assumed to underlie many processes and can be expected to operate in any suitable circumstances. The apex is classical mechanics whereby objects remain on fixed paths which may be predicted exactly from certain basic principles. (This does not, however, imply that probabilistic explanations do not also play an important role in these fields, as we already saw above.) The random variability in empirical studies, in such cases, may be assumed to arise purely from the production and observation process and, hence, may be termed *production* and *measurement errors.*

Example
Theoretically, a chemical reaction should always yield the same results, if the same proportions of chemicals are used under the same conditions. Production and measurement errors will arise from impurities, imprecision in weighing, slight changes in temperature or air pressure, and so on. □

On the other hand, in studies involving living organisms, and especially human beings who have language, memory, and a will of their own, each individual has an enormous amount of built-in variability, due to genetic, environmental, and other factors. Thus, here the systematic 'laws' to be studied must include a random part or be *stochastic.* Measurement error is proportionally minor as compared to other sources of variation.

Example
Young rats have a pattern of growth which can be studied. However, one cannot expect that two rats, even bred to be as identical as is possible, would follow exactly the same growth curve. In addition to small genetic differences, slight differences in their treatment during an experiment can never be avoided: the amount of light and heat, the exact time of feeding each animal, and so on. □

Thus, many scientific problems can only be answered by stochastic models.

Example
The question as to the chance that a natural population of living things, such as a species of rodent on a small isolated island, dies out in the course of time cannot be answered by a deterministic model. □

Unfortunately, much of the thinking about statistical inference has been premised on the first situation, deterministic models with production and measurement error. This is deeply embedded in statistical practice and terminology. For example,

- a probability distribution is often called an 'error distribution';
- an 'expected value' is what an observation would be (according to the model) if it were not for measurement error.

In the study of deterministic data generating mechanisms, increasing the number of observations should reduce the overall error in the parameters of interest, and, with enough observations, one should obtain the true values of the parameters in the model. Such circumstances are only available in tightly controlled laboratory conditions, and often not even then. However, more often in parametric statistical inference, we are faced with a dynamic situation, which can never be exactly reproduced.

Example
The level of unemployment varies greatly among countries and over the years. One has no hope of the same situation ever being reproduced and yet statistical models can be used to describe the important features of how unemployment is being generated. □

In scientific work, it is important to distinguish two types of material available prior to beginning a new study. *Prior theoretical knowledge* is embodied in the models constructed to describe and explain some phenomenon. *Prior empirical information* is supplied by previous empirical observation of the phenomenon of interest. Obviously, there must be an intimate relationship of interplay between the two, in which statistical inference has an important role to play. Only models which are believed to be pertinent will be considered, although the completely unexpected is central to scientific discovery and must be allowed for. Current information supplements prior information in ranking the models and improving theoretical knowledge. But, new empirical information can always place in question, and even overturn, prior theoretical knowledge.

We should also note that scientific *theories*, including the accompanying models, are generally developed as if measurements could be made to infinite precision. They do not take into account the actual precision of empirical observations related to them. In contrast, statistical inference is concerned with the process of obtaining information from such empirical measurements, and must take into account their precision.

Models serve to summarize, explain, and transmit available theoretical knowledge within the scientific community, making it unnecessary for the whole body of empirical observation to be directly available to every member. If a model can be completely specified beforehand, which is rarely the case, then data collection will serve only to check its validity. More usually, some parameters in the model have unknown values and data will also serve to provide information about them. Some idea about these values may be available but this is generally extremely difficult to quantify and rarely has the general agreement of the scientific community in the way that choice of model does. Thus, scientific data collection concentrates on obtaining empirical information about validity of models and about values of unknown parameters.

A stochastic model must

(1) be in accordance with previous theoretical knowledge and empirical information;
(2) specify the probability distribution of all observations included in the data, taken as a typical sample;
(3) include, as parameters, all unknown constants to be estimated;
(4) not be contradicted in any relevant way by the newly collected data.

(Fisher, 1955)

1.2 Probability models

The first step in the model building procedure will involve the selection of one or more appropriate probability distributions to describe the random generation of some *response variable*, say Y, to be observed. These will be chosen to describe, in an informative way, some underlying data generating mechanism which is of interest. A probabilistic data generating mechanism is a process producing a result which cannot be (completely) determined in advance, although the set from which it comes will usually be known. The process need not be repeatable, although scientific study usually is concerned with the repeatable aspects of processes. In the simplest cases, considered in this section, each observation can be assumed to be made independently of all others.

Example
For the toss of a coin, the response is a head or tail. Description of the data generating mechanism may take into account the fact that a coin has two sides and a narrow edge, the way a person normally throws a coin, and the surface on which it will land. In contrast, a machine could be constructed which tossed coins necessarily coming up heads, or alternating heads and tails in some fixed sequence. What makes the first data generating mechanism probabilistic, and the second not, is that the outcome is unknown for any given toss, not the long run variability. □

Thus, a model is some more or less reasonable simple mathematical approximation to the reality of the data generating mechanism, which may help us in understanding the functioning of that mechanism. A model *smooths* the observed data a chosen amount, which may help us to see patterns in them and to understand them. *No model is true.*

Examples

1. If the coin is believed to be unbiased, it could be modelled by assigning equal probabilities of one-half to the two outcomes, heads and tails. In many ways, this is a simplification of the data generating mechanism, ignoring, among other things, the possibility of the coin landing on its edge and the direction in which the wind is blowing.

2. The normal distribution, with

$$f(y;\mu,\sigma^2)=\frac{1}{\sqrt{2\pi\sigma^2}}\mathrm{e}^{-\frac{(y-\mu)^2}{2\sigma^2}} \qquad -\infty<y,\mu<\infty, \quad \sigma^2>0 \quad (1.2)$$

is commonly used as a model for measurements, such as length or weight, which are known to be positive. It ascribes positive probability to negative observations which are impossible. □

When scientists speak (loosely) of a scientific law, or a model, as being 'correct', they mean that any known deviations from it are sufficiently minor so as to be irrelevant for all practical purposes. The best *available* law or model will be used for scientific prediction, even given such discrepancies. Statistical inference has an important role to play in determining which is currently best.

As the number of observations increases, more complex models may be fitted, perhaps closer to the data generating mechanism. Conversely, if enough observations are collected, any model with a fixed finite number of unknown parameters can usually be shown to be untenable. If only simple models are of interest, then the *sample size* should be kept reasonably small. It should be just large enough for the aspect of the phenomenon of interest to be detectable; otherwise, time and money will be wasted.

Example

The gamma distribution, with

$$f(y;\mu,\alpha)=\frac{\alpha^\alpha y^{\alpha-1}\mathrm{e}^{-\frac{\alpha y}{\mu}}}{\mu^\alpha\Gamma(\alpha)} \qquad y,\mu,\alpha>0 \qquad (1.3)$$

can be interpreted as describing the total time of a sum of α exponential periods, whereas the Weibull, with

$$f(y;\mu,\alpha)=\frac{\alpha y^{\alpha-1}\mathrm{e}^{-\frac{y^\alpha}{\mu}}}{\mu} \qquad y,\mu,\alpha>0 \qquad (1.4)$$

could describe the total time until the first of a set of competing events occurs. (In both cases, $\alpha = 1$ yields an exponential distribution.)

Suppose now that we are interested in modelling the length of marriage before divorce in some country. With a sample of about 1000 cases, we find that a gamma distribution, with α close to three, is more plausible than a Weibull. This supports the theoretical idea that marriage is composed of three different stages through which a couple passes. However, if our sample is about 2000 cases, the gamma distribution, although still better than the Weibull, is no longer plausible, primarily because a low level of divorces in the first year or two of marriage, with a subsequent jump, is now detectable. □

Obviously, if we are only interested in exploring, as profoundly as possible, the phenomenon in question, and not looking for a particular pattern, we should gather the maximum amount of data feasible. But the models must then be correspondingly complex, if they are to be expected to represent the data satisfactorily.

Following our discussion to now, we can distinguish three types of randomness in data generation:

(1) the stochastic process leading to the specific observed response(s) on an individual unit – this is the data generating mechanism which the scientist or engineer wishes to study;
(2) random production and measurement errors in generating and observing the response;
(3) randomness introduced into the design of the observation scheme to ensure that the desired data generating mechanism is actually being observed.

Consider point (3) first. Good study *designs* introduce randomness, in the selection of individuals to observe, which is independent of or 'orthogonal' to the data generating mechanism of interest. This ensures that the way individuals are chosen does not depend on that mechanism.

Examples

1. In a sample survey, respondents are randomly selected from a well-defined population, in a way independent of the assumed data generating mechanism, to ensure 'representativity'. If attitude to abortion were being studied, and only women attending abortion clinics were interviewed, the results would tell us little about women in general in the country.

2. In an experiment, after a similar random choice of living subjects, treatments are randomly allocated to them to ensure that such allocation is not linked with the data generating mechanism, perhaps in some unknown manner. This might occur if a doctor gives what (s)he considers to be the most suitable treatment to each patient, thus linking the choice with the gravity of the disease. □

Such randomization may not be necessary in many physical and chemical experiments, unless human differences among the ways treatments are handled are suspected. A properly randomized experimental design generates a 'null' model, that the treatments under study have no effect, to which the results observed can be compared.

In certain designs, called sequential, the total number of observations, n, will not be fixed in advance; it is a random variable, N, in addition to the response, Y, from the data generating mechanism. Moreover, different designs can provide more or less relevant information about the data generating mechanism, even with the same number of observations. The vast fields of experimental and survey design are concerned with optimal choice of observations.

In all cases, the design should highlight the data generating mechanism and not distort it, so that it could be argued that, if its role of guarantor in the data collection process has been correctly accomplished, it should play no subsequent part in the inference procedure.

Point (2) above, production and measurement errors, is an unavoidable side effect of all empirical observation. If this is judged to be important, it may need to be included in the model describing the data generating mechanism, which is our prime objective. Often such errors can be modelled by a normal distribution. However, in many contexts, outside of the strict measurement error situation, which will be described in more detail in Section 2.7.1, such variability will (hopefully!) be minor as compared to that created by the data generating mechanism.

Suppose that we are interested in some large, but finite, clearly definable *population* of individuals. One fruitful, but simplistic, way to think of the data generating mechanism in point (1) above is as some process which allocates each individual to one of a number of discrete classes defined by the values of some variable. This is the response to be observed, the data to be generated.

The categories can be assumed discrete, for the moment, because any existing measuring instrument has finite precision, called the *unit of measurement.* This does not, however, imply that models must be constructed uniquely in terms of discrete variables. On the contrary, many phenomena must be conceived of as continuous, and the appropriate models developed. In these cases, the discreteness only arises in the empirical observation of data to check how well the model describes the mechanism of interest. It is paradoxical that statisticians, so concerned with the precision of their estimates, have generally ignored this important source of imprecision.

Then, if all individuals in the population, however conceived, could be observed, a histogram of the population frequencies could be constructed for that variable. Owing to the random elements in this generation process, described above, no simple mathematical function could ever be expected to describe it exactly. In fact, the larger is the population, the more difficult the task would be! However, certain systematic features should be present. (Patterns always are, even in purely random data!) What features would be noticed will depend on the object of the study.

When a sample is collected from the population, a histogram can actually be constructed. In spite of the random variation, one hopes to be able to discern the same systematic features as would be present in the population histogram and to be able to describe them by a simple model, not depending on sample size. If the sample is too small, only random fluctuation will be present. But, as we have seen, increasing it beyond some reasonable size will not generally augment knowledge about those aspects of the data generating mechanism, i.e. the pattern, of interest. Rather, the uninteresting irregularities of the population will become more apparent. The simple model for the pattern of interest will not converge to some 'true' data generating mechanism; instead, the complexity of the model would need to be allowed to increase with the sample size for the data to be adequately described.

Often the model describing the data generating mechanism is only specified up to some unknown constants, called parameters, indexing a family of distributions. (When statisticians interact with scientists, this can be confusing because the latter generally refer to observable quantities as parameters: the statistician's variables.) In other words, one or more mathematical functions are proposed as descriptions of the pertinent features of the histogram.

Example
Count data are often modelled by a Poisson distribution, with

$$f(y;\mu) = \frac{\mathrm{e}^{-\mu}\mu^y}{y!} \qquad y = 0, 1, 2, \ldots, \qquad \mu > 0 \tag{1.5}$$

and unknown parameter, μ, the mean intensity of events. However, among other things, this implies that the variance is equal to the mean (Section 2.1.3). Often, there may be too much variability in the data for this condition to hold. This is known as *overdispersion.*

If the systematic reasons for such variation are not of direct interest, we may change models for one which allows more random variability. Overdispersion may arise because the intensity of events is, in fact, not identical for all individuals observed or because it is varying over time for each individual. □

Each probability distribution corresponds to some aspects of one or more simplistic data generating mechanisms. Thus, the appropriate distribution should be chosen in light of knowledge about this mechanism.

Example
If the data are measurements believed to be generated as the sum of a large number of unknown factors, by the central limit theorem (Section B.6.3), the normal distribution of Equation (1.2) may be appropriate. If they are the product of a large number of unknown factors, a log normal distribution with

$$f(y;\mu,\sigma^2) = \frac{1}{y\sqrt{2\pi\sigma^2}} e^{-\frac{[\log(y)-\mu]^2}{2\sigma^2}} \qquad 0 < y, \mu < \infty, \quad \sigma^2 > 0 \quad (1.6)$$

might be chosen. □

This simplified description of model building has limited usefulness because it supposes that the population is static, whereas it will often be in dynamic evolution, perhaps even during the time in which a sample can be collected. This will be discussed further in Section 1.4 below.

In the next chapter (and in Appendix B), we shall study in more detail this basic element of the statistical model building structure: the probability distribution.

1.3 Regression models

A data generating mechanism will generally not be homogeneous for all units observed. In the simple situation described in the previous section, the variability in response outcomes was accounted for by the randomness in the probability distribution. However, often, we are interested in *systematic* changes in variability among distinguishable subgroups of the population, the latter described by *explanatory variables.* We are, then, concerned with how the probability distribution of the random response variable changes under these different conditions, a conditional distribution. In terms of our histograms above, we wish to see how they change form under the different conditions.

Example
In a simple scientific experiment, we randomly select some subjects as controls, while the others are given some special treatment. Interest will centre on how the distribution of responses differs in the two circumstances. □

Models are intended to influence our future expectations about observations not already made. A scientific study must be designed to provide information about such generalizations. Obviously, if enough conditions were specified, each individual in the population would be uniquely identified, i.e. would have a unique data generating mechanism. The model would be deterministic, and of little use. It would mean that the observed individuals were uninformative about those not observed and generalization could not be made. Probability statements never refer to individuals, except as typical members of some population.

In a stochastic model, any variability which is not of direct interest should be allowed to be random. Thus, explanatory variables are only used to describe the systematic effects of interest, within the data generating mechanism, while the random or stochastic part of the model, the probability distribution, must describe the remaining variability.

In simple cases, we can denote the conditions by a matrix, say $\mathbf{X}$, called the *design matrix*, where each column, indexed by j, specifies a type of different fixed condition, and each row, indexed by i, a different individual. Thus, an entry in the matrix gives the situation of the ith individual for the jth type of condition. This matrix defines a *design space* describing the conditions in which the study was conducted. Stating that conditions are fixed is arbitrary, but is done because, in our model, we are not interested in how they arise, only in how they affect the data generating mechanism, and hence the response.

Most often, we make the further simplifying assumption that only one parameter, say μ, in the probability distribution, describing the mean response, or some function of it, changes with the conditions of the individual. Thus, only the *location* of the histogram is changing. Then, we might choose functions such that

$$g(\mu_i) = h(\mathbf{x}_i, \boldsymbol{\beta}) \tag{1.7}$$

Here, the function $h(\mathbf{x}_i, \boldsymbol{\beta})$ is called the systematic component of the model, because the conditions specified by $\mathbf{X}$ are taken to be fixed. Because the function $g(\cdot)$ links this component to the random or stochastic component, the probability distribution, through the mean value, it is called the *link function.* The result is a *general regression model.*

Example
If we are interested in how the average weight, μ_t, of young mice changes over time, t, we might use an exponential growth model,

$$\mu_t = \alpha e^{\beta t} \tag{1.8}$$

Here, $g(\cdot)$ might be taken to be the identity function and $h(\cdot)$ an exponential function, or $g(\cdot)$ as a logarithmic link and $h(\cdot)$ as a linear function. □

Note, however, that there is no reason why only the mean parameter should change with the conditions. As suggested by Lindsey (1974b), other parameters, such as the variance, can also be made dependent on the explanatory variables.

Example
In the industrial production of foam rubber, the control of variability in hardness is at least as important as that of average hardness. If two people sit down on adjacent cushions of a sofa, one should not sink away down and the other stay up! □

In many situations, the systematic component, $h(\cdot)$, is assumed, for mathematical simplicity, to be a linear function of the unknown parameters, $\boldsymbol{\beta}$,

$$\begin{aligned} h(\mathbf{x}_i, \boldsymbol{\beta}) &= \boldsymbol{\beta}^{\mathrm{T}} \mathbf{x}_i \\ &= \eta_i \end{aligned} \tag{1.9}$$

called the *linear structure* or *linear predictor*. (Notice that this emphasis and terminology contrasts with the interests of scientists and engineers who are concerned with the linearity of $\mathbf{X}$.) When this linear function is equated to a link function for the mean, we have a *linear regression model*. We shall see, in Chapter 2, that one important family of such models is called the *generalized linear model*.

Example
One possible generalized linear regression model for frequency and count data is log linear regression,

$$\log(\mu_i) = \boldsymbol{\beta}^{\mathrm{T}} \mathbf{x}_i \tag{1.10}$$

with μ_i the mean of a Poisson distribution. □

We saw above that parameters index a family of models. Often they have no intrinsic interest, serving primarily to determine the shape of the probability distribution and of the systematic part. Thus, they usually can be specified in a number of different, but equivalent, ways.

Example
The additive two-way analysis of variance model can be written in the form,

$$\mathrm{E}[Y_{ij}] = \mu + \alpha_i + \beta_j \tag{1.11}$$

where $\{Y_{ij}\}$ is a set of independent random variables having a normal distribution with these means and constant variance. The 'conventional' constraints impose $\sum_i \alpha_1 = 0$ and $\sum_j \beta_j = 0$, while 'baseline' constraints may have, say, $\alpha_1 = 0$ and $\beta_1 = 0$. The values of the parameters will be quite different in the two cases, although they index the same models. □

We may now summarize some of the possible simplifying assumptions in model building suggested up until this point:

- observations can be made independently of each other;
- the data generating mechanism is stable enough over time (and space) so that a small selection of observations from a long series will provide useful information about patterns of interest;
- the random part of the process can be described usefully by some simple probability function;
- this function has the same form under all conditions, with only (some of) its parameter values changing;
- the systematic part involves only changes in the mean of that distribution; and
- these changes in the mean (or some function of it) are linearly (in the parameters) related to the conditions of interest.

As we have already indicated, most of these are simplistic but useful approximations which should be abandoned when necessary in any particular setting. That which is most important is evolution over time, which we shall now look at in a little more detail.

1.4 Stochastic processes

As we have seen, most empirical observations are a snapshot, through a small window, of some ongoing process. In the models described above, the aspect of the process taken to be of interest was assumed to be constant enough over time for it to be meaningful to study it in this way. However, in many situations, the dynamic random evolution over time itself may be of direct interest. This series of response values, $\{y_t\}$, at times t, defines the *state* of the process at each moment, and the corresponding model describing its probabilistic evolution is called a *stochastic process.*

Thus, time is an extremely important variable in statistical modelling. Another is spatial position. However, because of the symmetric (time does not go backwards) and three-dimensional nature of spatial dependence, it is much more difficult to model than time. What is happening in one spot may be influenced by everything around it (as well as what occurred previously in time!).

Here, our image of a fixed population, with a histogram of proportions for a response variable, may no longer be very useful. No concrete population will often be definable at all. The essential role of probability in modelling can be extended, however, still providing a description of variability in outcomes which are unknown before observation.

Example

Suppose that the response of interest is the number of unemployed in a given country. This can vary from zero to the total working population.

However, it will change over time, observable perhaps every month. Some complex data generating mechanism is producing the series of values of the variable and statistical models can be constructed to attempt to describe it. However, there is no clear 'population' of all 'individuals' that might be observed. All months? For how long? □

In addition, when these aspects of a data generating mechanism are to be studied, the assumption of independent observations must be abandoned. The same individual unit(s) will be observed repeatedly over time and the successive responses will be linked. A model can then consist of a series of conditional probability distributions for the sequence of observations, $y_1, y_2, y_2, \ldots$,

$$P(y_1)P(y_2|y_1)P(y_3|y_1, y_2)\cdots \tag{1.12}$$

Obviously, such a model is of little use as it stands in describing patterns of interest, because it tries to take into account the influence of all previous variability on the new response at each step. Again, a useful model requires simplifying assumptions. Thus, the dependence is usually assumed to be only short term, going back only a very few periods into the past, called a *Markov* assumption.

Example
A simple autoregression model allows the mean response from a normal distribution to depend on the response actually observed at the previous time point:

$$\mu_t = \beta_0 + \beta_1 y_{t-1} \tag{1.13}$$

Thus, in this way, a regression model can be used to introduce dependence among responses. □

Once the stochastic dependence has been incorporated, the model building techniques of the previous sections can still be applied in this new context. In particular, variation in the stochastic processes under differing conditions can be described by explanatory variables in regression models. Because dependence is within units, parameters describing it will also often be of direct interest; they describe one way in which successive responses are related.

When we follow a number of individuals over time, recording a series of responses on each, the data generating mechanism will have (at least) two inter-related aspects. For various reasons, individuals came to be different in the first place (their heterogeneous frailty or proneness to the phenomenon under study). Then, the processes under study will evolve to their current responses. Of course, these latter processes may operate to accentuate the former.

Example
For environmental and genetic reasons, some people are born weaker, and more susceptible to illness, than others. But then their history of childhood illnesses may further aggravate this situation. □

Models for stochastic processes in time and space are especially important because they often permit us to understand better how patterns of interest are produced, rather than only studying the final result of the process. Measures of time and space are truly continuous and realistic models will usually have to take this into account. At the same time, as with all empirical measurements, responses can only actually be observed at discrete points in time.

1.5 Inferences about models

In mathematics, specific conclusions are deduced from a general set of postulates; only those necessary need be used in deriving any given result. In contrast, in statistical inference, general conclusions are drawn from specific observations; here, *all* relevant available information must be used because arbitrary exclusion of certain available elements would permit equally arbitrary conclusions to be drawn. Deduction reveals already existing knowledge in the body of postulates. Inference adds new elements to theoretical knowledge.

The strength of conclusions will depend critically on the way in which data are obtained. A *causal* conclusion operationally means that modifying one (explanatory) variable will result in a changed distribution of another (response) variable; for such a conclusion, the former variable must have been appropriately manipulated, in an experiment, to produce the data. Otherwise, the conclusions can only be *observational*, describing the unmodified state of the phenomenon, even if it is an evolving process. This can be confusing, because the same models may be used in the two cases; more sophisticated models cannot compensate for the way in which the data were obtained. The way data are collected, not the model used, determines the strength of conclusions from those data; no empirical information about causality will be available if the appropriate variables have not been modified by the researcher.

One or more families of models will be chosen, according to prior theoretical knowledge, often supported by prior empirical information. Once (new) observations are made, the models must be *fitted* to these data. In other words, given the data and several chosen functional forms, i.e. one or more specific models indexed by their parameter values, as a first step, we perform a *model selection* process. Next, within a chosen family, one will be interested in which models, as indexed by their parameter values, are most plausible in light of the observations; these are obtained by *estimation.*

Examples

1. For our histogram, corresponding to categories of a variable, a natural guess at the proportion in each class in the population, called an estimate, is the corresponding observed proportion in the sample. In Chapter 3, we shall see that this also has more theoretical justifications.

2. In the divorce example of Section 1.2 above, we wished to choose between the gamma and Weibull models. Once the former was found to be more plausible, we required an estimate of the appropriate parameter to see if the three-stage model was plausible. □

Because of the random variability in the data, it will almost never be possible to limit selection to one model adequate to describe the aspects of the data generating mechanism interesting us. Instead, a subset of plausible parameter values must be specified for any functional model. These general procedures will be first discussed in Chapter 3.

Some statisticians claim that scientific practitioners, when given any such subset of models, will interpret it as containing the 'true' model with some given probability. A more reasonable interpretation, in light of our discussion of modelling, would be that they consider the proposed subset to contain plausible models, or models likely to be informative, given the observed data. Other statisticians propose methods for constructing subsets which will contain the 'true' model with a certain frequency in repetitions of the study, with little clear indication of what the observed data can actually tell us about the models under consideration. This approach is most appropriate for modelling such situations as the repetitiveness of industrial quality control systems.

Many statisticians reason primarily in terms of what would happen if the sample became extremely large. In most concrete situations, this is a mathematical fiction of limited usefulness, because populations studied are finite, or not even usefully definable, as in stochastic processes. In addition, if too many observations are made (except in a pure measurement error situation), none of the simple models under consideration may adequately describe the data: as we saw above, they would need to be made more complex to take into account the accumulating variability, which is, in fact, of no interest.

Each choice of a family of models can usually be checked by embedding it in a more complex model, containing extra parameters. That of interest will then be a special case, and one can look to see if the extra parameters are really necessary.

Examples
1. The normal distribution might be generalized by taking a power transformation, Y^{α}, of the response (Box and Cox, 1964), and then checking to see if the transformation differs from unity, i.e. if $\alpha = 1$.
2. As we have already seen, the exponential distribution, with

$$f(y;\mu) = \frac{e^{-\frac{y}{\mu}}}{\mu} \qquad 0 < y, \mu < \infty \tag{1.14}$$

can be embedded in either the gamma or Weibull distributions, both of which it is a special case, when $\alpha = 1$ in Equations (1.3) and (1.4). We would, then, usually also like to compare the fit of these latter two models, as in the divorce example. □

A model may also be placed in question if it fails to account for certain parts of the data, such as extreme observations. Model checking will be studied in detail in Chapter 4; it opens the possibility of scientific discovery.

One important practical aspect of empirical data collection is that unplanned circumstances will often arise. These may need to be taken into account, either in constructing the models or in making the inferences, or both.

Examples
1. Many studies involve repeated measurements of the same individuals. These may not all be made at the same time points, or some may, unavoidably, but unpredictably, be missing.
2. In the study of durations, such as survival of patients with some disease, observation must terminate at some point in time in order for results to be analyzed. Not all durations may then be complete, called censoring, i.e. many patients will, hopefully, still be alive. Others may be incomplete because the individual dropped out of or disappeared from the study early for some reason. □

Choice of a model to describe some phenomenon is a decision. In a scientific context, this decision is taken in order to obtain the best knowledge of the phenomenon under study. No specific action follows from it and, indeed, it can later be reversed. However, statistics is also widely used in other fields, primarily industrial, where decisions lead to actions, with costs and benefits, measurable or not. If these must be taken into account when the statistical choice of models is being made, then a model of human behaviour will also be necessary. Most such models are extremely simplistic, assuming, for example, that a study can be repeated infinitely often, or that individuals can have infinite knowledge. In some senses, these formalisms are models of deductive, not inductive, behaviour, because they assume the global statistical model to be true, making minor adjustments (to parameter values) for the data at hand and applying the general theory to decision-making in particular contexts. We shall look at some models of human decision-making behaviour connected with statistics in Part IV.

1.6 Characteristics of modelling

As we have seen, the empirical study of some phenomenon through statistical models has a number of important characteristics:

- No model is true.
- Features of interest can be described by some relatively simple mathematical function, helping to understand and to explain how they were generated.
- Patterns of interest are described by the systematic part of the model; all other variation is relegated to the random part.
- Observations can only be made to finite precision, and often this is rather rough, especially in the biological and human sciences.
- A reasonably small sample size may often be preferable to avoid wasting time and money and to prevent the random variation from swamping the simple patterns of interest.

Once one or more suitable models have been constructed from available theoretical knowledge, we shall require several things in order to make inferences:

- procedures for choosing how many observations to make, and in what way, including rules for stopping a study when sufficient, but not too much, information has accumulated: *sample size calculations*;
- principles for the selection of the individuals to observe in order to obtain maximum information efficiently: *experimental and survey design*;
- ways of choosing among competing models, especially if we believe that several different functional forms might be appropriate: *model selection criteria*;
- once a suitable functional form is selected, a method for deciding what subset of parameter values is plausible in light of the data: *interval estimation*;
- means of knowing how much smoothing a model applies to the data and criteria for judging globally how close each model is to the data: *goodness of fit*;
- methods for checking if individual (groups of) responses are extreme, in some sense, given the model: *diagnostics*.

The modelling approach has a number of advantages, as compared to the use of strictly descriptive statistics or of the closely related non-parametric statistics which do not assume a known functional form for the patterns in the data.

- Models incorporate previous theoretical knowledge.
- They summarize data in a parsimonious way, by smoothing uninteresting stochastic irregularities.
- They allow new data to be combined with previous empirical information.
- They provide some understanding of the underlying data generating mechanism.
- They are applicable to small sets of observations, but can become more complex as the information grows.
- They make explicit the assumptions being made in analyzing the data, assumptions that are always necessary but which would otherwise remain implicit.
- They allow measures of goodness of fit to the data, i.e. rigorous checking of the assumptions being made.
- They permit the detection of extreme observations, which may lead to modifications of the theory upon which the model was based.
- If not too complex, they are reproducible in replications of a study.
- They provide means of predicting future observations.

Descriptive and non-parametric methods may be useful for detecting and summarizing patterns in data; parametric models introduce the possibility of explaining them.

Further reading

Most applied probability texts provide examples using known probability models to predict future events; see, for example, Tijms (1994). For elementary introductions to statistical modelling, the reader may like to look at Derman *et al.* (1973) for probability models, Ross (1989) for applications in stochastic processes, and Dobson (1990) for the use of regression techniques called generalized linear models, the latter from a frequentist perspective. More general discussions of modelling include Berkson (1951), Cox (1990), Appleton (1995), and Bartholomew (1995), and, for model selection, Burnham *et al.* (1995). For more advanced modelling, the reader is advised to consult a standard text in the special area of interest.

At the other end of the scale of complexity, statistical models play a central role in pattern detection, in fields as diverse as image recognition and linguistics; see, for example, Grenander and Miller (1994).

Much of the original impetus for statistical modelling arose from the paper of Nelder and Wedderburn (1972), which demonstrated the unity of a set of apparently disparate models, and from the development of the interactive computer software, GLIM, shortly thereafter by the Royal Statistical Society. The latter allowed practitioners to experiment with such models in a wide variety of contexts.

The field of study design was launched by the path-breaking book by Fisher (1960). One of the most important instruments for obtaining information is the questionnaire. Much sociolinguistic, social psychological, and sociological research has been devoted to proper questionnaire construction, although this has had little effect within statistics; see, for example, Schwarz (1995).

The problems of communication between statisticians and scientists, in the modelling context, are ably discussed by Blyth (1970), McPherson (1989), and Hand (1994).

1.7 Exercises

(1) Choose some situation with which you are familiar and for which you could observe some response variable.
  (a) Suggest at least two data generating mechanisms which might be operating.
  (b) Develop possible simple statistical models which might allow you to distinguish among them.

(2) Find at least two types of non-biological phenomena for which a production and measurement error model would not be adequate because of the amount of non-stable variability.

(3) A study of the fuel efficiency of two models of cars is conducted in continental Europe and in Britain. Europeans measure fuel consumption in litres per 100 kilometres, while the British use miles per gallon (or equivalently, kilometres per litre). The results for two cars of each model are

	Model A			Model B		
	1	2	Mean	1	2	Mean
Britain (km/l)	8.3	33.3	20.8	16.7	16.7	16.7
Europe (l/100 km)	12.0	3.0	7.5	6.0	6.0	6.0

  (a) Which model of car would you choose to buy? Why?
  (b) How do you explain the difference in results between the two regions?
  (c) Can you develop an appropriate model to resolve the problem? Allow for variability among cars as well as average performance.

(Hand, 1994)

(4) (a) List the common probability distributions with which you are familiar.
  (b) For each, describe at least one data generating mechanism to which it might be applied.

(c) What departures from each model might be expected in practical circumstances?

(5) Give reasons why representation of the conditions under which a response is made by a matrix, **X**, will be of limited usefulness in many contexts.

(6) The field of statistics is dominated by the linear model of Equation (1.9). In most situations, this is unrealistic, especially for extreme conditions. Think of the exponential growth model of Equation (1.8) – which can, however, be written in a linear form.

(a) How would this model need to be modified if rats were followed from birth to old age?

(b) What would be the disadvantages of trying to construct a model for such data based on a polynomial function of time?

(c) Would it be reasonable to extend application of such a model function, found to be applicable to rats, but with different parameter values, to describe the growth of other living things, such as human beings?

(7) For what reasons may observed responses be extreme, given some model? Consider both the data generating mechanism and the data collection process.

(8) (a) List a number of ways in which the total number of observations actually made, the sample size, n, might arise as a random variable.

(b) Which of these depend on the design and which on the data generating mechanism?

(c) What are the implications of each for any model which you might construct?

(9) Data from studies of primary education in Third World countries indicate a wide range of ages for children in a given class. Often, this seems to arise, not because children fail, but because they have been absent from school. In certain cases, the ages can adequately be described by a Poisson distribution, indicating that they may be distributed at random, i.e. that absences occur randomly during the child's previous schooling history. Discuss the type of data which would need to be collected to study these hypotheses directly and the problems that might be encountered in drawing conclusions.

(10) A white noise generator produces signals with unknown mean, say μ, and unit variance. The number of signals randomly generated per unit time, say ρ, is also unknown. An instrument which only records positive signals is used to take one full day's sample. Only one signal, say y, is observed. Develop a model to describe this data generating and observation process.
(Irony, 1993)

2

Exponential family of probability distributions

Many of the models commonly used in statistics are based on probability distributions which have special characteristics in common. These can be grouped into a number of families, the most important of which is called the *exponential family*, originally introduced by Fisher (1934). As well, this family has a number of important statistical properties which have given it a central role in statistical inference.

Most of this chapter is concerned with this family, the fundamental results being presented in the first two sections. (For those interested in generalized linear models, Section 2.5.1 is also important.) The chapter concludes with shorter presentations of two other important families, those based on transformations and on stable distributions. The existence of such families means that we can structure a large variety of models in the same basic ways, and that the same type of procedures can be used for drawing conclusions in all of these cases.

2.1 General theory

Many of the most commonly used probability distributions are members of one special family, called the *exponential family.* These distributions are not necessarily the most useful or realistic ones in many situations. They have gained their prominence primarily through their nice statistical properties, especially their facility of mathematical manipulation.

First, we shall require two definitions:

- The *domain*, $\boldsymbol{\Theta}$, of the parameter vector is that set of values for which the normalizing constant, say $a_0(\boldsymbol{\theta})$, of the probability density is strictly positive.
- The *support*, $\mathcal{Y}$, of the distribution is that set of observable values, y, for which the density function is positive for all possible parameter values.

In an appendix to this chapter, the reader will find more rigorous discussion of some of the properties to be described in this section than is possible in the main body of the text.

2.1.1 EXPONENTIAL REPRESENTATION

For the moment, we shall only consider a single observation. Any probability density which can be written as the non-negative function,

$$f(y;\boldsymbol{\theta}) \stackrel{\text{def}}{=} a_0(\boldsymbol{\theta})t_0(y)e^{\boldsymbol{\theta}^{\mathrm{T}}\mathbf{t}(y)} \tag{2.1}$$

is said to have the *exponential representation*. Here, $\boldsymbol{\theta} \in \boldsymbol{\Theta}$ is the k-dimensional parameter vector indexing the distribution and $y \in \mathcal{Y}$ is a fixed or observed value of a random variable, Y, with $\mathbf{t}(y)$ some function calculable from the observed data alone, called a *statistic*. (Notice the distinction between capital letters for random variables, or functions of them, and small letters for observed values of them.) Finally, $a_0(\boldsymbol{\theta})$ is a normalizing constant, such that the integral or sum of all possible probabilities is unity:

$$a_0^{-1}(\boldsymbol{\theta}) = \int_{\mathcal{Y}} t_0(u)e^{\boldsymbol{\theta}^{\mathrm{T}}\mathbf{t}(u)}du \tag{2.2}$$

Theoretically, Y can take either real or integral values, corresponding to continuous or discrete distributions. In measure theory (Appendix A), these are said to have, respectively, Lebesgue and counting measure. The function $t_0(y)$ could be absorbed into this measure, but this would unnecessarily complicate matters, because the latter would no longer be counting or Lebesgue measure. Thus, $f(y;\boldsymbol{\theta})$ is indeed a density, although it is often called a probability mass function when y is discrete. Hence, we shall use $f(\cdot)$ generically to represent any such density, with counting or Lebesgue measure, not explicitly distinguishing between discrete and continuous variables. Thus, measure theory can provide a unified approach to both discrete and continuous distributions, simplifying the mathematics.

Unfortunately, results derived in this way are only a rough approximation to those applicable to what can actually be empirically measured or observed (in spite of the name!). Empirically, any random variable, Y, can only take discrete values, defined by the unit of measurement. In fact, all results for continuous distributions in this family, being based on the density function, assume that observations can be made with infinite precision. As it stands, a continuous density can give the probability of no observable event. Rather, the probability for such an event from a continuous distribution will be given by

$$P(Y = y;\boldsymbol{\theta}) = \int_{y-\Delta/2}^{y+\Delta/2} f(y;\boldsymbol{\theta})dy \tag{2.3}$$

where Δ is the unit of measurement, which generally does not have the exponential representation. Thus, the results of this chapter generally hold only for densities, not for the probabilities of observable values from continuous variables.

If k is the smallest integer such that the probability distribution is representable as in Equation (2.1), the family is said to be of *order* k and the representation is said to be *minimal.* This order must be less than or equal to the dimension of $\boldsymbol{\theta}$. This minimal representation is not unique, because it can be transformed to a different, but equivalent, representation by applying linked non-singular linear transformations to $\mathbf{t}(y)$ and $\boldsymbol{\theta}$.

Examples

1. The Poisson distribution is a discrete distribution with $k = 1$. Its density can be written, as in Equation (1.5),

$$\begin{aligned} f(y;\mu) &= \frac{\mathrm{e}^{-\mu}\mu^y}{y!} \qquad y = 0,1,2,\ldots, \quad \mu > 0 \\ &= \frac{\mathrm{e}^{-\mu}}{y!}\mathrm{e}^{y\log(\mu)} \end{aligned} \tag{2.4}$$

so that the normalizing constant is $a_0(\theta) = \mathrm{e}^{-\mu}$, while $t_0(y) = 1/y!$, $\theta = \log(\mu)$, and $t_1(y) = y$.

2. The density for the normal distribution, a continuous distribution with $k = 2$, can be written, as in Equation (1.2),

$$\begin{aligned} f(y;\mu,\sigma^2) &= \frac{1}{\sqrt{2\pi\sigma^2}}\mathrm{e}^{-\frac{(y-\mu)^2}{2\sigma^2}} \qquad -\infty < y,\mu < \infty, \quad \sigma^2 > 0 \\ &= \frac{\mathrm{e}^{-\frac{\mu^2}{2\sigma^2}}}{\sqrt{2\pi\sigma^2}}\mathrm{e}^{-y^2\frac{1}{2\sigma^2}+y\frac{\mu}{\sigma^2}} \end{aligned} \tag{2.5}$$

so that the normalizing constant is $a_0(\boldsymbol{\theta}) = \exp[-\mu^2/(2\sigma^2)]/\sqrt{2\pi\sigma^2}$, while $t_0(y) = 1$, $\boldsymbol{\theta} = (\mu/\sigma^2, -1/2\sigma^2)^{\mathrm{T}}$, and $\mathbf{t}(y) = (y, y^2)^{\mathrm{T}}$. □

Putting $\kappa(\boldsymbol{\theta}) = -\log[a_0(\boldsymbol{\theta})]$ and $t_0^*(y) = -\log[t_0(y)]$, we can note that

$$\log[f(y;\boldsymbol{\theta})] = \boldsymbol{\theta}^{\mathrm{T}}\mathbf{t}(y) - \kappa(\boldsymbol{\theta}) - t_0^*(y) \tag{2.6}$$

is additive in the functions and symmetric in $\boldsymbol{\theta}$ and $\mathbf{t}(y)$. We shall soon see that $\kappa(\boldsymbol{\theta})$ is an important function for understanding the properties of this family.

If we take the ratio of the densities of y for two members of the family with the same function, $\kappa(\cdot)$, but for two arbitrary values of $\boldsymbol{\theta}$, then

$$\frac{f(y;\boldsymbol{\theta}_1)}{f(y;\boldsymbol{\theta}_0)} = \frac{a_0(\boldsymbol{\theta}_1)}{a_0(\boldsymbol{\theta}_0)}\mathrm{e}^{[\boldsymbol{\theta}_1-\boldsymbol{\theta}_0]^{\mathrm{T}}\mathbf{t}(y)} \tag{2.7}$$

In Chapter 3, we shall call this a likelihood ratio. Only the information in $\mathbf{t}(y)$, and not that in $t_0(y)$, is needed to compare probability models in this way. Thus, the statistic, $\mathbf{t}(y)$, is said to be *sufficient* for $\boldsymbol{\theta}$ (Fisher, 1925), because it summarizes all of the available information in y about the parameter, given the functional form, $f(\cdot)$ (Section 3.4.1). If the elements of $\boldsymbol{\theta}$ are linearly independent, then $\mathbf{t}$ is called *minimal sufficient* and we say that the representation is (k, k), because both the parameter and the statistic have this dimension.

As we shall see in Chapter 3, such minimal sufficient statistics generally only exist for continuous distributions if we assume that responses can be measured with infinite precision, i.e. if we use the density, instead of the actual probability of observable data.

2.1.2 FUNDAMENTAL PARAMETRIZATIONS

Canonical parameters Obviously, members of the exponential family are not always, or even often, written in terms of the parameter $\boldsymbol{\theta}$. When they are, it is called the *natural* or *canonical parametrization* of the family indexed by that parameter and the statistic, $\mathbf{t}(y)$, is called the *natural* or *canonical statistic.* The family, itself, is said to be a *linear exponential family* if $\mathbf{t}(\mathbf{y}) = \mathbf{y}$, because of the term $\boldsymbol{\theta}^{\mathrm{T}}\mathbf{y}$ in the log density. It is minimal canonical if the representation is minimal.

Examples

1. The canonical representation of the Poisson distribution has

$$\theta = \log(\mu) \tag{2.8}$$

2. The canonical representation of the normal distribution has

$$\boldsymbol{\theta} = \left(\frac{\mu}{\sigma^2}, -\frac{1}{2\sigma^2}\right)^{\mathrm{T}} \tag{2.9}$$

□

Mean value parameters Now, let us define a new parameter vector which may be more familiar: $\boldsymbol{\mu}(\boldsymbol{\theta}) = \mathrm{E}[\mathbf{T}]$. This is called the *mean value parametrization.* Each element of this vector is a strictly increasing one-to-one function of the corresponding element of $\boldsymbol{\theta}$. Thus, all parameters in models from the exponential family have an interpretation as (functions of) expected values of observable quantities. This is not generally true with all parameters in other distributions.

The domain of the canonical parameter is convex, while, at least for continuous distributions, that of the mean value parameter is usually identical to the support of the family.

Examples

1. The mean value representation of the Poisson distribution has

$$\mu(\theta) = e^{\theta} \tag{2.10}$$

2. The mean value representation of the normal distribution has

$$\begin{aligned} \boldsymbol{\mu}(\boldsymbol{\theta}) &= \left(-\frac{\theta_1}{2\theta_2}, \frac{\theta_1^2 - 2\theta_2}{4\theta_2^2}\right)^{\mathrm{T}} \\ &= (\mu, \sigma^2 + \mu^2)^{\mathrm{T}} \end{aligned} \tag{2.11}$$

These are the expected values of $T_1 = Y$ and $T_2 = Y^2$, respectively. □

Mixed parameters A parametrization may also be *mixed* canonical and mean value, say $(\boldsymbol{\mu}_1, \boldsymbol{\theta}_2)$, with corresponding partition of the statistic, $(\mathbf{t}_1, \mathbf{t}_2)$. For a regular (see Section 2.1.5 below) family of probability distributions, the components of a mixed parameter representation, $(\boldsymbol{\mu}_1, \boldsymbol{\theta}_2)$, are *variation independent*, i.e. the parameter space is the Cartesian product, $M_1 \times \Theta_2$.

Example

A mixed representation for the normal distribution is $(\mu, \theta_2) = (\mu, -\frac{1}{2\sigma^2})$, so that μ and σ^2 are variation independent. One example is plotted in Figure 2.1. For any fixed positive value of the variance, the mean can take any value on the real line. □

2.1.3 GENERATING FUNCTIONS

From Equation (2.2), we know that $a_0^{-1}(\boldsymbol{\theta})$ is the normalizing constant for the probability density. We can see that this has a form similar to a Laplace transform. Then, for any $\boldsymbol{\theta} \in \boldsymbol{\Theta}$, it is easy to show that the moment generating function of the family is

$$M_T(\mathbf{u}) = \frac{a_0(\boldsymbol{\theta})}{a_0(\mathbf{u} + \boldsymbol{\theta})} \tag{2.12}$$

proof:

$$\begin{aligned} \mathrm{E}\left[e^{\mathbf{u}^{\mathrm{T}}\mathbf{T}}\right] &= \int e^{\mathbf{u}^{\mathrm{T}}\mathbf{t}} a_0(\boldsymbol{\theta}) t_0(y) e^{\boldsymbol{\theta}^{\mathrm{T}}\mathbf{t}} dt \\ &= a_0(\boldsymbol{\theta}) \int t_0(y) e^{(\mathbf{u}+\boldsymbol{\theta})^{\mathrm{T}}\mathbf{t}} dt \end{aligned}$$

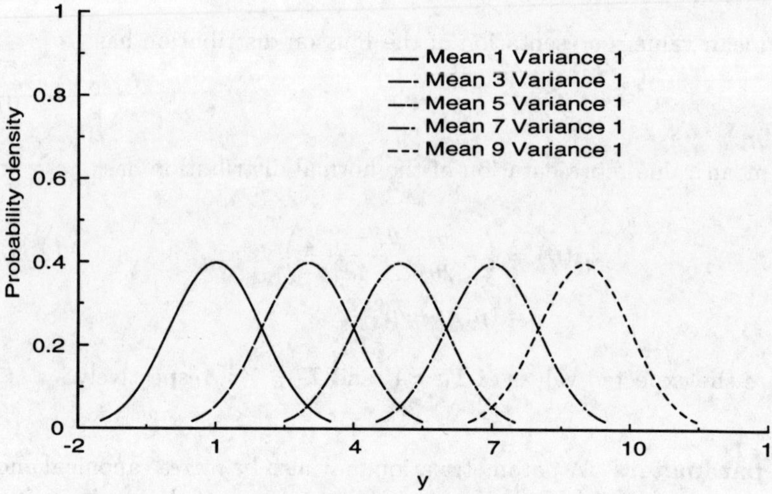

Fig. 2.1. Normal distributions with $\sigma^2 = 1$ and various means.

$$= \frac{a_0(\boldsymbol{\theta})}{a_0(\mathbf{u}+\boldsymbol{\theta})}$$

□

For values of $\boldsymbol{\theta}$ not on the boundary of $\boldsymbol{\Theta}$, the statistic, $\mathbf{t}$, will have moments of all orders.

The cumulants are obtained by taking derivatives of the cumulant generating function, i.e. of the logarithm of the moment generating function, at $u_i = 0$. However, because

$$\frac{\partial}{\partial u_i} \log[a_0(\mathbf{u}+\boldsymbol{\theta})]\bigg|_{u_i=0} = \frac{\partial}{\partial \theta_i} \log[a_0(\boldsymbol{\theta})] \tag{2.13}$$

the derivatives of $\kappa(\boldsymbol{\theta}) = -\log[a_0(\boldsymbol{\theta})]$ with respect to the canonical parameters are the cumulants of $\mathbf{t}$ under $f(y;\boldsymbol{\theta})$. We shall call this convex function the *cumulant function.* Thus, this important function is directly available as a term in the log density for this family. In particular, we have

$$\begin{aligned} \mathrm{E}_\theta[\mathbf{T}] &= \frac{\partial \kappa(\boldsymbol{\theta})}{\partial \boldsymbol{\theta}} &(2.14)\\ &= \boldsymbol{\mu}(\boldsymbol{\theta}) \\ \mathrm{var}_\theta[\mathbf{T}] &= \frac{\partial^2 \kappa(\boldsymbol{\theta})}{\partial \boldsymbol{\theta} \partial \boldsymbol{\theta}^{\mathrm{T}}} &(2.15)\\ &= \frac{\partial \boldsymbol{\mu}(\boldsymbol{\theta})}{\partial \boldsymbol{\theta}} \end{aligned}$$

These two derivatives of the cumulant function are sufficient completely to characterize any member of the linear exponential family (Tweedie, 1947).

Examples

1. For the Poisson distribution, recall that

$$\begin{aligned} a_0(\theta) &= \mathrm{e}^{-\mu} \\ &= \exp(-\mathrm{e}^{\theta}) \end{aligned} \tag{2.16}$$

so that

$$M_T(u) = \frac{\exp(-\mathrm{e}^{\theta})}{\exp(-\mathrm{e}^{\theta+u})} \tag{2.17}$$

with $T(Y) = Y$, and

$$\begin{aligned} \kappa(\theta) &= \mu \\ &= \mathrm{e}^{\theta} \end{aligned} \tag{2.18}$$

Then,

$$\begin{aligned} \mathrm{E}[Y] &= \mathrm{e}^{\theta} \\ &= \mu \end{aligned} \tag{2.19}$$

$$\begin{aligned} \mathrm{var}[Y] &= \mathrm{e}^{\theta} \\ &= \mu \end{aligned} \tag{2.20}$$

All cumulants are equal to the mean, an important characteristic of the Poisson distribution.

2. For the normal distribution, recall that

$$a_0(\boldsymbol{\theta}) = \frac{\exp(-\mu^2/2\sigma^2)}{\sqrt{2\pi\sigma^2}} \tag{2.21}$$

so that

$$\begin{aligned} \kappa(\boldsymbol{\theta}) &= \frac{\mu^2}{2\sigma^2} + \frac{1}{2}\log(2\pi\sigma^2) \\ &= -\frac{\theta_1^2}{4\theta_2} - \frac{1}{2}\log\left(-\frac{\theta_2}{\pi}\right) \end{aligned} \tag{2.22}$$

Then,

$$\begin{aligned} \mathrm{E}[T_1(Y)] &= \mathrm{E}[Y] \\ &= -\frac{\theta_1}{2\theta_2} \end{aligned}$$

$$= \mu \tag{2.23}$$

$$\text{var}[Y] = -\frac{1}{2\theta_2}$$
$$= \sigma^2 \tag{2.24}$$

as might be expected. The variance of Y does not depend on the mean and all cumulants for θ_1 after the second are zero.

As well,

$$\text{E}[T_2(Y)] = \text{E}[Y^2]$$
$$= \frac{\theta_1^2}{4\theta_2^2} - \frac{1}{2\theta_2}$$
$$= \mu^2 + \sigma^2 \tag{2.25}$$
$$\text{var}[Y^2] = -\frac{\theta_1^2}{2\theta_2^3} + \frac{1}{2\theta_2^2}$$
$$= 2\sigma^2(\sigma^2 + 2\mu^2) \tag{2.26}$$
$$\text{cov}[Y, Y^2] = \frac{\theta_1}{2\theta_2^2}$$
$$= 2\mu\sigma^2 \tag{2.27}$$

□

2.1.4 ONE-PARAMETER FAMILIES

Some important common members of the exponential family, such as the binomial and Poisson distributions, have order $k = 1$, with only a mean parameter. As members of the linear exponential family, $t(y)$ is the identity function. From Equation (2.15), the variance,

$$\tau^2 \stackrel{\text{def}}{=} \frac{\partial^2 \kappa(\theta)}{\partial \theta^2} \tag{2.28}$$
$$= \text{var}[Y]$$

will be a function of the mean, called the *variance function.*

Example

As we have already seen, for the Poisson distribution,

$$\text{var}[Y] = \tau^2$$
$$= \mu$$

We plot this distribution for various values of the mean in Figure 2.2. This

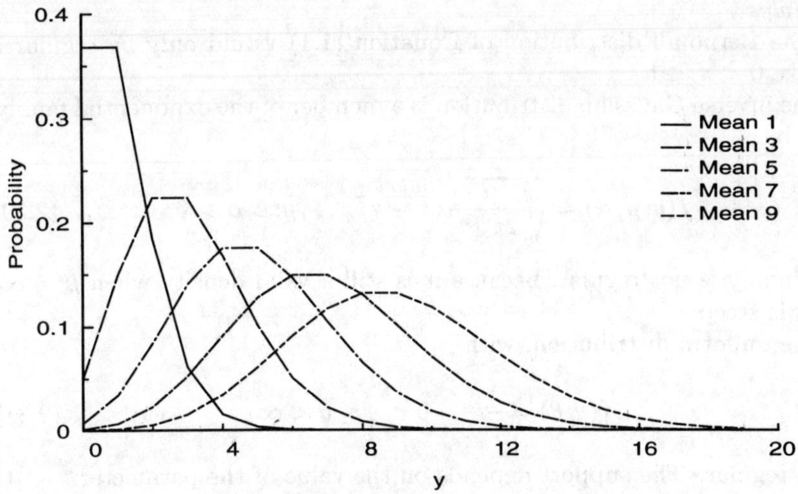

Fig. 2.2. Probability polygons for Poisson distributions with various means.

clearly shows how the variability increases with the mean. Contrast it with the normal distributions of Figure 2.1. □

2.1.5 REGULARITY CONDITIONS

Regularity conditions for a set of probability distributions, indexed by some parameter, concern discontinuities in the set, i.e. the possibility of integration and differentiation. In most common cases, they primarily involve the existence of values of that parameter lying on the boundary of the parameter space. Note that this is not necessarily a question of the values possibly being infinite. Any parameter with domain in **R**, i.e. $-\infty \le \psi \le \infty$, can be transformed to have only finite values, for example by using

$$\psi^* = \frac{e^{\psi}}{1 + e^{\psi}} \tag{2.29}$$

which has domain $[0, 1]$. The existence of discontinuities, such as at boundary values, and the behaviour of the density function at them, is only of real concern for inference in asymptotic statistical theory, discussed in Chapter 5.

That set of models with parameters encompassing the domain is called the *full* exponential family. If the set $\boldsymbol{\Theta}$ is open, the family will be *regular*. (A weaker condition on the boundary of the set is *steepness*, which is discussed in the appendix to this chapter, along with further details of all of these conditions.)

Examples

1. The Bernoulli distribution of Equation (1.1) would only be regular if we take $0 < \pi < 1$.

2. The inverse Gaussian distribution is a member of the exponential family, given by

$$f(y;\mu,\alpha) = \sqrt{\frac{\alpha}{2\pi y^3}}\mathrm{e}^{-\frac{\alpha(y-\mu)^2}{2\mu^2 y}} \qquad y,\mu,\alpha > 0 \tag{2.30}$$

This family is not regular, because it is still a valid density when $\mu = \infty$, but it is steep.

3. The uniform distribution, with

$$f(y;\theta) = \frac{1}{\theta} \qquad 0 < y \leq \theta < \infty \tag{2.31}$$

is not regular. The support depends on the value of the parameter. □

To summarize,

- a full exponential family has a parameter space consisting of all values for which the norming constant, $a_0^{-1}(\cdot)$, is finite, i.e. Equation (2.108);
- a steep exponential family satisfies Equation (2.110);
- a regular exponential family, in addition, is full with an open parameter space.

2.2 Distributions for random samples

Up until now, we have only looked at theory involving the probability of one observation, y, from the exponential family. In most practical situations, a sample of n observations will be available, so that we shall require results for this more general case.

2.2.1 PRODUCT FAMILIES

Suppose that the random variable is observed, independently, n times from a member of the exponential family, so that the Y_i are independent and identically distributed (i.i.d.). For the moment, we are not concerned about how the value, n, was determined.

The resulting product family

$$f(\mathbf{y};\boldsymbol{\theta}) = a_0^n(\boldsymbol{\theta}) \prod_{i=1}^{n} [t_0(y_i)]\, \mathrm{e}^{\mathbf{a}(\boldsymbol{\theta})^{\mathrm{T}} \Sigma \mathbf{t}(y_i)} \tag{2.32}$$

is also exponential of the same order and with the same minimal canonical parametrization as $f(y;\boldsymbol{\theta})$. The minimal sufficient canonical statistic is now $\sum \mathbf{t}(y_i)$, conditional on n being fixed and known. To simplify notation, we shall also call this $\mathbf{t}$: $\mathbf{t}(\mathbf{y})$.

Examples
1. The minimal sufficient canonical statistic for the Poisson distribution is $\sum y_i = y_\bullet$.
2. Now take n independent Poisson random variables, $Y_1, \ldots, Y_n$, having means $\mu_1, \ldots, \mu_n$ and $\theta_i = \log(\mu_i)$, an (n,n) model. Suppose that the linear function is $\theta_i(\boldsymbol{\beta}) = \beta_0 + \beta_1 x_i$ for known x_i, the explanatory variable. This is a two-parameter exponential family model, an example of a linear regression model, and a special case of a generalized linear model. It is still in the linear exponential family, now $(2,2)$, because the sufficient statistic is $(\sum y_i, \sum x_i y_i)$.
3. The minimal sufficient canonical statistic for the normal distribution is given by $(\sum y_i, \sum y_i^2)$. □

Generally, $n > k$, so that this implies that available information in $\mathbf{y}$, a vector of dimension n, about the parameter can be reduced to dimension k, that of the unknown parameter. Thus, no matter how many independent observations are available, all of this information can be reduced to a vector of fixed dimension, k. This is an important property in many aspects of statistical inference, one which, many statisticians believe, makes this family so important. Notice, once again, however, that such reduction generally only occurs for continuous distributions if infinite measurement precision is assumed.

The sufficient statistics can also be written $\bar{\mathbf{t}}(\mathbf{y}) = \sum \mathbf{t}(y_i)/n$, with canonical parameter, $\boldsymbol{\theta}_n = n\boldsymbol{\theta}$, so that the cumulant function becomes

$$\kappa_n(\boldsymbol{\theta}_n) = n\kappa(\boldsymbol{\theta}_n/n) \tag{2.33}$$

Note that it depends on n, which is assumed to be a fixed constant.

Example
For n i.i.d. observations from a Poisson distribution, $\theta_n = n\theta$, and

$$\begin{aligned}\kappa_n(\theta_n) &= n e^{\theta_n/n} \\ &= n\kappa(\theta_n/n)\end{aligned} \tag{2.34}$$

□

This latter parametrization is related to the mean value representation: μ is the expected value or theoretical mean, while $\bar{\mathbf{t}}(\mathbf{y})$ is the observed empirical mean, of the canonical statistic.

2.2.2 MARGINALITY

As we have just seen, the marginal density of $\mathbf{T}$, derived from a set of n i.i.d. observations, $\mathbf{y}$, is in the exponential family, as in Equation (2.32). However, the marginal density of a component of $\mathbf{T}$, say $\mathbf{t}_1$,

$$f(\mathbf{t}_1;\boldsymbol{\theta}) = a_0(\boldsymbol{\theta})t_0(\mathbf{t}_1,\boldsymbol{\theta}_2)\mathrm{e}^{\boldsymbol{\theta}_1^{\mathrm{T}}\mathbf{t}_1(\mathbf{y})} \tag{2.35}$$

is generally not in the family. Here, $t_0(\mathbf{t}_1,\boldsymbol{\theta}_2)$ can depend on $\boldsymbol{\theta}_2$ and $\mathbf{t}_1$ in a complicated way (the log density will generally not be a linear function). Thus, the marginal density of $\mathbf{T}_1$ depends on the full vector of parameters, $\boldsymbol{\theta}$. On the other hand, each of its subfamilies, obtained by fixing $\boldsymbol{\theta}_2$, is exponential.

Example
For n i.i.d. observations from a normal distribution, the marginal density of $t_1 = \sum y_i$ is also a normal distribution,

$$f(t_1;\mu,\sigma^2) = \frac{\mathrm{e}^{-\frac{(t_1-n\mu)^2}{2n\sigma^2}}}{\sqrt{2n\pi\sigma^2}} \tag{2.36}$$

so that

$$a_0(\boldsymbol{\theta}) = \frac{\mathrm{e}^{-\frac{n\mu^2}{2\sigma^2}}}{\sqrt{2n\pi\sigma^2}}$$

and

$$t_0(t_1,\theta_2) = \mathrm{e}^{-\frac{t_1^2}{2n\sigma^2}}$$

Because $t_0(t_1,\theta_2)$ has the exponential form, the marginal density is a member of the exponential family. Thus, it is a partial exception to the general rule, as is often the case for the normal distribution. □

2.2.3 CONDITIONALITY

The conditional density of $\mathbf{T}_2$, given $\mathbf{t}_1$, is

$$f(\mathbf{t}_2|\mathbf{t}_1;\boldsymbol{\theta}_2) = a_0(\boldsymbol{\theta}_2|\mathbf{t}_1)t_0(\mathbf{y})\mathrm{e}^{\boldsymbol{\theta}_2^{\mathrm{T}}\mathbf{t}_2(\mathbf{y})} \tag{2.37}$$

where $a_0^{-1}(\boldsymbol{\theta}_2|\mathbf{t}_1) = t_0(\mathbf{t}_1,\boldsymbol{\theta}_2)$, from Equation (2.35). It depends on $\boldsymbol{\theta}$ only through $\boldsymbol{\theta}_2$, so that the conditional density of $\mathbf{T}_2$ only involves the corresponding parameter vector, $\boldsymbol{\theta}_2$. Thus, the family of conditional densities is in the exponential family with $\mathbf{t}_2$ and $\boldsymbol{\theta}_2$ canonical. This result, that parameters can be eliminated by conditioning, will be important for certain aspects of inference in the following chapters.

Example
For n observations from a normal distribution, the conditional density, given $t_1 = \sum y_i$, is

$$f(y_1, \ldots, y_n | t_1; \sigma^2) = \frac{e^{-\frac{1}{2\sigma^2}\Sigma(y_i - t_1/n)^2}}{\sqrt{n}(2\pi\sigma^2)^{\frac{n-1}{2}}} \tag{2.38}$$

from which the *Chi-squared distribution* (Section C.1) may be obtained as the conditional distribution of $t_2 = \sum(y_i - t_1/n)^2$ given t_1. This is a special case of a gamma distribution and a member of the exponential family. Notice how the mean parameter, μ, has disappeared in this expression. □

2.3 Special cases

2.3.1 TRUNCATED DISTRIBUTIONS

A *truncated* distribution, with density $f^t(y; \boldsymbol{\theta})$, from the exponential family, has the domain of Y restricted to a subset, say $\mathcal{Y}^t$, of that usually considered. It will still be in the exponential family with

$$f^t(\mathbf{y}; \boldsymbol{\theta}) \stackrel{\text{def}}{=} a_0(\boldsymbol{\theta}; \mathbf{y}) t_0(\mathbf{y}) e^{\boldsymbol{\theta}^\mathrm{T}\mathbf{t}(\mathbf{y})} \tag{2.39}$$

where

$$a_0^{-1}(\boldsymbol{\theta}; \mathbf{y}) = \int_{\mathcal{Y}^t} e^{\boldsymbol{\theta}^\mathrm{T}\mathbf{t}(\mathbf{u})} d\mathbf{u} \tag{2.40}$$

The order of $f^t(\mathbf{y}; \boldsymbol{\theta})$ may be less than that of $f(\mathbf{y}; \boldsymbol{\theta})$ because the restriction of $\mathbf{t}$ to a subset may make the components dependent.

Example

Suppose that we have count data following a Poisson distribution, but that zero counts cannot be observed, so that $Y > 0$. Because

$$f(Y = 0; \mu) = e^{-\mu}$$

by renormalization, the truncated Poisson density is

$$f^t(y; \mu) = \frac{\mu^y e^{-\mu}}{(1 - e^{-\mu})y!} \tag{2.41}$$

which is still a member of the exponential family. □

When observed data are truncated, we do not know how many values exceed the truncation point (nor their actual values). This contrasts with *censored* data, where the number exceeding the censoring point is known and only their values are missing.

2.3.2 BIVARIATE DISTRIBUTIONS

A multivariate distribution is a joint probability distribution of two or more random variables. Such distributions are useful for describing the simultaneous variability of several random response variables, and thus the relationships among them. In general, the (univariate) marginal and conditional distributions of a multivariate distribution will not have the same form.

Suppose that we require a bivariate distribution, with both conditional distributions being members of the one-parameter exponential family. Then, our required density will be given by

$$\log[f(y_1, y_2; \theta_1, \theta_2, \rho)] \stackrel{\text{def}}{=} y_1\theta_1 + y_2\theta_2 + \rho y_1 y_2 \\ -t^*_{01}(y_1) - t^*_{02}(y_2) - \kappa(\theta_1, \theta_2, \rho) \qquad (2.42)$$

where, as usual, $\kappa(\theta_1, \theta_2, \rho)$ is a normalizing constant so that the probability integrates to one. The parameter ρ measures the dependence between the two random variables; when it is zero, they are independent (Arnold and Strauss, 1991). These bivariate distributions are members of the linear exponential family.

Example

A bivariate density with Poisson conditionals is

$$f(y_1, y_2; \mu_1, \mu_2, \rho) = \frac{\mu_1^{y_1} \mu_2^{y_2} \mathrm{e}^{\rho y_1 y_2}}{y_1! y_2! \mathrm{e}^{\kappa(\mu_1, \mu_2, \rho)}} \qquad (2.43)$$

The marginal density of, say, Y_1 can easily be obtained as

$$f(y_1; \mu_1, \mu_2, \rho) = \frac{\mu_1^{y_1} \mathrm{e}^{\mu_2 \mathrm{e}^{\rho y_1}}}{y_1! \mathrm{e}^{\kappa(\mu_1, \mu_2, \rho)}} \qquad (2.44)$$

by summation. This is not a Poisson distribution. Then, the conditional density, obtained by division, is

$$f(y_2|y_1; \mu_1, \mu_2, \rho) = \frac{\mu_2^{y_2} \mathrm{e}^{\rho y_1 y_2} \mathrm{e}^{-\mu_2 \mathrm{e}^{\rho y_1}}}{y_2!} \\ = \frac{[\mu_2 \mathrm{e}^{\rho y_1}]^{y_2} \mathrm{e}^{-\mu_2 \mathrm{e}^{\rho y_1}}}{y_2!} \qquad (2.45)$$

which is Poisson with mean $\mu_2 \mathrm{e}^{\rho y_1}$. □

Such bivariate densities can easily be generalized to the more complex multivariate case. Unfortunately, the normalizing constants for these densities usually have very complex forms, so that they have so far proved to be of limited practical usefulness (see, however, Lindsey, 1995b, pp. 132–135).

2.3.3 DOUBLE EXPONENTIAL FAMILY

The double exponential family is defined by

$$f(y; \theta, \psi) \stackrel{\text{def}}{=} c(\theta, \psi)\sqrt{\psi}[f(y; \theta)]^{\psi}[f(y; y)]^{1-\psi} \tag{2.46}$$

where $f(y; \theta)$ is some member of the one-parameter exponential family. It can easily be seen that the resulting density is a member of the two-parameter exponential family (Efron, 1986b).

Examples

1. The double Poisson density is given by

$$f(y; \mu, \psi) = c(\mu, \psi)\frac{\sqrt{\psi}}{\mathrm{e}^{\psi\mu}y!}\left(\frac{y}{\mathrm{e}}\right)^{y}\left(\frac{\mu\mathrm{e}}{y}\right)^{y\psi} \tag{2.47}$$

2. The double binomial density is

$$f(y; \pi, \psi) = c(\pi, \psi)\binom{n}{y}\frac{y^{y}(n-y)^{n-y}}{n^{n}}\frac{n^{n\psi}}{y^{y\psi}(n-y)^{(n-y)\psi}}\pi^{y\psi}(1-\pi)^{(n-y)\psi} \tag{2.48}$$

In neither case is $c(\cdot)$ analytically simple. □

These distributions can be useful when the strict relationship between the mean and the variance of a random variable, defined by the Poisson or binomial distribution, and given in Equation (2.28), does not hold. Such a case is known as under- or overdispersion (Section 1.2). Again, this family has been little used because the normalizing constant, $c(\theta, \psi)$, is generally very complex (see, however, Lindsey, 1995b, pp. 130–132).

2.3.4 CONDITIONAL EXPONENTIAL FAMILY

Suppose that $Y_1, \ldots, Y_n$ is a series of identically distributed random variables to be observed, as a stochastic process, over equally spaced discrete points in time, $t = 1, \ldots, n$. As we saw in Section 1.4, if Y_t, given y_{t-1}, is independent of y_k for all $k < t-1$, we say that the series is a *Markov process*.

The *conditional exponential family* of Markov processes is given by

$$f(y_t|y_{t-1}; \boldsymbol{\theta}) \stackrel{\text{def}}{=} a_0(\boldsymbol{\theta}, y_{t-1})t_0(y_t, y_{t-1})\mathrm{e}^{\boldsymbol{\theta}^{\mathrm{T}}\mathbf{t}(y_t, y_{t-1})} \tag{2.49}$$

If, as well, $\log[a_0(\boldsymbol{\theta}, y_{t-1})]$ factors,

$$\log[a_0(\boldsymbol{\theta}, y_{t-1})] = c(\boldsymbol{\theta})d(y_{t-1}) \tag{2.50}$$

it is called a conditionally additive exponential family, because the cumulant has the form of an additive process (Feigin, 1981). Because this is a member of the exponential family, we have

$$\mathrm{E}[\mathbf{T}(Y_t, Y_{t-1})|y_1, y_2, \ldots, y_{t-2}] = \frac{\partial\{-\log[a_0(\boldsymbol{\theta}, y_{t-1})]\}}{\partial\boldsymbol{\theta}} \tag{2.51}$$

Example

Suppose that the conditional distribution of Y_t, given y_{t-1}, is gamma, from Equation (1.3), with parameters $\rho = \mu/\alpha$ and $\alpha = y_{t-1}$, the latter known when Y_t is to be observed. Then,

$$f(y_t|y_{t-1};\rho) = \frac{y_t^{y_{t-1}-1}\mathrm{e}^{-\frac{y_t}{\rho}}}{\rho^{y_{t-1}}\Gamma(y_{t-1})} \tag{2.52}$$

so that

$$\mathrm{E}[Y_t|y_{t-1}] = \rho y_{t-1} \tag{2.53}$$

a type of autoregression model. □

2.3.5 CURVED EXPONENTIAL FAMILY

For the *full* linear exponential family of order k, an r-dimensional submodel of this ($r < k$) is obtained when the vector, $\boldsymbol{\theta}$, is a function of an r-dimensional parameter, $\boldsymbol{\beta}$ as $\boldsymbol{\theta}(\boldsymbol{\beta})$, an r-dimensional differentiable manifold. If the function is non-linear in the new parameters, so that the sufficient statistics still have dimension k, this is called a *curved exponential family* (Efron, 1975), a (k, r) model, because it has a k-dimensional sufficient statistic, but an r-dimensional parameter vector. It is embedded in the (k, k) family.

Examples

1. Consider a normal distribution with mean μ and variance $c\mu^2$, where $c > 0$ is a known constant. This is a $(2, 1)$ model.
2. Let Y_1 and Y_2 have normal distributions with means μ and $c\mu^2$ and unit variance, where $c \neq 0$ is a known constant. Again, this is a $(2, 1)$ model.
3. In Section 2.2.1, we saw that a linear Poisson regression, with $\theta_i(\boldsymbol{\beta}) = \beta_0 + \beta_1 x_i$, was a (2,2) model. Now, instead, suppose that the function is $\theta_i(\boldsymbol{\beta}) = \beta_0 \mathrm{e}^{\beta_1 x_i}$, again for known x_i. Although no longer a linear exponential family model, it is a member of the $(n, 2)$ curved exponential family. This is a *non-linear regression model.* □

We, thus, see that, for this family, the minimal sufficient statistic will be of larger dimension than the parameter vector and may provide no reduction in dimension at all with respect to the observed data.

2.4 Parametrizations

In Section 2.1.2, we studied the two fundamental parametrizations for the exponential family, the canonical and mean value. We now need to look a little more closely at ways in which the parameters of a family of distributions can be specified.

The primary role of a parameter is to index a family of statistical models. We have also seen that parameters for the exponential family have an interpretation in terms of (functions of) expected values of observable quantities. In some situations, they can also be viewed as limits of sequences of observable quantities (see Chapter 5), although this is a mathematical attribute, not statistical.

2.4.1 TYPES OF PARAMETERS

Any one-to-one transformation, $\xi(\mu)$, of the mean parameter, $\mu = \mathrm{E}[Y]$, will be called a *location parameter.* It specifies the position of the distribution with respect to the support. An alternative definition of a measure of location is any parameter such that adding a constant to the random variable changes the parameter by the same amount. In a certain sense, this is considerably less general, because it only allows $\xi(\mu)$ to be a linear function of the mean. At the same time, it is less restrictive in allowing other measures of location than the mean.

Any parameter, ψ_2, such that $\theta_1 = \xi(\mu)/s(\psi_2)$, where $\xi(\mu)$ is a location parameter, and $s(\cdot)$ is an arbitrary function, will be called a *dispersion parameter.* It describes the variability of the random variable or the spread of the distribution. Here, an alternative definition of spread is any parameter such that a linear transformation of the random variable, $a + bY$, changes the parameter by a factor $|b|^c$ for some constant, c. This is a measure of *scale* when $c = 1$. Again, in a sense, this second definition is less general.

Other parameters are generally known as *shape parameters.*

Example
For the normal distribution, the mean, μ, is a location parameter, the variance, σ^2, a dispersion parameter, and the standard deviation, σ, a scale parameter. □

2.4.2 PARAMETER TRANSFORMATIONS

In specific contexts, special parametrizations are important, either as an aid in interpretation or in making inferences. Thus, we might require a transformation of a parameter which makes the distribution of a statistic look as much like a normal distribution as possible, or which makes its variance reasonably constant.

For the one-parameter linear exponential family, a useful family of transformations of θ, indexed by δ, is defined by

$$g(\theta,\delta) \stackrel{\text{def}}{=} \int_{\theta_0}^{\theta} \left[\frac{\partial^2 \kappa(u)}{\partial u^2} \right]^{\delta} du \tag{2.54}$$

where κ is the cumulant function. Then,

$\delta = 0$: canonical parametrization,
$\delta = \frac{1}{3}$: log density approximately quadratic in the parameter,
$\delta = \frac{1}{2}$: stable variance,
$\delta = 1$: mean value parametrization.

This may be generalized to the one-parameter curved exponential family, with parameter, β. Then, the differential equation,

$$\frac{\frac{\partial^2 g}{\partial \beta^2}}{\frac{\partial g}{\partial \beta}} = \frac{\delta \frac{\partial^3 \kappa}{\partial \theta^3} \frac{d\theta}{d\beta} + \left(\frac{d^2 \theta}{d\beta^2} \right)^{\mathrm{T}} \frac{\partial^2 \kappa}{\partial \theta^2} \frac{d\theta}{d\beta}}{\left(\frac{d\theta}{d\beta} \right)^{\mathrm{T}} \frac{\partial^2 \kappa}{\partial \theta^2} \frac{d\theta}{d\beta}} \tag{2.55}$$

must be solved (Hougaard, 1982).

Examples

1. For the Poisson distribution, with

$$f(y;\theta) = \frac{\mathrm{e}^{-\mathrm{e}^{\theta}}}{y!} \mathrm{e}^{y\theta}$$

with $\kappa(\theta) = \mathrm{e}^{\theta}$ so that

$$g(\theta,\delta) = \begin{cases} \mathrm{e}^{\theta\delta} & \text{when } \delta \neq 0 \\ \theta & \text{when } \delta = 0 \end{cases}$$

The variance stabilizing transformation is $\mathrm{e}^{\theta/2} = \sqrt{\mu}$.

2. For the normal distribution with unit variance, $\kappa(\theta) = \frac{1}{2}\theta^2$ so that all choices of δ give the same result. □

Scale parameters can be transformed, by taking logarithms, so that they have some of the characteristics of a location parameter. Thus, if Y/σ is such that σ is a scale parameter, then $\log(\sigma)$ is a location parameter for $\log(Y)$.

We also saw in Section 2.1.5 that any parameter with domain containing infinite values can always be transformed to one with only finite values.

2.5 Related families

A number of important probability distributions would be members of the exponential family except for one parameter.

2.5.1 EXPONENTIAL DISPERSION FAMILY

Dispersion family A multivariate *dispersion family* is defined by

$$\log[f(\mathbf{y};\boldsymbol{\mu},\sigma^2)] \stackrel{\text{def}}{=} \frac{t(\mathbf{y},\boldsymbol{\mu})}{\sigma^2} - a_0^*(\sigma^2, y) \tag{2.56}$$

with $(\boldsymbol{\mu},\sigma^2) \in M \times \Sigma$, where σ^2 is called the *dispersion parameter* and its reciprocal, the *precision parameter* (Jørgensen, 1987a).

Dispersion models are generally characterized by the clear separation between a location parameter, $\boldsymbol{\mu}$, and the dispersion parameter, σ^2. As we shall see subsequently in examples, this will be useful in regression-type situations, where $\boldsymbol{\mu}$ is assumed to change under various observed conditions while σ^2 is assumed to remain constant.

The family is a *proper dispersion family* if the normalizing constant, a_0^*, can be written as a sum of two terms, one containing only the dispersion parameter, and the other only the random variable:

$$a_0^*(\sigma^2,\mathbf{y}) = a^\ddagger(\sigma^2) + b(\mathbf{y}) \tag{2.57}$$

In this case $\int \exp[t(\mathbf{y},\boldsymbol{\mu})/\sigma^2 - b(\mathbf{y})]d\mathbf{y}$ does not depend on $\boldsymbol{\mu}$ and, for $\boldsymbol{\mu}$ known, the family is a one-parameter linear exponential family with $\theta = 1/\sigma^2$ and canonical statistic, $t(\mathbf{y},\boldsymbol{\mu})$. Thus, the quantity, $T(\mathbf{y},\boldsymbol{\mu})$, has the special property that, for fixed σ^2, its distribution does not depend on $\boldsymbol{\mu}$.

Exponential dispersion family When

$$\log[f(\mathbf{y};\boldsymbol{\xi},\sigma^2)] \stackrel{\text{def}}{=} \frac{\mathbf{y}^{\mathrm{T}}\boldsymbol{\xi} - \kappa^*(\boldsymbol{\xi})}{\sigma^2} - a_0^*(\sigma^2,\mathbf{y}) \tag{2.58}$$

where $\boldsymbol{\xi}$ is a function of $\boldsymbol{\mu}$, we have a special dispersion model called the *exponential dispersion family.* Often $a_0^*(\sigma^2,\mathbf{y})$ is additive so that we have a proper exponential dispersion family (Tweedie, 1947; Jørgensen, 1986, 1987a, 1992). If the mean value parametrization is used, it is indexed by $\boldsymbol{\mu}$ instead of $\boldsymbol{\xi}(\boldsymbol{\mu})$.

The variance of $\mathbf{Y}$ is given by

$$\begin{aligned}\text{var}[\mathbf{Y}] &= \sigma^2\boldsymbol{\tau}^2 \\ &= \sigma^2\frac{\partial^2\kappa^*(\boldsymbol{\xi})}{\partial\boldsymbol{\xi}\partial\boldsymbol{\xi}^{\mathrm{T}}}\end{aligned} \tag{2.59}$$

a product of the dispersion parameter and $\boldsymbol{\tau}^2$ (with slight abuse of notation), a matrix of elements defined as in Equation (2.28), a function only of the mean. We may write this as $\mathbf{Y} \sim \text{ED}(\boldsymbol{\mu}, \sigma^2\boldsymbol{\tau}^2)$. Here, $\boldsymbol{\mu}$ and σ^2 are variation independent.

Sets of independent random variables, $\mathbf{Y}_1, \ldots, \mathbf{Y}_n$, with exponential dispersion distributions, $\text{ED}(\boldsymbol{\mu}, \boldsymbol{\tau}^2\sigma^2/w_i)$, have a *convolution property*:

$$\frac{\sum_{i=1}^n w_i \mathbf{Y}_i}{\sum_{i=1}^n w_i} \sim \text{ED}\left(\boldsymbol{\mu}, \frac{\sigma^2}{\sum_{i=1}^n w_i}\boldsymbol{\tau}^2\right) \tag{2.60}$$

In particular, if they are identically distributed,

$$\overline{\mathbf{Y}}_{\bullet} \sim \text{ED}\left(\boldsymbol{\mu}, \frac{\sigma^2}{n}\boldsymbol{\tau}^2\right) \tag{2.61}$$

One special case of an exponential dispersion family is a generalization of the bivariate distribution given in Section 2.3.2:

$$\begin{aligned}\log[f(y_1, y_2; \xi_1, \xi_2, \rho)] = {} & \frac{y_1\xi_1 + y_2\xi_2 + \rho y_1 y_2}{\sigma^2} - \kappa(\xi_1, \xi_2, \rho) \\ & - a_0^*(y_1, y_2, \sigma^2)\end{aligned} \tag{2.62}$$

Example

The multivariate Poisson distribution of Equation (2.43) is a multivariate exponential dispersion model with $\sigma^2 = 1$. □

Univariate case If y is a scalar in Equation (2.58), we have an important special case, the univariate (one-dimensional) exponential dispersion family, upon which generalized linear models (Nelder and Wedderburn, 1972) have been constructed. This may be written

$$\log[f(y; \xi, \sigma^2)] \stackrel{\text{def}}{=} \frac{y\xi - \kappa^*(\xi)}{\sigma^2} - a_0^*(\sigma^2, y) \tag{2.63}$$

where σ^2 is the dispersion parameter, while $\xi(\mu)$ is the canonical location parameter for this representation. Because, for fixed value of σ^2, we have a one-parameter linear exponential family, that family belongs to the exponential dispersion family with $\sigma^2 = 1$.

Example

The normal distribution has

$$\begin{aligned}\xi &= \mu \\ \kappa^*(\xi) &= \mu^2/2\end{aligned}$$

$$a_0^*(\sigma^2, y) = \frac{1}{2}\log(2\pi\sigma^2) + \frac{y^2}{2\sigma^2}$$

□

The canonical linear regression model is then

$$\xi_i = \boldsymbol{\beta}^{\mathrm{T}}\mathbf{x}_i \tag{2.64}$$

while the mean value linear regression model is

$$\mu_i = \boldsymbol{\beta}^{\mathrm{T}}\mathbf{x}_i \tag{2.65}$$

where $\mathbf{x}_i$ is a vector of explanatory variables, a row of the design matrix, $\mathbf{X}$, and $\xi(\mu)$ is the canonical link function. In a certain sense, the first is simpler because it retains the linearity of Equation (2.63) so that, for example, the sufficient statistics for $\boldsymbol{\beta}$ are $\{\mathbf{y}^{\mathrm{T}}\mathbf{x}_j\}$, where $\mathbf{x}_j$ is a column of the design matrix. However, other transformations of μ may make more sense in modelling a given problem. Note that the question of the form of this function of the mean only arises when the systematic component is linear in the parameters; see Equation (1.7).

The cumulants can be obtained from

$$\kappa(\theta) = \frac{\kappa^*(\xi)}{\sigma^2} \tag{2.66}$$

with $\theta = \xi/\sigma^2$. The second derivative of κ^* will be a function only of the mean; it is called the *variance function*,

$$\tau^2 \stackrel{\text{def}}{=} \frac{\partial^2\kappa^*(\xi)}{\partial\xi^2} \tag{2.67}$$

Then, the variance of Y is given by

$$\mathrm{var}[Y] = \tau^2\sigma^2 \tag{2.68}$$

Notice how this is a product of the variance function and the dispersion parameter, with only the former depending on the mean.

Example
For the normal distribution, $\tau^2 = 1$. As we already saw above, μ and σ^2 are variation independent; the variance is not a function of the mean. □

Within the univariate exponential dispersion family, distributions are uniquely defined by their variance functions. An important special case occurs when the variance function is a known power of the mean:

$$\tau^2 = \mu^c \tag{2.69}$$

Example

The family of univariate exponential dispersion models with power variance functions includes the normal ($c = 0$), Poisson ($c = 1$), gamma ($c = 2$), and inverse Gaussian ($c = 3$) distributions, as well as the stable distributions ($c \leq 0$ and $c > 2$; see Section 2.7.2 below) and some compound Poisson distributions ($1 < c < 2$).

These distributions are closed under scale transformations. If $Y \sim \text{ED}(\mu, \sigma^2\tau^2)$, then $cY \sim \text{ED}(c\mu, c^{2-c}\sigma^2\tau^2)$ for any positive constant, $c \in \mathbf{R}^+$. □

Several important common members of the exponential family, such as the normal, gamma, and inverse Gaussian distributions, have order $k = 2$, with a mean value or location and a scale or dispersion parameter. These are also exponential dispersion models. More generally, if the only additive term in $\kappa(\boldsymbol{\theta}) = -\log[a_0(\boldsymbol{\theta})]$ containing the location parameter can be factored, as $\kappa^*(\xi)/\sigma^2$, then the two-parameter exponential family is an exponential dispersion model, where ξ is some function of μ_1 such that $\theta_1 = \xi/\sigma^2$ and $-\theta_2 t_2(y) + \kappa(\boldsymbol{\theta}) + t_0^*(y) = a_0^*(\sigma^2, y) + \kappa^*(\xi)/\sigma^2$. The only three densities which have this form are the normal, gamma, and inverse Gaussian (Blaesild and Jensen, 1985).

2.5.2 EXTRA PARAMETERS

When a parameter is added to a model, the old model is said to be embedded in the new one. One common case occurs when a power transformation is applied to the random variable:

$$Y \rightarrow Y^\alpha \tag{2.70}$$

or, for better limit properties at $\alpha = 0$,

$$Y \rightarrow \frac{Y^\alpha - 1}{\alpha} \tag{2.71}$$

The density must be adjusted by the Jacobian.

Examples

1. For the normal distribution, this is called the Box–Cox transformation:

$$f(y; \mu, \sigma^2, \alpha) = \frac{\alpha y^{\alpha-1}}{\sqrt{2\pi\sigma^2}} \mathrm{e}^{-\frac{(y^\alpha - \mu)^2}{2\sigma^2}} \tag{2.72}$$

2. For the exponential distribution of Equation (1.14), with

$$f(y; \mu) = \frac{\mathrm{e}^{-\frac{y}{\mu}}}{\mu} \qquad y, \mu > 0$$

we obtain the Weibull distribution of Equation (1.4),

$$f(y;\mu,\alpha) = \frac{\alpha y^{\alpha-1} e^{-\frac{y^\alpha}{\mu}}}{\mu} \qquad y,\mu,\alpha > 0$$

□

Embedding is often useful for checking the adequacy of a model (Section 4.1.1).

2.6 Conjugate distributions

2.6.1 EXPONENTIAL FAMILY

Recall, for an exponential family in canonical form, from Equation (2.6), that

$$\log[f(y;\boldsymbol{\theta})] = \boldsymbol{\theta}^{\mathrm{T}}\mathbf{t}(y) - \kappa(\boldsymbol{\theta}) - t_0^*(y)$$

Let us now suppose that the parameter $\boldsymbol{\theta}$ is itself a random variable. We can imagine at least two ways in which this might occur. The population for which we wish to construct a model may be heterogeneous. In our model building process, some parameter, such as a mean, might be assumed to vary among subgroups in an unknown way not of direct interest (Section 6.4.1). Or, we may have some model for human behaviour which gives relative weights representing knowledge about the possible 'true' value of $\boldsymbol{\theta}$ before making observations and which could be described by a probability distribution (Chapter 8).

Now let us consider Bayes' formula, from Equation (B.38):

$$p(\boldsymbol{\theta}|y) = \frac{f(y|\boldsymbol{\theta})p(\boldsymbol{\theta})}{f(y)} \qquad (2.73)$$

Here, $p(\boldsymbol{\theta})$ is called the *prior density* for $\boldsymbol{\theta}$, because it is applicable before y is observed, and $p(\boldsymbol{\theta}|y)$ its *posterior density*, conditional on observing y. (Densities of observable quantities will be denoted, as usual, by $f(\cdot)$, and those for parameters by $p(\cdot)$.)

If $p(\boldsymbol{\theta}|y)$ and $p(\boldsymbol{\theta})$ have the same form, $p(\cdot)$ is the *conjugate prior* for $f(y;\boldsymbol{\theta})$. It is said to be *closed under sampling*, because the density has the same form before and after observing the sample data (Barnard, 1954; Wetherill, 1961). This implies that the marginal density of Y, called a *compound distribution*,

$$f(y) = \int_{\Theta} f(y|\boldsymbol{\theta})p(\boldsymbol{\theta})d\boldsymbol{\theta} \qquad (2.74)$$

must have an analytically tractable form.

In order for the prior and posterior to have the same form, within the exponential family, it is easy to demonstrate that the conjugate prior must have the general form,

$$\log[p(\boldsymbol{\theta};\boldsymbol{\zeta},\gamma)] = \boldsymbol{\theta}^{\mathrm{T}}\boldsymbol{\zeta} - \gamma\kappa(\boldsymbol{\theta}) + s(\boldsymbol{\zeta},\gamma) \tag{2.75}$$

where $\boldsymbol{\zeta}$ and γ are new parameters, with $\gamma > 0$ and $s(\boldsymbol{\zeta},\gamma)$ is not a function of $\boldsymbol{\theta}$. For a given k-order exponential family, the conjugate prior is an exponential family of order $k+1$, with canonical statistics, $\boldsymbol{\theta}$ and $\kappa(\boldsymbol{\theta})$. Thus, for one observation, $\boldsymbol{\zeta}$ and γ in the prior become $\boldsymbol{\zeta}+\mathbf{t}(\mathbf{y})$ and $\gamma+1$ in the posterior distribution. Then, the marginal compound density is given by

$$\log[f(y;\boldsymbol{\zeta},\gamma)] = s(\boldsymbol{\zeta},\gamma) - t_0^*(\mathbf{y}) - s[\boldsymbol{\zeta}+\mathbf{t}(y),\gamma+1] \tag{2.76}$$

Although the prior and posterior for $\boldsymbol{\theta}$ are members of the exponential family, the compound distribution for Y generally will not be.

Example
The Poisson density, with $\xi = \log(\mu)$,

$$f(y|\mu) = \exp[y\log(\mu) - \mu - \log(y!)] \tag{2.77}$$

is a member of both the linear exponential and exponential dispersion families. The conjugate prior density for μ is

$$p(\mu;\zeta,\gamma) = \exp[\zeta\log(\mu) - \mu\gamma + s(\zeta,\gamma)] \tag{2.78}$$

which is a gamma distribution of Equation (1.3), with

$$s(\zeta,\gamma) = (\zeta+1)\log(\gamma) - \log[\Gamma(\zeta+1)] \tag{2.79}$$

This pair of distributions might be interpreted, for example, as applying to a heterogeneous population where events are generated according to a Poisson distribution, but where not every individual has the same value of μ. Then, with this model, the values of μ would be distributed in the population according to a gamma distribution.

The resulting marginal compound density is

$$\begin{aligned} f(y;\zeta,\gamma) &= \exp[s(\zeta,\gamma) - \log(y!) - s(\zeta+y,\gamma+1)] \\ &= \frac{\Gamma(\zeta+y+1)\gamma^{\zeta+1}}{\Gamma(\zeta+1)y!(\gamma+1)^{\zeta+y+1}} \end{aligned} \tag{2.80}$$

which is a negative binomial distribution. It is commonly used as a model when the equality relationship between the mean and variance of the Poisson distribution does not hold, so that there is over- or underdispersion of the random variable (Section 1.2).

For n observations, the posterior density will be

$$p(\mu|\mathbf{y};\zeta,\gamma) = \exp\{(\zeta + y_\bullet)\log(\mu) - \mu(\gamma+n) \\ +(\zeta + y_\bullet + 1)\log(\gamma+n) - \log[\Gamma(\zeta + y_\bullet + 1)]\} \quad (2.81)$$

again a gamma distribution. □

Other examples will be given in Section 8.3.5.

These results can be directly extended to finite mixtures of K conjugate densities

$$p_K(\boldsymbol{\theta};\boldsymbol{\pi},\boldsymbol{\zeta},\boldsymbol{\gamma}) = \sum_{k=1}^{K} \pi_k p(\boldsymbol{\theta};\zeta_k,\gamma_k) \quad (2.82)$$

with $\sum \pi_k = 1$, $\pi_k > 0\ \forall k$, which are also closed under sampling (Dalal and Hall, 1983).

2.6.2 EXPONENTIAL DISPERSION FAMILY

Consider now the univariate exponential dispersion family of Equation (2.63):

$$\log[f(y;\xi,\sigma^2)] = \frac{y\xi - \kappa^*(\xi)}{\sigma^2} - a_0^*(\sigma^2, y)$$

Because, for fixed σ^2, this is a member of the exponential family, a conjugate prior, for ξ alone, can be written in the general form

$$\log[p(\xi;\zeta,\gamma)] = \xi\zeta - \gamma\kappa^*(\xi) + s(\zeta,\gamma) \quad (2.83)$$

where again $s(\zeta,\gamma)$ is not a function of ξ. Then, the marginal density is given by

$$\log[f(y;\zeta,\gamma,\sigma^2)] = s(\zeta,\gamma) - a_0^*(\sigma^2,y) - s\left(\zeta + \frac{y}{\sigma^2}, \gamma + \frac{1}{\sigma^2}\right) \quad (2.84)$$

which will generally not be an exponential dispersion model.

Example

Because the normal density, with $\xi = \mu$, is given by

$$f(y|\mu;\sigma^2) = \exp\left[\frac{y\mu - \frac{\mu^2}{2}}{\sigma^2} - \frac{y^2}{2\sigma^2} - \frac{1}{2}\log(2\pi\sigma^2)\right] \quad (2.85)$$

the conjugate prior density for μ is

$$p(\mu;\zeta,\gamma) = \exp\left[\zeta\mu - \frac{\gamma\mu^2}{2} + s(\zeta,\gamma)\right] \tag{2.86}$$

which is, itself, a normal distribution, with mean ζ/γ, variance $1/\gamma$, and

$$s(\zeta,\gamma) = -\frac{\zeta^2}{2\gamma} - \frac{\log(2\pi/\gamma)}{2} \tag{2.87}$$

Thus, in a heterogeneous population, under this model, the mean, μ, could have a normal distribution over its members.

The resulting marginal compound density is also normal

$$\begin{aligned} f(y;\zeta,\gamma,\sigma^2) &= \exp\left\{ s[\zeta,\gamma] - \frac{y^2}{2\sigma^2} - \frac{1}{2}\log(2\pi\sigma^2) \right. \\ &\qquad \left. -s\left[\zeta + \frac{y}{\sigma^2}, \gamma + \frac{1}{\sigma^2}\right]\right\} \\ &= \exp\left\{ \frac{\zeta y}{\gamma\sigma^2+1} - \frac{\zeta^2}{2\gamma(\gamma\sigma^2+1)} \right. \\ &\qquad \left. - \frac{\gamma y^2}{2(\gamma\sigma^2+1)} - \frac{\log[2\pi(\gamma\sigma^2+1)/\gamma]}{2} \right\} \end{aligned} \tag{2.88}$$

with mean ζ/γ and variance $\sigma^2 + 1/\gamma$. For n observations, this becomes a multivariate normal distribution with mean and variance as given, and constant covariance $1/\gamma$. Models based on the idea of letting a parameter have a random distribution are called random effects models (Section 6.4.1).

For n observations, the posterior density will be

$$\begin{aligned} p(\mu|\mathbf{y};\zeta,\gamma,\sigma^2) = \exp&\left\{ \left(\zeta + \frac{y_\bullet}{\sigma^2}\right)\mu - \frac{(\gamma + n/\sigma^2)\mu^2}{2} \right. \\ &\left. - \frac{(\zeta + y_\bullet/\sigma^2)^2}{2(\gamma+n/\sigma^2)} - \frac{\log[2\pi/(\gamma+n/\sigma^2)]}{2} \right\} \end{aligned} \tag{2.89}$$

a normal distribution, as expected. □

2.7 Other families

The exponential family may be thought of as one natural generalization of the normal distribution. Two other important extensions will now be studied more briefly. The transformation family is primarily useful in production and measurement error models, while stable distributions often find their place when extreme responses are generated, in robust inference (Section 4.1.2) and in stochastic processes.

2.7.1 TRANSFORMATION FAMILIES

In many situations, the response variable studied is some kind of measurement. In certain cases, especially in the natural sciences, deterministic phenomena are of interest, such as chemical reactions or physical processes, which should always give the same results under the same conditions. A model may take the form of some deterministic response equation, such as

$$y = \alpha e^{\beta x}$$

Nevertheless, variability can arise either from small differences in the results given by the measuring instrument, aptly called measurement error, or from small differences among objects observed, owing to production errors. The former may be observational or experimental error. Thus, probability distributions describing such phenomena are often called error distributions. This variability, say ε, in response can be included in the model as, for example,

$$y = \alpha e^{\beta x} \varepsilon$$

if the random error is multiplicative, where $f(\varepsilon)$ will be the density of the error distribution. It does not depend on the parameter values of the response equation (although it may contain a separate dispersion parameter), so that ε is a *pivotal quantity* (Section 3.5.4) for the former parameters.

When the data generating mechanism is stochastic, the parameters of the probability distribution determine its location and shape, but have relatively minor inherent interest in themselves, simply indexing possible models. In contrast, we have, here, a situation where parameters are of primary interest, without regard to the probability distribution. This occurs because the data generating mechanism is deterministic and the random variation is only a nuisance factor. It can usually be modelled by some simple symmetric distribution. Thus, one will be wanting to calculate the parameter value and an estimate of the precision with which it was obtained.

Certain classes of models are particularly suitable in such contexts. A transformation family is defined by a group of transformations on the sample space generating distributions of the same form but with different parameter values (Fraser, 1968, Ch. 1). The most important case, the *location–scale family* of distributions, introduced by Fisher (1934) at the same time as the exponential family, is defined by

$$f(y;\mu,\sigma) = \frac{1}{\sigma} f\left(\frac{y-\mu}{\sigma};0,1\right) \tag{2.90}$$

with $f(\cdot)$ a known density function. In this family, the two parameters correspond to changes in location and scale of the measurements (Section

2.4.1). Thus, a pivotal quantity is

$$\varepsilon = \frac{y-\mu}{\sigma}$$

whose distribution does not depend on either the parameters or the random variable. It corresponds to a simple production and measurement error model,

$$y = \mu + \sigma\varepsilon$$

Here, μ gives the parameter value of interest and σ a measure of its precision. The most common member of this family is the normal distribution, so that there is overlap between this and the exponential family.

For the location–scale family, the group of transformations corresponds to change of origin and scale:

$$\begin{aligned} Y_i &\to aY_i + b \\ \mu &\to a\mu + b \qquad (2.91) \\ \sigma &\to a\sigma \end{aligned}$$

with $a > 0$ and b arbitrary real numbers. This family of transformations partitions both the sample space and the parameter space into a collection of orbits. Thus, for example,

- each point in the sample space is on only one orbit;
- any point on an orbit can be transformed onto any other point on the same orbit by choice of a and b;
- points on different orbits cannot be so transformed to each other.

If there is only one transformation taking one point into another, then y can be represented as $(\mathbf{u}, \mathbf{v})$ where $\mathbf{u}$ specifies the orbit on which y is found and $\mathbf{v}$ gives the position on the orbit. The distribution of $\mathbf{u}$ does not depend on the parameters; only the conditional distribution of the observations given $\mathbf{u}$ does. The statistic, $\mathbf{u}$, is said to be *ancillary* (Fisher, 1925; see Section 3.4.2).

In many situations, the response is related in a deterministic manner to certain other variables, to which the error variability is added. This can be written as the *Gauss linear family*, defined by

$$Y_i = \boldsymbol{\beta}^{\mathrm{T}}\mathbf{x}_i + \sigma\varepsilon_i \qquad (2.92)$$

where $\mathbf{x}_i$ is a vector of these explanatory variables which, when combined with the parameter vector, defines the location, μ_i. Again, ε_i is the random error, scaled by σ. This error, then, has some arbitrary, but specified, distribution, usually from a location–scale family.

Much of classical statistics is dominated by the terminology of production and measurement error models. Thus, for example, the probability distribution was often called the error distribution. However, in fields outside physics, chemistry, and engineering, most variability in observations is not due to measurement or production errors, but to the stochastic nature of the data generating mechanism.

2.7.2 STABLE DISTRIBUTIONS

The family of stable distributions is an extremely important, if neglected, group of models. Much of the reason for this neglect arises from the absence of simple forms for the density functions in almost all cases. However, modern computing power allows this handicap to be overcome, permitting wide ranging applications.

A distribution is stable if the sum of independent random variables from the distribution, appropriately normalized, follows the same distribution: $cY = \sum c_i Y_i$ with $c^\alpha = \sum c_i^\alpha$ $(0 < \alpha \leq 2)$. In other words, the convolution of the distribution of these random variables has this same distribution.

The characteristic function, $C_Y(u)$ (Section B.5.4), is defined by four parameters,

- the index of stability or characteristic exponent, α,
- the skewness, β,
- the scale, σ, and
- the location (or shift), μ,

such that

$$\begin{aligned}\log[C_Y(u)] &= i\mu u - \sigma^\alpha |u|^\alpha \left[1 - i\beta \operatorname{sign}(u) \tan\left(\frac{\alpha\pi}{2}\right)\right] \qquad (2.93)\\ &\qquad 0 < \alpha < 1 \text{ and } 1 < \alpha \leq 2\\ &= i\mu u - \sigma|u| \left[1 + i\beta\frac{2}{\pi}\operatorname{sign}(u)\log(u)\right]\\ &\qquad \alpha = 1\end{aligned}$$

with $-1 \leq \beta \leq 1$ and $\sigma > 0$.

The support of stable distributions is the real line except if $\alpha < 1$ and $|\beta| = 1$, in which case it is $(-\infty, 0)$ for $\beta = 1$ and $(0, \infty)$ for $\beta = -1$.

If Y has density $f_\alpha(y; \mu, \sigma, \beta)$, then $Z = (Y - \mu)/\sigma$ will have density $f_\alpha(z; 0, 1, \beta)/\sigma$, in an obvious notation. Thus, this family is the subclass of location–scale distributions which are closed under convolution.

Stable distributions are continuous, with smooth unimodal densities. When $\alpha < 2$, the tails of the density function decay like a power function, thus exhibiting more variability than the normal distribution. The smaller is α, the slower the decay and the heavier the tails, so that the parameter determines the total probability contained in the extreme tails of the density. Those members with $\beta = 0$ are symmetric about μ, while those with

$|\beta| = 1$ are called extreme stable distributions. Distributions with $\beta > 0$ are left skewed and those with $\beta < 0$ right skewed.

Examples
1. For $\beta = 0$, the normal distribution has $\alpha = 2$. Note that, in this notation (which is, unfortunately, standard), the variance is $2\sigma^2$.
2. The Cauchy distribution, with

$$f(y;\mu) = \frac{\sigma}{\pi[\sigma^2 + (y-\mu)^2]} \qquad -\infty < y, \mu < \infty, \quad \sigma > 0 \qquad (2.94)$$

has $\alpha = 1$ and $\beta = 0$. This can be generalized to the Student t distribution (Section C.2).
3. For $\alpha = 1/2$ and $\beta = 1$, we have the Lévy distribution, with

$$f(y;\mu,\sigma) = \sqrt{\frac{\sigma}{2\pi(y-\mu)^3}}\mathrm{e}^{-\frac{\sigma}{2(y-\mu)}} \qquad (2.95)$$
$$-\infty < \mu < y < \infty, \quad \sigma > 0$$

Note that the support is $Y > \mu$.
4. A Lévy process, $\{Y_t\}$, has independent increments, $Y_{t+s} - Y_t$, for all $s, t > 0$. Then, a stochastic process with stable Lévy motion has $Y_0 = 0$ and independent increments such that $Y_t - Y_s$ has a stable distribution with $\mu = 0$ and $\sigma = (t-s)^{1/\alpha}$ for $0 \le s < t < \infty$. When $\alpha = 2$, this is called Brownian motion. □

When $0 < \alpha < 2$, we have

$$\begin{aligned} \mathrm{E}[|Y|^i] &< \infty \qquad 0 < i < \alpha \\ &= \infty \qquad i \ge \alpha \end{aligned} \qquad (2.96)$$

Thus, for $\alpha < 2$, the variance is infinite, so that it only is finite for the normal distribution; for $\alpha \le 1$, the mean is infinite as well. When $1 < \alpha \le 2$, the location parameter is the mean: $\mathrm{E}[Y] = \mu$.

Example
For $\alpha < 1$, the mean of n i.i.d. random variables is likely to be much larger than any given individual value. This is possible because the maximum value will be exceedingly large and will dominate their sum. □

Thus, such distributions are useful when some very large observations may be expected and may dominate sums of these random variables. It is not valid to treat such values as outliers (Section 4.3.3), because excluding them takes away much of the specificity of the data. Precisely these observations may be of greatest interest in understanding the data generating mechanism.

Generally, the stable family is stated not to have a known explicit density. Nevertheless, in most cases, it can be expressed in terms of known functions. Thus, the characteristic function can be inverted using Equation (B.90), which can be expressed in terms of incomplete hypergeometric functions (Hoffmann-Jørgensen, 1994, I, pp. 406–411). The latter are generally only defined as the sum of infinite series, which, however, can be numerically approximated. Thus, the normal, Cauchy, and Lévy distributions are usually said to be the only non-degenerate ones with known closed forms.

Another possibility is to consider the (standardized) bivariate distribution ($\alpha \neq 1$),

$$f_\alpha(z,u;0,1,\beta) = \frac{\alpha}{|\alpha-1|}\left|\frac{z}{s(u)}\right|^{\frac{\alpha}{\alpha-1}}\frac{1}{|z|}\exp\left[-\left|\frac{z}{s(u)}\right|^{\frac{\alpha}{\alpha-1}}\right] \quad (2.97)$$

where

$$s(u) = \left[\frac{\sin(\pi\alpha u+\zeta)}{\cos(\pi u)}\right]\left[\frac{\cos(\pi u)}{\cos[\pi(\alpha-1)u+\zeta]}\right]^{\frac{\alpha-1}{\alpha}}$$

$\zeta = \pi\beta\inf(\alpha, 2-\alpha)/2$, and $u \in (-1/2, 1/2)$. From this, numerical integration, for example by Monte Carlo methods (Section D.2.2), can be applied to obtain the marginal stable distribution of Z (Buckle, 1995).

If the Y_i have parameters σ_i, β_i, μ_i, and common α, called α-stable distributions, then $Y_1 + Y_2$ has parameters

$$\begin{aligned}\sigma &= (\sigma_1^\alpha + \sigma_2^\alpha)^{\frac{1}{\alpha}}\\ \beta &= \frac{\beta_1\sigma_1^\alpha + \beta_2\sigma_2^\alpha}{\sigma_1^\alpha + \sigma_2^\alpha}\\ \mu &= \mu_1 + \mu_2\end{aligned} \quad (2.98)$$

The *codifference* of two jointly α-stable random variables, Y_1 and Y_2, with scale parameters σ_i, is defined by

$$\rho_{12} = \sigma_1^\alpha + \sigma_2^\alpha - \sigma_{12}^\alpha \quad (2.99)$$

where σ_{12} is the scale parameter for $Y_1 - Y_2$. For $\alpha = 2$, this gives the covariance.

The distribution of i.i.d. random variables, Y_i, is said to belong to the domain of attraction of the distribution of Y if

$$\lim_{n\to\infty} d_n \sum_{i=1}^{n} Y_i + e_n = Y \quad (2.100)$$

where d_n and e_n are sequences of positive real numbers, most generally with $d_n = n^{-1/\alpha}$. A random variable has a stable distribution iff it has such a domain of attraction. If $e_n = 0$, it is strictly stable. Thus, stable distributions provide the only possible class of limiting distributions for sums of i.i.d. random variables. This is a generalization of the central limit theorem to cases where the variance, and perhaps even the mean, is undefined.

Infinite divisibility If a random variable, Y, can be represented as a sum of n independent random variables, Y_i, with common distribution, for all n, it is said to be *infinitely divisible.* Stable distributions are infinitely divisible, such that the common distribution of the Y_i is also stable. However, this property holds more widely.

Example
Because of the convolution property of Equation (2.60), members of the exponential dispersion family might be thought to be infinitely divisible. However, this is true iff all cumulants are non-negative (Jørgensen, 1986).

By the convolution property, an exponential dispersion model yields a stochastic process with stationary and independent increments. If it is infinitely divisible, it is a Lévy process (Jørgensen, 1992), although not generally a member of the stable family. □

2.8 Appendix: technical results

This appendix provides a more detailed account of the general theory of exponential families than was possible in Section 2.1. Reading it will require familiarity with Appendices A and B.

For a probability space $(\Omega, \mathcal{F}, P)$, let us call $(\Omega, \mathcal{F}, \mathcal{P})$ a *statistical field,* where $\mathcal{P}$ is a family of possible probability measures on $(\Omega, \mathcal{F})$ dominated by a measure, ν, usually Lebesgue or counting measure. A *statistic* is any real-valued measurable function of a random variable, Y, while a *parameter* is a real-valued function on the probability measure, P.

2.8.1 EXPONENTIAL REPRESENTATION

Suppose that $a_0 \geq 0$ and $\mathbf{a}$ are a scalar and a k-dimensional ($k > 0$) column vector of real-valued functions on P. They are the parameters, indexing the probability measure. Also, let $t_0 \geq 0$ and $\mathbf{t}$ be a scalar and a k-dimensional column vector of real-valued measurable functions on Ω, i.e. statistics. Suppose that $\mathbf{t} \in \mathcal{T}$, a convex set.

The *exponential representation* of the probability measure is given by

$$\frac{dP}{d\nu}(y) \stackrel{\text{def}}{=} a_0(P) t_0(y) \mathrm{e}^{\mathbf{a}(P)^\mathsf{T} \mathbf{t}(y)} \tag{2.101}$$

where $P \in \mathcal{P}$, $y \in \mathcal{Y}$ is a fixed or observed value of a random variable, Y, on the field, and ν is either Lebesgue or counting measure on Y. In other words, we are using Lebesgue–Stieltjes measures, so that the exponential family contains both discrete and continuous members. The dominating measure is

$$\nu(dy) = t_0(y)d\nu \tag{2.102}$$

We may immediately note that

$$a_0^{-1}(P) = \int t_0(y)\mathrm{e}^{\mathbf{a}(P)^{\mathrm{T}}\mathbf{t}(y)}d\nu \tag{2.103}$$

a normalizing constant. It is the Laplace transform of the dominating measure, $\nu(dy) = t_0(y)d\nu$, and, hence, its moment generating function.

If k is the smallest integer such that the probability measures are representable in this way, $\mathcal{P}$ is said to be of *order* k and the representation is said to be *minimal.* This order must be less than or equal to the minimum of the dimension of $\mathbf{a}(P)$ and that of the convex support of ν. The components of $\mathbf{t}(y)$ are not linearly dependent, with probability one. This minimal representation is not unique, because it can be transformed to a different, but equivalent, representation by linearly transforming $\mathbf{t}(y)$ and $\mathbf{a}(P)$ with linked non-singular affine transformations.

Suppose that $P_i \in \mathcal{P}$ is an arbitrary probability measure. Then,

$$\frac{dP}{dP_i} = \frac{a_0(P)}{a_0(P_i)}\mathrm{e}^{[\mathbf{a}(P)-\mathbf{a}(P_i)]^{\mathrm{T}}\mathbf{t}(y)} \tag{2.104}$$

In this way, P is generated from P_i, some arbitrary origin, by what is called exponential tilting. We know that this ratio exists because of the Radon–Nikodým theorem (Section B.2.2). We can see that only the information in $\mathbf{t}(y)$, and not that in $t_0(y)$, is needed to compare probability models. The statistic, $\mathbf{t}(y)$, is said to be *sufficient* for $\mathbf{a}(P)$. If the elements of $\mathbf{a}$ are affinely independent, then $\mathbf{t}$ is called *minimal sufficient* and we say that the representation is (k, k).

2.8.2 FUNDAMENTAL PARAMETRIZATIONS

Let $\Theta = \mathbf{a}(\mathcal{P})$, which is a one-to-one mapping, $P \to \mathbf{a}(P)$ for $P \in \mathcal{P}$. Then, $\{P_\theta : \boldsymbol{\theta} \in \Theta\}$, where $P_\theta = \mathbf{a}^{-1}(\boldsymbol{\theta})$, is a *natural* or *canonical parametrization* of $\mathcal{P}$ and $\boldsymbol{\theta}$ now indexes the members of the family. The mapping $\mathbf{t}(y)$ is called the *natural* or *canonical statistic.* If this mapping is the identity, $\mathcal{P}$ is said to be a *linear exponential family.* Then,

$$\frac{dP_\theta}{d\nu}(y) = a_0(\boldsymbol{\theta})t_0(y)\mathrm{e}^{\boldsymbol{\theta}^{\mathrm{T}}\mathbf{t}(y)} \tag{2.105}$$

is the *canonical representation* of the exponential family. It is minimal canonical if the representation is minimal. Here, $\kappa(\boldsymbol{\theta}) = -\log[a_0(\boldsymbol{\theta})]$ is a convex function.

2.8.3 GENERATING FUNCTIONS

From Equation (2.2), we have

$$a_0^{-1}(\boldsymbol{\theta}) = \int t_0(u)e^{\boldsymbol{\theta}^{\mathrm{T}}\mathbf{t}(u)}du$$

for $\boldsymbol{\theta} \in \Theta$. Then, for any $\boldsymbol{\theta} \in \Theta$, the Laplace transform, and hence the moment generating function, of the family is

$$M_T(u) = \frac{a_0(\boldsymbol{\theta})}{a_0(u+\boldsymbol{\theta})} \tag{2.106}$$

Thus, $a_0(\boldsymbol{\theta})$ is regular enough to be differentiated under the integral sign. If $\boldsymbol{\theta}$ is not on the boundary of Θ ($\boldsymbol{\theta} \in \text{int}\Theta$), the statistic, $\mathbf{t}$, has moments of all orders with respect to P_θ, and, hence, $a_0(\boldsymbol{\theta})$ is analytic for such values of $\boldsymbol{\theta}$.

The derivatives of $\kappa(\boldsymbol{\theta}) = -\log[a_0(\boldsymbol{\theta})]$ are the cumulants of $\mathbf{t}$ under P_θ. We call this the *cumulant function.* It is the cumulant generating function of the dominating measure, $\nu(dy)$, and is a strictly convex analytic function on $\boldsymbol{\theta} \in \text{int}\Theta$.

2.8.4 REGULARITY CONDITIONS

If $0 \in \Theta$ and $a_0(0) = 1$, then

$$\frac{dP_\theta}{dP_0} = a_0(\boldsymbol{\theta})e^{\boldsymbol{\theta}^{\mathrm{T}}\mathbf{t}(y)} \tag{2.107}$$

This is called the *standard representation* of the exponential family. The *domain* of the parameter is

$$\Theta = \left\{\boldsymbol{\theta} : a_0^{-1}(\boldsymbol{\theta}) = \int t_0(u)e^{\boldsymbol{\theta}^{\mathrm{T}}\mathbf{t}(u)}du < \infty\right\} \tag{2.108}$$

while the *support* of the family is

$$\mathcal{T} = \left\{\mathbf{T} : \frac{dP_\theta}{dP_0} > 0 \quad \forall\boldsymbol{\theta}\right\} \tag{2.109}$$

The domain of a mean value parameter is contained within the closed convex hull of the support of the corresponding canonical statistic. The domain of the canonical parameter is *convex* and $\mathcal{P}$ is said to be *full.* If the $\mathcal{P}$ is full and the set Θ is open, $\mathcal{P}$ is said to be *regular.* Any full family having finite support is regular because $\Theta = \mathbf{R}^k$.

A family of probability distributions is said to be *complete* if the statistic $T(\mathbf{y}) \in \mathcal{L}^1$ with $\int T(\mathbf{y})dF(\mathbf{y}) = 0 \;\forall\; \boldsymbol{\theta} \in \Theta \Rightarrow T(\mathbf{y}) = 0 \;\forall\; \boldsymbol{\theta} \in \Theta$. In the exponential family, if the interior of Θ, int$\Theta \neq \emptyset$, i.e. if Θ has members not on its boundary, then the family is complete.

Now, let $h : \mathbf{R}^k \to (-\infty, \infty)$ be a convex function and take $\Theta = \{\boldsymbol{\theta} \in \mathbf{R}^k : h(\boldsymbol{\theta}) < \infty\}$. Assume h to be continuously differentiable in intΘ. Suppose that $\boldsymbol{\theta}_0 \in \text{int}\Theta$ and $\boldsymbol{\theta}_1 \in \Theta \setminus \text{int}\Theta$. Then, if

$$\lim_{\alpha \uparrow 1} \frac{\partial}{\partial \alpha} h[\boldsymbol{\theta}_0 + \alpha(\boldsymbol{\theta}_1 - \boldsymbol{\theta}_0)] = \infty \quad 0 < \alpha < 1 \tag{2.110}$$

h is *steep.* An exponential family is said to be steep if its cumulant function, κ, is steep. Note that, if the family is regular, it is steep, because, then, Θ is open. The converse is not true.

For a minimal steep canonical exponential family, $\boldsymbol{\mu}(\boldsymbol{\theta}) = \text{E}[\mathbf{Y}]$ defines a homeomorphism mapping between intΘ and the interior of the convex support of ν.

Further reading

Statistical models have been around for a long time. However, little distinction was made between such a theoretical model and empirical observations connected with it. Thus, Fisher (1922a) was the first to introduce a clear difference, by means of the terms 'parameter' and 'statistic'.

Ample details about exponential families, to complete the outline given in this chapter, can be obtained in Barndorff-Nielsen (1978), from a frequentist point of view, or in Brown (1986), from a Bayesian point of view. The original article of Fisher (1934), on both the exponential and location–scale families, is definitely worth consulting; he already introduced the concept of sufficient statistic in Fisher (1920). The characterization of exponential families by their variance function was first noticed by Tweedie (1947); see also Morris (1982, 1983a) and Letac and Mora (1990). Barnard (1954) first developed the idea of conjugate distributions.

Generalized linear models were introduced by Nelder and Wedderburn (1972). Jørgensen (1987a), following the ideas of Tweedie (1947), later generalized them to the exponential dispersion family.

See Feller (1966), Fisz (1962), and Samorodnitsky and Taqqu (1994) for stable distributions. Interesting applications to modelling survival times are provided by Hougaard (1986), to stochastic processes by Janicki and Weron (1994), and to Bayesian inference by Buckle (1995) and Damien *et al.* (1995).

2.9 Exercises

(1) Redo the examples in this chapter, replacing the Poisson distribution by the binomial, with

$$f(y;\pi) = \binom{n}{y}\pi^y(1-\pi)^{n-y} \qquad (2.111)$$

$$y = 0, 1, 2, \cdots, n, \quad 0 \leq \pi \leq 1$$

a special case of which, the Bernoulli, was given in Equation (1.1), the normal distribution by the gamma of Equation (1.3) and the inverse Gaussian of Equation (2.30).

(2) Suggest why the exponential family might also be called the family of power series distributions.

(3) (a) Show that the negative binomial distribution, with

$$f(y;\pi,\alpha) = \frac{\Gamma(y+\alpha)}{\Gamma(y+1)\Gamma(\alpha)}(1-\pi)^\alpha \pi^y \qquad (2.112)$$

$$y = 0, 1, 2, \ldots, \qquad \alpha > 0, 0 < \pi < 1$$

is a one-parameter exponential family for fixed index, α.

(b) Is it a two-parameter family for unknown α?

(c) Is it in the exponential dispersion family?

(4) A canonical one-parameter exponential family with var$[Y]$ being a quadratic function of the mean, E$[Y]$, is called a *quadratic variance function family.* Show that the following six distributions have this property:

(a) Poisson distribution;

(b) binomial distribution;

(c) normal distribution with known mean;

(d) gamma distribution with known mean;

(e) negative binomial distribution with known index, α;

(f) hyperbolic secant distribution

$$f(y;\theta) = \frac{\cos(\theta)\mathrm{e}^{\theta y}}{2\cosh(\pi y/2)} \qquad (2.113)$$

These are the only quadratic variance function distributions. (Morris, 1982, 1983a).

(5) In a branching process, we have n independent observations following sequentially over time, each having a Poisson distribution with mean

$$\mu(\alpha^i - \alpha^{i-1}) \quad \alpha > 1, \mu > 0 \quad i = 1, \ldots, n$$

(a) Write down the joint density for the n observations.
(b) Show that this distribution is a member of the exponential family.
(c) What are the canonical parameters?
(d) What are the canonical statistics?
(e) Give mean value and mixed representations.
(f) What is the conjugate distribution for the location parameter?
(Basawa and Scott, 1983, pp. 105–106)

(6) Suppose that $Y_1, \ldots, Y_n$ are independently distributed in a random sample from the density,

$$\begin{aligned} f(y;\theta) &= \theta(2y)^{\theta-1} & y \leq \tfrac{1}{2} \\ &= \theta[2(1-y)]^{\theta-1} & y \geq \tfrac{1}{2} \end{aligned} \tag{2.114}$$

where $\theta > 0$, $0 \leq y \leq 1$.
(a) Write down the joint density for the n observations using an indicator function.
(b) Show that this distribution is a member of the exponential family.
(c) What is the canonical parameter?
(d) What is the canonical statistic?
(e) Is this distribution regular?
(Bernardo and Smith, 1994, p. 313)

(7) Consider a series of random variables, $Y_1, \ldots, Y_n$, from a normal distribution. Define $\mu_t = \beta_0 + \beta_1 x_t$ for explanatory variables x_t, observed at the same time, t, as y_t. Suppose that Y_1 has mean μ_1 and variance $\sigma^2/(1-\rho^2)$, while the rest are conditionally independent with means $\mu_t + \rho(y_{t-1} - \mu_{t-1})$ and variance σ^2. For $|\rho| < 1$, this is a stationary first-order normal autoregressive process.
(a) Write down the joint density for the n observations.
(b) Show that this distribution is a member of the exponential family.
(c) What are the canonical parameters?
(d) What are the canonical statistics?
(e) What changes if the first observation is taken to be a known fixed constant?
(f) What changes if, instead, the means of the last $n-1$ values are $\mu_t + \rho y_{t-1}$?
(Lindsey, 1993, pp. 54–56, 99–112)

(8) Let $Y_1, \ldots, Y_n$ be a series of n random variables from an exponential autoregressive process, with known initial value, y_0. Thus (conditionally) $Y_t - \rho y_{t-1}$ have independent exponential distributions, Equation (1.14), with canonical parameter, θ.
   (a) Write down the joint density for the n observations.
   (b) Show that this distribution is a member of the exponential family.
   (c) What are the canonical parameters?
   (d) What are the canonical statistics?
   (e) Derive a conjugate distribution for the two parameters.
   (Girón *et al.*, 1994)

(9) For the multinomial distribution, with

$$f(\mathbf{y}; \boldsymbol{\pi}) = \binom{n}{y_1 \cdots y_k} \prod \pi_i^{y_i} \tag{2.115}$$

with fixed $n = \sum y_i$, find
   (a) the order;
   (b) the minimal sufficient statistics;
   (c) the minimal canonical parameters.

(10) Consider a 2×2 contingency table of frequencies, y_{ij}, having a multinomial distribution with parameters π_{ij}.
   (a) Find the conditional density of $\mathbf{Y}$ given the observed values of $y_{i\bullet} = \sum_j y_{ij}$ and $y_{\bullet j} = \sum_i y_{ij}$.
   (b) Show that the resulting distribution is a one-parameter exponential family.
   (c) What is the canonical parameter?
   When $\pi_{ij} = \pi_{i\bullet}\pi_{\bullet j}$, we have independence in the contingency table and the resulting conditional distribution is called the hypergeometric.

(11) (a) Is the gamma distribution of Equation (1.3) steep?
   (b) Is it regular?
   (c) What conclusions can we draw about the exponential distribution, i.e. when $\alpha = 1$?

(12) Show that the inverse Gaussian distribution of Equation (2.30) is steep but not regular.
   (Barndorff-Nielsen, 1978, p. 117; Brown, 1986, p. 72)

(13) For n i.i.d. observations from a normal distribution, find
   (a) the marginal density of $t_2 = \sum y_i^2$;
   (b) the conditional density, given t_2.

(14) (a) For the conditional density of Equation (2.37), show that

$$a_0^{-1}(\boldsymbol{\theta}_2|\mathbf{t}_1) = f(\mathbf{t}_1;\boldsymbol{\theta})e^{\kappa(\boldsymbol{\theta})-\boldsymbol{\theta}_1^{\mathrm{T}}\mathbf{t}_1} \tag{2.116}$$

(b) For the inverse Gaussian distribution of Equation (2.30),
  i. What are the sufficient statistics?
  ii. What is the cumulant function?
  iii. Derive a conditional distribution which only involves the mean, μ.

(Bar-Lev, 1994)

(15) In a Behrens–Fisher model, two independent normal populations have common mean, μ, but different variances, σ_1^2 and σ_2^2.
  (a) Write down the probability distribution.
  (b) To what family does this model belong?
  (c) What are the canonical parameters?
  (d) What are the mean value parameters?

(Barndorff-Nielsen and Cox, 1994, p. 68)

(16) Suppose that we have n i.i.d. observations of durations over time following the exponential distribution of Equation (1.14), but that some of the times are incomplete. These are called censored times (Section 2.3.1); for them, the only information we have is in their cumulative distribution function, up to the time of censoring. Assume that the censoring mechanism is non-informative and thus ignorable, i.e. independent of the durations and of μ (see Section 6.4.3).
  (a) Write down the resulting distribution for the n observations, where c are censored.
  (b) Show that it is a curved exponential family.
  (c) How can the parameter, μ, be interpreted?
  (d) Suggest a transformation of this parameter which might be useful in certain contexts.

(17) Let Y_1 and Y_2 have independent exponential distributions, from Equation (1.14), with means μ_1 and μ_2, where μ_1 is a function of μ_2 such that

$$\frac{e^{\frac{1}{\mu_2}}}{\mu_1(\mu_2)\mu_2} = 1$$

  (a) Write down the probability density.
  (b) To what family does this model belong?
  (c) What are the canonical parameters?
  (d) What are the mean value parameters?

(Barndorff-Nielsen and Cox, 1994, p. 229)

(18) The Poisson, binomial, normal, gamma, and inverse Gaussian distributions are all members of both the linear exponential and exponential dispersion families.

(a) Demonstrate the latter fact.
(b) For each, list the canonical regression models, $\xi_i = \boldsymbol{\beta}^{\mathrm{T}}\mathbf{x}_i$ as a function of the mean.
(c) Do any of these link functions provide particular advantages?
(d) What might be a more useful link function for some of these distributions?
(e) If a non-canonical link function is used, to what family do they now belong?

(19) The gamma distribution of Equation (1.3) is a member of the exponential dispersion family.
(a) Use its cumulant function to show the convolution property.
(b) The exponential distribution, with $\alpha = 1$, is a special case. What is the distribution of a sum of p independent exponential random variables with common mean, μ?
(c) Another special case of this distribution is the Chi-squared distribution (Section C.1). What is the distribution of a sum of two independent Chi-squared random variables with, respectively, p_1 and p_2 degrees of freedom?

(20) Suppose that we observe independent random variables, Y_1 and Y_2, with gamma distributions of Equation (1.3), having parameters α_i and μ_i, the former being known values.
(a) What is the marginal density of $Z = Y_1/Y_2$?
(b) What is the conditional density of Y_2 given Z?
(c) If the means are unknown but equal, to what family does this model belong?

(Barndorff-Nielsen, 1978, p. 53; Barndorff-Nielsen and Cox, 1994, pp. 68–69)

(21) Consider a random variable, Z, with density $f_Z(z)$. Transform linearly to a new random variable, $Y = a + bZ$, with density $f_Y(y)$.
(a) Show that

$$\int h(a+bz)f_Z(z)dz = \int h(y)f_Y(y)dy \tag{2.117}$$

for all functions $h(\cdot)$.
(b) From this, demonstrate that we must have

$$f_Y(y) = \frac{1}{|b|}f_Z\left(\frac{y-a}{b}\right) \tag{2.118}$$

the location–scale family.
(c) Find at least two density functions, other than the normal, which fulfil these conditions.

(d) Under what conditions can this be generalized to vector variables?

(Whittle, 1992, pp. 133, 140)

(22) Suppose that we have i.i.d. random variables, Y_i, with a sequence of normalizing constants, d_n, such that $d_n \sum_1^n Y_i$ has the same characteristic function as Y_i.

(a) Show that the characteristic function must necessarily have the form $\exp[-c|u|^\alpha]$.

(b) To what family of distributions does this correspond?

(c) Give two members of the family.

(d) What are the values of α for these two?

(e) In general, what are the limits to the value of α?

(Whittle, 1992, p. 134)

Part II

Inference

3
Likelihood

In the preceding two chapters, we have looked at various aspects of probability theory and of statistical model building. In that context, we assume that we have a reasonable model, and explore what predictions can be made about the possible events which might occur. This model formulation is the first step in solving a problem, as well as an an ultimate goal of any model construction process. However, in order to develop such models, we must also be able to integrate information from the real world: which models describe well what we have observed and what can we learn from them. Thus, statistical inference is classically concerned with the reverse process as compared to Chapter 2 and Appendix B: instead of assuming a model and predicting events, we know that certain events have occurred and attempt to discover what can be said about the probabilistic process which might have produced them. As we proceed to study ways of making such inferences, we shall also introduce, as examples, some of the more common statistical models which are based on this probability theory, and, in particular, on the families that we studied in Chapter 2.

Measure theory is important in the development of statistical models, as we can see from Appendix B and Section 2.8. Nevertheless, it will play little role in our presentation of statistical inference, because the observable values of all random variables are discrete (see, however, Chapter 8).

3.1 Statistical inference

3.1.1 THE PROBLEM

Measurement precision Consider a data generating mechanism which yields observable events. If it could be completely known, these events could be predicted, i.e. precise probabilities could be assigned to them. Models will describe this mechanism in some simplistic but hopefully useful way. In a model, continuous variables, such as time, should be treated as such, where feasible; they generally are in scientific theories. However, only discrete events can actually be observed. Thus, these events will be recorded as values of the random variable, Y, with some *unit of measurement*, $\Delta = y_2 - y_1$, determined by the *precision* of the measuring instrument, such that the corresponding probability of observing the event is $P(y_1 < Y \leq y_2)$. This unit of measurement is always finite in any real observable situation (see, for example, Fisher, 1922, or de Finetti, 1974a, II, pp. 292–298).

If empirical observations were truly continuous, then, with probability one, no two observed responses would ever be identical. Indeed, use of continuous distributions can lead to 'paradoxes' in many situations. One involves the use of conditioning, which is important in many statistical contexts (see Section B.3.5). Thus, statistical inference, the science of obtaining information from empirical observations subject to random variability, must take into account this measurement precision.

Statistical models Consider now the case with which statistical inference is classically concerned. Given a clearly formulated problem, we construct one or more sets of possible models which may help us to understand some data generating mechanism. These represent the available prior theoretical knowledge. This is the first crucial step, perhaps the most difficult: model formulation by the choice of a set of suitable model functions which might reasonably describe the mechanism. It generally involves close collaboration between the statistician and the people who formulated the problem. Once this is accomplished, we observe a suitable *sample* of one or more values, $\mathbf{y}$, of the random variable, Y.

This set of models under consideration can be conveniently written as a *parametric statistical model*, $(\mathcal{Y}, \mathcal{P}, \Psi)$, where

- $\mathcal{Y}$ is the *sample space*, as measured through the random variable.
- $\mathcal{P}$ is the family of *model functions*, P_ψ.
- Ψ is the *parameter space* indexing the models.

In many situations, further simplification is made. For example, the set of possible models may contain only one functional form. Then, the parameter index will usually be a vector of unknown real values, defined by the function.

Sample space For one observation, the sample space will be clearly defined for any family of models, $\mathcal{P}$, to describe a data generating mechanism under study. But, most often, several observations will be made according to some sample design, and the space will often depend on which scheme is chosen.

Example

The data generating mechanism under study is a Bernoulli trial to determine the bias of a coin, with $\mathcal{Y} = \{0, 1\}$, the number of heads. A sample design might specify continuing for a fixed total of n trials, with random variable $Y \in \{0, 1, \ldots, n\}$, or until a fixed number of tails is observed, with $Y \in \{0, 1, \ldots\}$. Here, we have one finite and one infinite sample space, depending on the rule for stopping the study, both to obtain information about the same phenomenon. □

Obviously, the choice of sample design should be made to aid in elucidating the data generating mechanism. We shall soon discover that certain designs produce more relevant information about the data generating mechanism than others. However, as we shall also see, considerable controversy has arisen over the effect of such different schemes on the subsequent inferences to be made.

Inference When the observations have been obtained, we finally come to the basic question of statistical inference:

Once values of the random variable have been observed, what can be said about which members of $\mathcal{P}$ best describe how it was generated?

Answering this question involves a number of steps. The principal ones are

(1) selecting an appropriate model function from some set of models of interest,
(2) given a model function, determining the set of plausible parameter values, and
(3) checking the goodness of fit of the model function (for these parameter values),

all in the light of the observed data (eventually) combined with any prior empirical data. Of course, these steps may need to be iterated several times before a suitable model function and set of parameter values are selected. The first two points are closely related and, in fact, involve an arbitrary distinction between families of models indexed by a parameter and those having different functional forms (Section 4.1.1). They will be studied in this chapter. The third point usually involves more *ad hoc* techniques and will be discussed in Chapter 4.

3.1.2 INFERENCE AND DECISION-MAKING

As we have seen in Chapter 1, statistical inference and decision-making are concerned with drawing conclusions from empirical observations on probabilistic or stochastic processes. We may distinguish three cases:

(1) how observations can be interpreted as containing information about the process;
(2) how they can be used to make decisions about the process; and
(3) how beliefs about the process are modified in the presence of new observations.

Thus, the need to make statistical inferences or decisions can arise in a number of different situations. These include:

(1) A scientific experiment or sample survey is performed. By definition, this is repeatable, but not usually repeated, at least not in exactly the same circumstances. Prior theoretical knowledge is available permitting the formulation of one or more (families of) statistical models. What information is provided from this, and only this study, about the data generating mechanism, as described by the models?

(2) A study is performed, as in point (1), but additional previous theoretical knowledge or empirical information is available. This may yield relative preferences for the various models, and in particular, about their parameter values, and must be incorporated in drawing conclusions.

(3) An evaluation of costs and benefits, given the decision to choose each model, is available and must be combined with the information in the data at hand.

(4) Samples are continually being drawn from a production process to test for quality. A bad sample indicates that the batch must be rejected. Wrong decisions must be minimized over the long run.

(5) An evolving process is periodically observed, and the information accumulated, with the goal, at each observation period, of predicting future values. Bad prediction may be costly.

Point (1) may be considered the minimal case for the other situations. In order to accomplish the other goals, the maximum relevant information must first be obtained about the data generating mechanism from the available observations. However, here, the idea is not to make some irrevocable decision or choice among the possible models, but rather to give them some (provisional) weighting or ordering in light of the present data. Information is to be obtained from data without consideration as to the use to which it will be put.

Points (2) and (5), and perhaps also (3) and (4), require the inclusion of past knowledge. Points (3), (4), and (5) all involve making decisions and, in addition, can incur costs in the face of wrong decisions. They will require a model of human decision-making behaviour when faced with resulting costs and benefits. Point (4) perhaps relies least on the information in present observations, being more concerned with the repetitive nature of the situation. It is more important to be correct often in the long run than on any individual test.

In this and the next three chapters, we shall primarily be concerned with the minimal situation of extracting maximum relevant information about a model (hopefully) based on prior theoretical considerations, from present observations of random variables, and incorporating prior empirical information. Then, the other, more complex goals will be related to this in Chapters 7 and 8.

3.1.3 A LIKELIHOOD PRINCIPLE

Suppose that we have observed one or more values, $\mathbf{y}$, of the random variable, Y, of interest. Then, for each of the possible fixed models $P_\psi \in \mathcal{P}$, the probability of these observations can be calculated. We can now *postulate* the

Likelihood principle
A model, P_1, is more plausible or *likely* than another, P_2, in the light only of the given observed data, written $\mathrm{L}(P_1) > \mathrm{L}(P_2)$, if it makes those data more probable. □

Thus, the *likelihood* (Fisher, 1921) of a model is the probability of the observed data given that model. It provides a comparative measure of how well various models can predict the observations which were actually made. It is a measure of relative support for models, regardless of the fact that the observations were certainly not actually generated according to one of the models.

Observability implies a finite unit of measurement. The probability of observing a point is always zero; thus, the probabilities to be used must always be with respect to the unit of measurement of the empirical observations.

What we have done is to use the random variability in possible outcomes of the data generating mechanism, specifically, the uncertainty that the observed outcome would have occurred, to provide a measure of uncertainty about the parameters of the model used to describe that mechanism. This should be distinguishable from any uncertainty about the choice of model (function) due to lack of (theoretical) information, which does not appear in the likelihood. The latter assumes the model family to be given. In this perspective, inference from likelihood is especially adapted to small samples; indeed, we shall see as we proceed that, with the approximations customarily used, inferences may not even be valid for very large samples!

This likelihood principle does not state that statistical inference should *only* be based on this principle. In addition, it is a *relative* statement, comparing models, but providing no absolute knowledge about any one possible model. This is logical, if no model can be true. We only want the best model among those currently available to help us in understanding how the data might have been produced.

If it were felt necessary to formalize the procedure for making a statistical inference from a given data set in the light of prior theoretical knowledge, this likelihood principle would be the only axiom required.

Comparing sets of models This likelihood principle does not allow us easily to compare arbitrary *sets* of models. One may agree that it follows from this principle that set $\mathcal{P}_1 \subset \mathcal{P}$ is more likely than $\mathcal{P}_2 \subset \mathcal{P}$ with $\mathcal{P}_1 \bigcap \mathcal{P}_2 = \emptyset$ if

$$\inf_i \mathrm{L}(P_{1i}) > \sup_j \mathrm{L}(P_{2j})$$

where $P_{1i} \in \mathcal{P}_1$ and $P_{2j} \in \mathcal{P}_2$. However, it is not reasonable to conclude that the data support set 1 over set 2 simply if, for example, the sum of the likelihoods for the first is greater than that of the second.

Example
Consider two finite sets of models with, for simplicity, the same number in each. Suppose that set 1 contains one very likely model, say P_{11}, and a number of extremely unlikely ones, such that

$$\mathrm{L}(P_{11}) > \sup_{i \neq 1} \mathrm{L}(P_{1i})$$

In contrast, set 2 contains models which are all more likely than those in set 1, with the exception of P_{11}, so that

$$\inf_j \mathrm{L}(P_{2j}) > \sup_{i \neq 1} \mathrm{L}(P_{1i})$$

but with a sum such that

$$\sum_i \mathrm{L}(P_{1i}) > \sum_j \mathrm{L}(P_{2j})$$

All models in set 2 are more likely than all those, except P_{11}, in set 1, but the sum would indicate set 1 to be more 'likely' than set 2. □

The same argument applies to any arbitrary weighted sum of likelihoods. Likelihood is a point function on the parameter space; the operations of addition and averaging are not defined without further major assumptions (Chapter 8).

Model complexity Because we are interested in the best means (for our purposes) of representing a data generating mechanism, a second, possibly more reasonable, way to compare sets of models is to represent each set by the model in it which is most likely. In this way, each set is given its best 'chance' to explain the observations. However, this immediately raises the problem of model complexity. Generally, a more complex model will inherently tend to make the observations more probable. (For example, it is always possible to construct a model for which the probability of the observations is unity!) Consider a set of models indexed by the parameter vector, ψ. Members of this set will always be at least as likely as those of the subset where we fix one element, say $\psi_1 = 0$.

Thus, this likelihood principle is not sufficient in itself to order satisfactorily all possible models. It does not take into account the relative complexity of the different candidate models. The likelihood principle only indicates if the models *fit* relatively closely to the data, called goodness of fit. However, a model which is selected to fit a given sample of data extremely well generally will have a poor chance of also fitting a second equivalent set well. It has not isolated the patterns of interest in the data, but follows the random fluctuations of that particular random sample too closely. In other words, models must *smooth* the data to some extent, to make apparent the patterns of interest.

A measure of complexity of a set of models is the minimum dimension, p, of the *unknown* parameter vector indexing the model. The number of *degrees of freedom* (Fisher, 1922b) is the amount by which the dimension of this parameter vector could still be increased, before the model becomes *saturated*, i.e. with as many parameters as observations. If there are n observations and p (estimated) parameters, the degrees of freedom will usually be $n - p$. Thus, when the members of $\mathcal{P}$ do not all have the same minimum dimension, i.e. the same number of unknown parameters, closeness of fit and complexity must, somehow, be balanced in weighing the relative merits of candidate models. This is the essence of what is called the model selection problem, which, in fact, refers to selection of an appropriate model function. We shall come back to this in Section 3.6.

3.2 Likelihood function

3.2.1 DEFINITION

In general, the probability of the observed values, $\mathbf{y}$, of the random variable, Y, is given by the joint probability distribution, $P_\psi(y_{11} < Y_1 \leq y_{21}, \ldots, y_{1n} < Y_n \leq y_{2n})$. Here, the observed value, y_i, of the random variable is only known to lie within the unit of measurement, $y_{1i} < y_i \leq y_{2i}$. For independent observations, this probability is

$$\begin{aligned} & P_\psi(y_{11} < Y_1 \leq y_{21}, \ldots, y_{1n} < Y_n \leq y_{2n}) \\ &= \prod_{i=1}^{n} P_\psi(y_{1i} < Y_i \leq y_{2i}) \qquad (3.1) \\ &= \prod_{i=1}^{n} [F(Y_i \leq y_{2i}) - F(Y_i \leq y_{1i})] \end{aligned}$$

If observations are dependent, but occur serially, in time, for example, the product of the appropriate conditional probabilities can be used, as in Equation (1.12).

In probability theory, the values of parameters are assumed fixed and known, while the random variable is unknown. However, now, because the random variable has been observed, the vector $\mathbf{y}$ is fixed, and no longer variable, while $\boldsymbol{\psi}$ is unknown.

The *likelihood function* is defined as the model taken as a function of the parameter vector, $\boldsymbol{\psi}$, for the fixed, given, observed value of $\mathbf{y}$:

$$\mathrm{L}(\boldsymbol{\psi};\mathbf{y}) \stackrel{\mathrm{def}}{=} P_{\psi}(y_{11} < Y_1 \leq y_{21}, \ldots, y_{1n} < Y_n \leq y_{2n}) \tag{3.2}$$

where $\mathbf{y}^{\mathrm{T}} = (y_1, \ldots, y_n)$ and y_i has unit of measurement $\Delta_i = y_{2i} - y_{1i}$, so that it is at the centre of the class interval so defined. The likelihood function is a point function (Fisher, 1922a, 1925); no measure has been defined on $\boldsymbol{\psi}$ (but see Chapter 8).

Example
The likelihood function for the binomial distribution of Equation (2.111) is

$$\mathrm{L}(\pi; y, n) = \binom{n}{y} \pi^y (1-\pi)^{n-y} \tag{3.3}$$

Because we have a discrete distribution, the unit of measurement is one.□

For independent continuous variables, from Equation (2.3), the likelihood function can be written

$$\mathrm{L}(\boldsymbol{\psi};\mathbf{y}) = \prod_{i=1}^{n} \int_{y_{1i}}^{y_{2i}} f(u_i;\boldsymbol{\psi}) du_i \tag{3.4}$$

This is based on the probability of empirically observing values from a continuous variable, thus providing the link between a theoretical model with such continuous variables and the discreteness of empirical data. The data have been coarsened in the observation and recording process. Inferences based on ignoring the grouping and substituting class centres, i.e. $\mathbf{y}$, are not in general correct (Heitjan and Rubin, 1991; see Section 6.4.3).

Because this function is analytically complex, it is often approximated by

$$\mathrm{L}(\boldsymbol{\psi};\mathbf{y}) \doteq \prod_{i=1}^{n} f(y_i;\boldsymbol{\psi})\Delta_i \tag{3.5}$$

which we shall call the density approximation to the likelihood function, where Δ_i is usually not a function of the parameters of interest (see Section 6.4.3).

Examples

1. An approximate likelihood for n observations from a normal distribution is

$$L(\mu, \sigma^2; \mathbf{y}) \doteq \frac{e^{-\frac{\Sigma(y_i-\mu)^2}{2\sigma^2}}}{(2\pi\sigma^2)^{\frac{n}{2}}} \prod \Delta_i \qquad (3.6)$$

2. Consider censored exponential survival times, based on Equation (1.14). If censoring occurs at the fixed time, t_c, the unit of measurement is (t_c, ∞) for events occurring after that time. Thus, Equation (3.5) will not provide a valid approximation and the exact factor in Equation (3.4), here $\exp(-t_c/\mu)$, must be used. □

Except in very special cases, the likelihood of Equation (3.4) will not be of the form of the exponential family, even when the density is of this form, so that many of the results for that family in Chapter 2 are only approximations where observed data are concerned. As we shall see in Chapter 5, the approximation in Equation (3.5) depends not only on the Δ_i being reasonably small, but also on the sample size being small. For fixed measurement precision, the quality of the density approximation to the likelihood function in Equation (3.5) degrades rather rapidly as n increases, because, for independent observations, it involves a growing product of small approximation errors (Sections 5.2.3 and 5.3.2).

Although a 'likelihood function' defined directly in terms of the probability density function may provide some mathematical conveniences, it loses its interpretation based on our likelihood principle and leads to a number of paradoxes (Section 3.6.6). It has no exact meaning for empirical data. However, as is the usual convention, the approximate character of these likelihood functions will not be noted in what follows.

Empirical likelihood If we are to judge models by how probable they make the data, we may ask what the upper bound to such probability is, i.e. the most likely possible model for the data. Obviously, there is a bound, the trivial model which gives the data probability one. Thus, the likelihood function, as defined in Equation (3.2), is always bounded above. This assertion would not be true if we were working with densities.

In most cases, however, we can obtain more useful upper bounds. Thus, if we can assume that observations are independent and identically distributed, the maximum value that the likelihood function can have is obtained by placing an arbitrary point probability mass at each observation, a multinomial distribution. Then, the maximum of this likelihood occurs when each probability is $1/n$, so that the likelihood function is

$$\sup \mathrm{L} = \prod_{i=1}^{p} \left(\frac{n_i}{n}\right)^{n_i} \tag{3.7}$$

where there are p distinct responses, with n_i in category i, defined by the unit of measurement. When all observed responses are distinct, this becomes

$$\sup \mathrm{L} = \left(\frac{1}{n}\right)^{n} \tag{3.8}$$

This maximum has been called the (maximized) *empirical likelihood.* When there are explanatory variables, as in a regression model, there will be a different multinomial likelihood for each distinct condition described by these variables. If all conditions are distinct, $\sup \mathrm{L} = 1$.

At least theoretically, such models can have an enormous number of parameters. Because a functional form is not specified, they are often called 'non-parametric'. In contrast, within a parametric family of models, defined by some smooth model function, the maximum will not normally be reached.

Likelihood ratios A likelihood function can provide no absolute statement about the validity of any candidate models, but only relative comparisons among them. Consider the simplest case, with only two possible models. For given data, $\mathbf{y}$, they can only be compared by looking at the probability of $\mathbf{y}$ given each model. In this comparison, any *irrelevant* event, say E, independent of $\mathbf{y}$, must not be taken into account. Because such an event is independent, the joint probability is $P(E)P(\mathbf{y}; \boldsymbol{\psi}_i)$. For the irrelevant information to disappear, we must take a ratio of probabilities under the two models. (A similar argument might be made for densities in terms of transforming the observations from $\mathbf{y}$ to, say, $\mathbf{z}$, where the Jacobian would disappear in the ratio.)

In drawing final (provisional) conclusions after any study, all *relevant* information should be included in the likelihood function (Section 1.5). However, this does not imply that it is all the relevant information in the data about the data generating mechanism!

Thus, when interest centres on comparing parameter values within some functional family of models, the likelihood can be taken as *proportional* to the joint probability of the observations. In such situations, the likelihood function need only be determined up to a multiplicative constant. In other words, in this context, the only relevant statistical statements when comparing the relative merits of the models, P_{ψ_1} and P_{ψ_2}, with different fixed values of the same parameter, are in the form of a *likelihood ratio*

$$\mathrm{LR}(\boldsymbol{\psi}_1, \boldsymbol{\psi}_2; \mathbf{y}) = \frac{\mathrm{L}(\boldsymbol{\psi}_1; \mathbf{y})}{\mathrm{L}(\boldsymbol{\psi}_2; \mathbf{y})} \tag{3.9}$$

for $\psi_1, \psi_2 \in \Psi$.

Example
The likelihood ratio, for the two different parameter values, π_1 and π_2, of a binomial distribution, has the form

$$\mathrm{LR}(\pi_1, \pi_2; y) = \left(\frac{\pi_1}{\pi_2}\right)^y \left(\frac{1-\pi_1}{1-\pi_2}\right)^{n-y} \tag{3.10}$$

Notice how the combinatorial has cancelled. □

However, when comparing different model functions, there will be no cancellation of constants. Thus, the complete likelihood function of Equation (3.2) allows comparison of models with different functional forms, or containing different parameters, because they are all based on the respective probabilities of the same observed data and are all comparable to the upper bound, given by the empirical likelihood (Lindsey, 1974a, b).

Examples
1. In a normal distribution linear multiple regression, the mean of the distribution varies with the explanatory variables. Thus, with two explanatory variables, we have

$$\mu_i = \beta_0 + \beta_1 x_{1i} + \beta_2 x_{2i} \tag{3.11}$$

The approximate likelihood function can be derived from Equation (3.6):

$$\mathrm{L}(\boldsymbol{\mu}, \sigma^2; \mathbf{y}) = \frac{\mathrm{e}^{-\frac{\Sigma(y_i-\mu_i)^2}{2\sigma^2}}}{(2\pi\sigma^2)^{\frac{n}{2}}} \prod \Delta_i \tag{3.12}$$

Suppose that we wish to compare the models,

$$\begin{array}{c} \mu_i^1 = \beta_0^1 + \beta_1^1 x_{1i} \\ \mu_i^2 = \beta_0^2 + \beta_2^2 x_{2i} \end{array} \tag{3.13}$$

Applying Equation (3.9), we obtain

$$\mathrm{LR}(\boldsymbol{\mu}^1, \boldsymbol{\mu}^2; \mathbf{y}) = \left[\frac{\sum_i (y_i - \beta_0^1 - \beta_1^1 x_{1i})^2}{\sum_i (y_i - \beta_0^2 - \beta_2^2 x_{2i})^2}\right]^{-\frac{n}{2}} \tag{3.14}$$

by letting $\sigma_j^2 = \sum_i (y_i - \mu_i^j)^2$, with the appropriate values of μ_i^j in the numerator and denominator. This ratio gives the relative probability of the data under the two models. Here, because we have two normal distributions, the constant, $\exp(-n/2)/(2\pi)^{n/2}$, has cancelled.

2. Suppose now that we believe that values of a response depend on an explanatory variable with the following response equation,

$$y = \frac{\alpha x}{\beta + x} \tag{3.15}$$

If Y has a normal distribution with mean μ_i, this might be written

$$\log(\mu_i^N) = \log(\alpha x_i) - \log(\beta + x_i) \tag{3.16}$$

This model has a log link function, from Equation (1.7).

If instead it has a log normal distribution, the regression equation might be

$$\mu_i^{LN} = \log(\alpha x_i) - \log(\beta + x_i) \tag{3.17}$$

where μ_i is now the mean of $\log(Y)$. This model has an identity link function. The approximate likelihood function for a log normal distribution, from Equation (1.6), is

$$\mathrm{L}(\boldsymbol{\mu}, \sigma^2; \mathbf{y}) = \frac{\mathrm{e}^{-\frac{\Sigma[\log(y_i)-\mu_i]^2}{2\sigma^2}}}{(2\pi\sigma^2)^{\frac{n}{2}}} \prod \frac{\Delta_i}{y_i} \tag{3.18}$$

Thus, the models may be compared using

$$\mathrm{LR}(\boldsymbol{\mu}^{LN}, \boldsymbol{\mu}^N; \mathbf{y}) = \prod_i \frac{1}{y_i} \left\{ \frac{\sum_i [\log(y_i) - \mu_i^{LN}]^2}{\sum_i [y_i - \mu_i^N]^2} \right\}^{-\frac{n}{2}} \tag{3.19}$$

with the same type of replacement for σ^2 as in the first example. □

These two examples illustrate how model selection can be carried out directly by likelihood ratios, at least if all models have the same number of fixed known parameters. Such ratios can provide positive or negative evidence about a given model, but only as compared to other models.

If the parameters are unknown, so that they index two sets of models, we might consider comparing the sets by maximizing the corresponding likelihood functions in the ratio over these parameters. In this way, we can determine which set can best fit the data. As already suggested above, we are giving the best chance to each set. Again, this will only be a fair comparison if all sets have the same number of unknown parameters.

3.2.2 MAXIMUM LIKELIHOOD ESTIMATE

The most likely model within the parametric family indexed by ψ, for the observed data, is given by

$$\hat{\psi} = \arg\left[\sup_{\psi \in \Psi} \mathrm{L}(\psi; \mathbf{y})\right] \tag{3.20}$$

i.e. that value of ψ which makes the observations most probable, given the family. The corresponding parameter vector, which we have denoted by $\hat{\psi}$, is called the *maximum likelihood estimate* (m.l.e.; Fisher, 1922a). When the density approximation of Equation (3.5) is used, this maximum will also be approximate.

Example
A coin tossed $n = 10$ times yields, say, $y = 3$ heads. Suppose that the family of models under consideration is the binomial distribution indexed by the parameter, π. If we maximize the likelihood function in Equation (3.3), we obtain $\hat{\pi} = y/n = 0.3$. □

Because the likelihood function is bounded above, the m.l.e. always exists. However, the likelihood function may not have a unique maximum or may have a maximum on the boundary of the parameter space. This is not a problem. In the first case, it tells us that several models are equally likely to describe how the data were generated, but many models will generally be almost as likely. In the second case, it shows that a set of plausible models all lie to only one side of the most likely one. In some cases, the estimate may be infinite, but, if found inconvenient, this can be avoided by transformation, as we saw in Section 2.1.5.

Example
A coin tossed 10 times yields 10 heads. We are interested in the log odds or logit function

$$\theta = \log\left(\frac{\pi}{1-\pi}\right) \tag{3.21}$$

Although $\hat{\pi} = 1.0$, $\hat{\theta} = \infty$. The binomial likelihood function for θ is plotted in Figure 3.1. It is relatively flat for values above about $\theta = 6$, these being all about equally likely. As well, it shows that values of θ less than, say, about two, or of π less than about 0.88, are implausible. □

A *point estimate* is a special kind of statistic which it is hoped will be close to the 'true' value of an unknown parameter. In a production and measurement error situation, such an estimate may be the essential information sought from the data, along with an indication of precision. Location–scale models (Section 2.7.1) are specifically designed to accommodate these goals. Another area where point estimation is important is in a decision context (Part IV). The maximum likelihood estimate can be

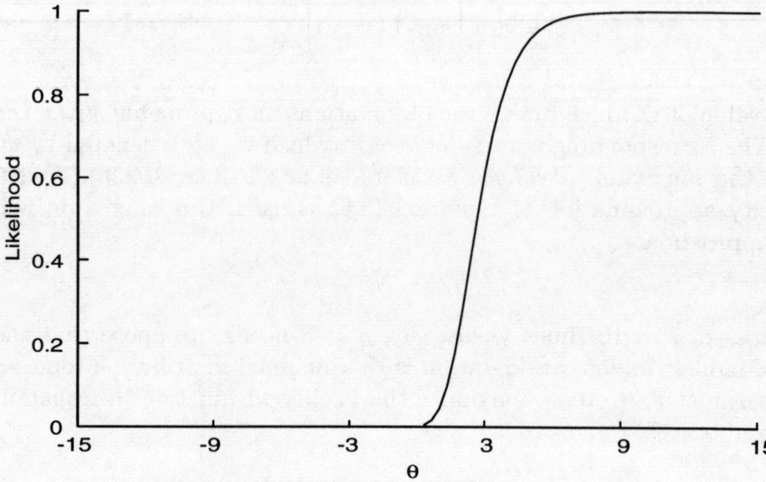

Fig. 3.1. Likelihood function for the logit function from a binomial experiment with 10 heads in 10 tosses.

looked upon as a point estimate. However, like any point estimate, in most contexts, outside of those just mentioned, it is often of little use because many other models will be almost as likely. We need to look at the form of the whole likelihood function, as we did in the example above.

3.2.3 NORMED LIKELIHOOD AND DEVIANCE

Normed likelihood When only one functional form is under consideration, it is often useful to compare all other models to the most likely model using a special likelihood ratio, the *relative* or *normed likelihood function*:

$$\mathrm{R}(\boldsymbol{\psi}; \mathbf{y}) \stackrel{\text{def}}{=} \frac{\mathrm{L}(\boldsymbol{\psi}; \mathbf{y})}{\mathrm{L}(\hat{\boldsymbol{\psi}}; \mathbf{y})} \tag{3.22}$$

which has a range of $[0, 1]$ (Fisher, 1959, pp. 68–69, 128–133). This normed function can be used to obtain *interval estimates* of the parameter, i.e. a set of plausible models. However, it is only easily interpretable if the dimension of the parameter vector is one or two, in which case it can be examined directly, for example by plotting it.

Example

For our coin, with 3 heads in 10 tosses, the normed binomial likelihood is

$$\mathrm{R}(\pi) = \left(\frac{\pi}{0.3}\right)^3 \left(\frac{1-\pi}{0.7}\right)^7$$

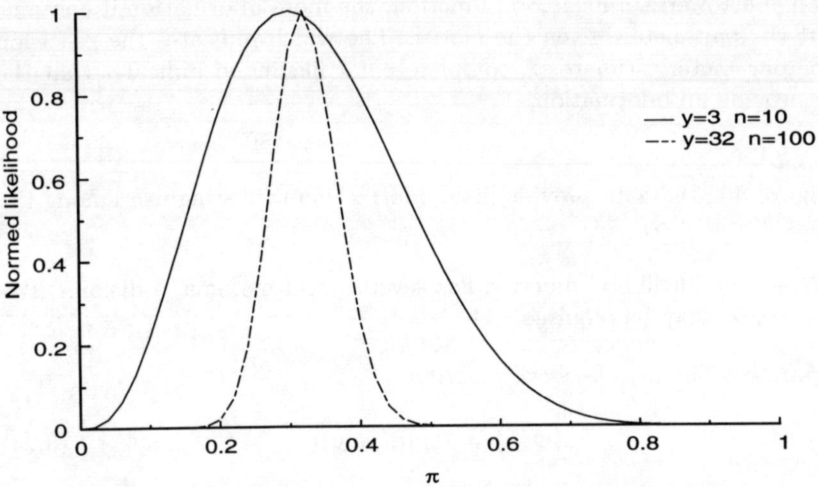

Fig. 3.2. Normed likelihood functions for two binomial experiments.

We may plot this, as with the solid curve in Figure 3.2. We see that any value of π in the interval $[0.10, 0.58]$ is at least one-fifth as likely as $\hat{\pi} = 0.3$. Note that this interval is not symmetric about the m.l.e. (although not as asymmetric as Figure 3.1!). □

As the number of observations increases, we would expect more information to become available about a parameter, if the given model remains valid. In other words, the *precision* of the estimate should increase. However, with increased information, the simple model will usually gradually lose its usefulness as more complexities of the data generating mechanism become apparent. Goodness of fit will need to be examined closely (Chapter 4).

Example
The same coin tossed 100 times yields, say, 32 heads. Now $\hat{\pi} = 0.32$ and the normed likelihood is

$$\mathrm{R}(\pi) = \left(\frac{\pi}{0.32}\right)^{32} \left(\frac{1-\pi}{0.68}\right)^{68} \tag{3.23}$$

This is the broken line in Figure 3.2. The 0.2 likelihood interval is now $[0.23, 0.41]$. As might be expected, with more information available, our interval becomes narrower (and more symmetric, like a normal distribution likelihood). But if we continue tossing too long, idiosyncrasies of the throws may become apparent or the coin will wear, introducing biases. □

The narrower is a likelihood function, the more information it contains about the parameter, given the model. The width indicates the precision of the parameter estimate. A completely flat likelihood indicates that the data provide no information.

Example
In Figure 3.1, the data provide little information to distinguish among the models, $6 < \theta < \infty$. □

When the likelihood function has several local maxima, a disjoint *likelihood region* may be required.

Deviance The *log likelihood function*

$$\mathrm{l}(\boldsymbol{\psi}; \mathbf{y}) \stackrel{\text{def}}{=} \log[\mathrm{L}(\boldsymbol{\psi}; \mathbf{y})] \tag{3.24}$$

is often convenient because it allows models to be compared additively instead of multiplicatively. For many applications, it need only be determined up to an additive constant.

Often the log normed likelihood is used, but because it is always non-positive, it can be multiplied by an arbitrary negative number (Fisher, 1950). When this number is minus two, it has come to be called the *deviance* (Nelder and Wedderburn, 1972):

$$\mathrm{D}(\boldsymbol{\psi}) \stackrel{\text{def}}{=} -2[\mathrm{l}(\boldsymbol{\psi}; \mathbf{y}) - \mathrm{l}(\hat{\boldsymbol{\psi}}; \mathbf{y})] \tag{3.25}$$

a non-negative number, which will be used in Section 5.4. The larger is the deviance, the further is the model under consideration from the most likely model, in the set under study, given the observed data.

Example
The same coin experiments can be plotted as log normed likelihoods or deviances, as is seen in Figure 3.3. □

Again, such plots allow the range of plausible values of the parameter, given the observed data, to be determined.

Because the deviance value gives no indication as to which direction a parameter of a model is situated in relation to the maximum likelihood estimate, for a scalar parameter, it is sometimes more useful to take the *directed deviance*,

$$\mathrm{D}_d(\psi) \stackrel{\text{def}}{=} \mathrm{sign}(\hat{\psi} - \psi)\sqrt{\mathrm{D}(\psi)} \tag{3.26}$$

When the model has a linear structure containing a parameter vector, as in Equation (1.9), this may be generalized, for the components of the sum

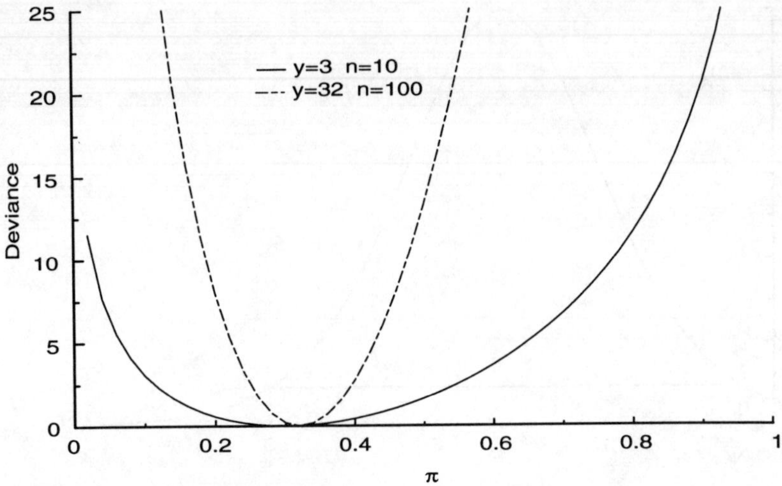

Fig. 3.3. Normed deviance functions for two binomial experiments.

Table 3.1. Guidelines for strength of evidence comparing two models, from Kass and Raftery (1995).

$R(\psi)$	$D(\psi)$	Evidence
0.37–1	0–2	Little
0.05–0.37	2–6	Positive
0.007–0.05	6–10	Strong
< 0.007	> 10	Very strong

making up the deviance, to

$$D_{di}(\boldsymbol{\beta}) \stackrel{\text{def}}{=} \text{sign}(\hat{\eta}_i - \eta_i)\sqrt{D_i(\boldsymbol{\beta})} \tag{3.27}$$

It will be used in Section 4.3.2 for goodness of fit.

Useful guidelines for judging the strength of evidence provided by various likelihood ratios or differences in deviance between models with a scalar parameter are given in Table 3.1.

3.3 Samples

As we saw in Section 1.2, a study design may involve a fixed or random number of observations. In the first case, an optimal sample size (in the sense discussed in Chapter 1 and further elaborated below) may need to be calculated before beginning the study; in the second, the way in which the study is stopped, the stopping rule, must be specified and may need

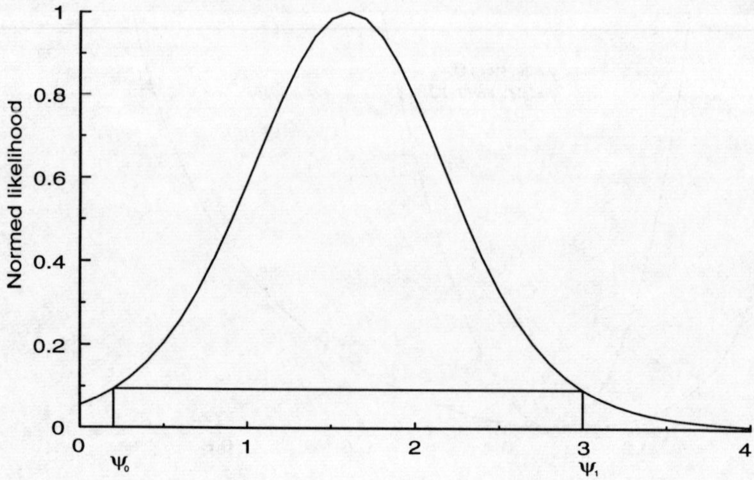

Fig. 3.4. The worst case likelihood function for distinguishing between two models.

to be taken into account in the likelihood function. The choice of design generally will determine what kind of relevant information is available so that goodness of fit of the models can be judged (Chapter 4).

3.3.1 SAMPLE SIZE CALCULATION

In planning a study, it is often important to know how large a sample will be required in order to detect some specified difference between two models, i.e. a given size of likelihood ratio or (difference in) deviance, without being too large, hence wasting time and money. For example, one is often interested in comparing a regression model with no effect of an explanatory variable to one with a prespecified difference. When the two models to be compared in a likelihood ratio or deviance are completely specified, then that ratio or (difference in) deviance is some function of the sample size, n, as well as of the m.l.e. In a worst case situation, when the m.l.e. is centred between the parameter values of the two models, this will require more observations to distinguish between them than in any other situation. Then, at worst, we would like each of them to be just at the limit of being unlikely by some criterion which we have adopted.

Suppose that we wish to be able to detect a specified difference between two values of some parameter, ψ, say between ψ_0 and ψ_1, where often $\psi_0 = 0$. This is illustrated in Figure 3.4, where a hypothetical normed likelihood function is plotted and the interval for a deviance of $-2\log(0.1) = 4.6$, our selected criterion, is indicated. Here, the curve does not arise from observed data, but is specifically chosen to lie between the two values of interest, ψ_0 and ψ_1, with equal normed likelihood of 10% for each of them. If $\hat{\psi}$ turns out to be different than that indicated by this hypothetical curve when the study is actually made, as it almost surely will, the observed curve will be displaced, either to the left or to the right. Then, one of ψ_0 and ψ_1 will have a larger normed likelihood than 10% and the other smaller so that we can choose between them.

Example
Consider the normal distribution, with unknown variance, where we want to be able to distinguish between μ_0 and μ_1. In the worst case, we shall have $\bar{y}_\bullet = \hat{\mu} = (\mu_0 + \mu_1)/2$. In this case, the deviance for, say, μ_0 would then be

$$\begin{aligned} \mathrm{D} &= \frac{1}{\sigma^2}\left(-2n\mu_0\bar{y}_\bullet + n\mu_0^2 + n\hat{\mu}\bar{y}_\bullet\right) \\ &= \frac{n}{\sigma^2}\frac{(\mu_0 - \mu_1)^2}{4} \end{aligned}$$

so that the required sample size is given by

$$n = \frac{4\mathrm{D}\sigma^2}{(\mu_0 - \mu_1)^2} \tag{3.28}$$

Because of the symmetry of the normal distribution, the answer will be the same if we used the deviance for μ_1.

The sample size is directly proportional to the variance in the model and inversely proportional to the square of the difference between the means. Suitable values of these, and of the required deviance, can be substituted into the formula to obtain the necessary sample size. □

When the likelihood function is not symmetric, calculations are slightly more complex, but can still easily be carried out, at least within the exponential family (see Section 3.7.4).

3.3.2 STOPPING RULES

Many sample designs do not fix the number of observations, n, in advance. If it is not fixed, the design is said to be *sequential*, because a decision must generally be made about stopping after each new set of observations, and perhaps after each one. Thus, some *stopping rule* must specify how to end observation. The likelihood function, for a given model function, will be the same for experiments studying the same data generating mechanism, even with different stopping rules, if these rules do not depend on that mechanism.

Example
In Section 3.1.1, we already outlined two possible ways of studying Bernoulli

trials. Thus, our family of models, $\mathcal{P}$, with constant unknown probability, π, will describe a series of (supposedly) independent Bernoulli trials which are to be performed. Let us first consider the two experiments already described, which differ only in the rule for stopping them.

Experiment 1, a fixed sample size design, consists in tossing a coin a fixed number of times, n, and recording the number of heads, y. The likelihood function is binomial, as in Equation (3.3),

$$\mathrm{L}(\pi; y, n) = \binom{n}{y} \pi^y (1-\pi)^{n-y}$$

and the sample space is $\mathcal{Y} = \{y : 0 \leq y \leq n\}$.

Experiment 2, a sequential design, consists in tossing a coin until c tails, a fixed number, have been observed, so that y is the number of heads recorded in a random total of $n = c + y$ tosses. A study, such as this, which only stops when certain conditions are met, instead of after a fixed number of observations, is said to yield a sequential sample. If the tosses are independent with constant probability of heads, the likelihood function for independent Bernoulli trials with constant probability is negative binomial,

$$\mathrm{L}(\pi; y, c) = \binom{c+y-1}{y-1} \pi^y (1-\pi)^c \tag{3.29}$$

and the sample space is $\mathcal{Y} = \{y : y \in \mathbf{N}\}$.

Note that this experiment differs in an important way from Experiment 1. Here, we must have available the result of each trial, as it occurs, in order to know whether to stop or not, and, hence, necessarily have the time-ordered sequence of results. Experiment 1 can be performed without this sequential knowledge being available.

Once the total number of tosses, $n = c + y$, and the number of heads, y, are known, both likelihood functions, for a model of independent events with constant probability, are proportional to

$$\pi^y (1-\pi)^{n-y} = \pi^y (1-\pi)^c \tag{3.30}$$

and y is the sufficient statistic for π (given n or c fixed). However, in specifying the second likelihood, we have discarded the relevant information necessarily available on the sequential order of results. □

Stopping rules must only use information available up until the point in time when stopping is to occur. Technically, this is called a martingale stopping time (Dawid, 1979; see Section B.4).

Example

In Experiment 3, suppose that the stopping rule for Bernoulli trials is 'if the fourth trial gives a head, keep the results of the first two trials only, otherwise keep the first three only'. This is not a martingale stopping time because one cannot know that the experiment is to be stopped at the second or third trial until after the fact; information is being lost. □

If the probability of stopping, given only the data so far observed, does not depend on the parameters, ψ, of the model of interest, and if that model indeed adequately describes the generation of each observation, then the stopping rule can be ignored in the likelihood function. However, studies with different stopping rules will provide different types of relevant supplementary information about the data generating mechanism which may be used to check if the model being used to describe that data generating mechanism is reasonable.

Example
The negative binomial Experiment 2 *necessarily* provides information on the order in which heads and tails occur, which may not be available from a binomial Experiment 1. This information can be used to check if the model with independence among trials is appropriate. In other words, it provides relevant information about π: whether it is constant or not over trials. However, this information is lost when the negative binomial likelihood is formed. Thus, it is the incorrect likelihood for this experiment unless prior theoretical knowledge excludes the possibility of dependence among trials, and unless this assumption does not need to be checked. □

In summary, to be adequate for complete inferences about model functions of interest, and their parameters, the likelihood function must contain the sample path of data collection in the study, when available. Reduction to sufficient statistics discards relevant information in the data, a fact which we shall study in more detail in the next section. Model checking of this kind, called goodness of fit, will be explored in more detail in Chapter 4.

3.4 Special statistics

3.4.1 SUFFICIENT STATISTICS

Because likelihood functions which are proportional are equivalent for inferences about a given model function, it is sometimes sufficient to summarize the information in the data about the parameters of that model by some function of the random variable, a statistic, say $\mathbf{T}(\mathbf{Y})$. Here, we assume, for the moment, that the number of observations is not random, but fixed at n by the sample design. To be useful, this statistic, $\mathbf{T}(\mathbf{Y})$, should generally be of lower dimension than $\mathbf{Y}$; it should provide a reduction in the data. We saw, in Section 2.2.1, that this occurs for the exponential family.

Then, any statistic, $\mathbf{T}$, such that the conditional distribution of $\mathbf{Y}$ given

$\mathbf{t}$ is the same for all $P_\psi \in \mathcal{P}$, no matter what $\mathbf{t}$ is observed, is said to be *sufficient* for ψ, given the model, P_ψ. It is *minimal sufficient* if any further reduction of dimension, beyond that of $\mathbf{T}$, results in a statistic which is not sufficient (Fisher, 1922a).

This is another situation where the discreteness of observed data is important. The conditional distribution of $\mathbf{Y}$ given $\mathbf{t}$ is not uniquely defined for continuous variables; see the example in Section B.3.5. But, on the other hand, Equation (3.4) cannot generally have the exponential family form for continuous distributions, so that reduction to canonical or sufficient statistics for densities within that family will not be possible without assuming infinite measurement precision. For continuous distributions, reduction of data to sufficient statistics is an approximation depending on the validity of Equation (3.5).

It is also important to note that a sufficient statistic only contains all of the information in the data about the parameter if the model is assumed to be adequate to describe the aspects of the data generating mechanism of interest. As we have seen in Section 3.3.2, reduction to sufficiency eliminates relevant information which would permit one to check that assumption, i.e. the suitability or goodness of fit of the model. Thus, potentially useful information, such as the order in which observations were made, is lost in reducing the data to a sufficient statistic.

Sufficient statistics have certain important characteristics:

(1) If $\mathbf{T}$ is sufficient, so is any one-to-one function of $\mathbf{T}$.
(2) If $\mathbf{T}$ is minimal sufficient, it is necessarily a function of all other possible sufficient statistics.
(3) If the statistic, $\mathbf{T}$, is sufficient, then $P(\mathbf{y}|\mathbf{t})$ does not depend on ψ with probability one. This conditional probability will be used to assess goodness of fit in Chapter 4.
(4) The observed $\mathbf{t}$ is a summary of $\mathbf{y}$ which contains all of the information about ψ in the data, under the given family of models. It divides the sample space, $\mathcal{Y}$, completely into disjoint equivalence classes, each class containing all possible observations, $\mathbf{y}$, with the same value, $\mathbf{t}$.

Note that the minimal sufficient statistic cannot be determined from inspection of an *observed* likelihood function. One also requires information as to which variables were random when the data were obtained, i.e. about the sample design.

Example

The likelihood function for n independent observations from a normal distribution with unit variance is proportional to $\exp(\mu \sum y_i - n\mu^2/2)$. If N is random, the sufficient statistic is $(n, \sum y_i)$, not just $\sum y_i$. □

Factorization theorem
A necessary and sufficient condition that $\mathbf{T}$ be sufficient for ψ in the family $\mathcal{P}$ (dominated by a σ-finite measure) is that the likelihood function factor

$$\mathrm{L}(\psi;\mathbf{y}) = \mathrm{L}_1(\psi;\mathbf{t})\mathrm{L}_2(\mathbf{y}) \tag{3.31}$$

for all $\psi \in \Psi$, where the second factor is not a function of ψ. (Fisher, 1922a, 1934; Halmos and Savage, 1949)

proof: If $\mathbf{T}$ is sufficient, by definition, $f(\mathbf{y}|\mathbf{t})$ does not depend on ψ, so that

$$\mathrm{L}(\psi;\mathbf{y}) = f(\mathbf{t};\psi)f(\mathbf{y}|\mathbf{t}) \tag{3.32}$$

Conversely, if the likelihood function factors, we have

$$\begin{aligned} f(\mathbf{y}|\mathbf{t};\psi) &= \frac{f(\mathbf{y};\psi)}{f(\mathbf{t};\psi)} \\ &= \frac{f(\mathbf{y};\psi)}{\int_{\mathbf{u}:\mathbf{t}(\mathbf{u})=\mathbf{t}} f(\mathbf{u};\psi)d\mathbf{u}} \end{aligned} \tag{3.33}$$

Substituting the factored likelihood, Equation (3.31), into the top and bottom, we see that, under suitable regularity conditions, $\mathrm{L}_1(\psi;\mathbf{t})$ cancels out so that the conditional distribution does not involve ψ. □

As a result of this theorem, certain principles for making statistical inferences have been proposed.

Sufficiency principle
Given the adequacy of the family of models, $\mathcal{P}$, identical conclusions should be drawn from all sets of observations with the same sufficient statistics. □

Equivalently, we have the

Weak likelihood principle
Data sets with proportional likelihood functions lead to identical conclusions, where the proportionality constant can contain functions of $\mathbf{y}$, but not of ψ. (Barnard, 1947, 1949; Birnbaum, 1962) □

Notice that the latter is not the same as our likelihood principle stated in Section 3.1.3; hence, to avoid confusion, we shall refer to it as the sufficiency principle. In contrast, a strong sufficiency principle would limit all inference to only such conclusions, drawn from the likelihood function alone.

The sufficiency principle implies that the same conclusions would be made from samples drawn from different probability spaces, i.e. with different sample designs, if the resulting likelihood functions are proportional.

Inference does not depend on $\mathcal{Y}$, but only on $\mathcal{P}$ and on the observed values, $\mathbf{y} \in \mathcal{Y}$.

Example
According to the sufficiency principle, the same conclusions about π would be drawn from the first two Bernoulli experiments in Section 3.3.2, although they are conducted in different ways. □

Inference procedures which use sums of probabilities over possible, but unobserved, events depending on the sample design will not obey the sufficiency principle.

Examples
1. A large observation, y, may be said to be a rare event under some model if $P(Y > y)$ is sufficiently small. However, this probability will depend on the sample space, so that an inference based on it will not obey the sufficiency principle. This probability will be different for binomial and negative binomial observations, even though the numbers of successes and failures in the Bernoulli trials are the same.
2. Any inference procedure which involves an expectation (over samples) will violate the sufficiency principle, because it will depend on an average over the whole sample space, instead of just taking that set of data, $\mathbf{y}$, which is observed. □

We saw, in Section 3.3.2, that the likelihood function does not necessarily contain all of the relevant information in the data about the data generating mechanism, or even about a parameter in the model. The weak likelihood principle is often interpreted to mean that proportional likelihood functions contain the same information about the parameter (Berger, 1985b, p. 28; Berger and Wolpert, 1988, p. 19). This is only true because some likelihood functions, for a given model function, discard different relevant information than others.

As we shall discover (Chapter 7), such a sufficiency principle is not acceptable to all statisticians. However, there is general agreement on a still weaker form, where observations are restricted to replications of the same type of experiment, and thus have the same sample space.

3.4.2 ANCILLARY STATISTICS

Certain useful statistics, calculable from the observed data, have distributions which do not depend on the parameter, ψ, of the model of interest. We shall call these *ancillary statistics*. The most important case occurs when $\mathbf{T}$ is minimal sufficient for ψ, but its dimension is greater than that of ψ. Sometimes, we can write $\mathbf{T} = (\mathbf{T}_1^{\mathrm{T}}, \mathbf{T}_2^{\mathrm{T}})^{\mathrm{T}}$, where $\mathbf{T}_2$ has a marginal distribution not depending on ψ. The distribution of $\mathbf{T}_2$ is the same for all $P_\psi \in \mathcal{P}$. Then, $\mathbf{T}_2$ is an ancillary statistic (Fisher, 1925, 1934) and $\mathbf{T}_1$ is *conditionally sufficient*, given $\mathbf{T}_2$:

$$\mathrm{L}(\psi; \mathbf{y}) \propto \mathrm{L}_1(\psi; \mathbf{t}_1|\mathbf{t}_2)\mathrm{L}_2(\mathbf{t}_2) \tag{3.34}$$

The statistic, $\mathbf{T}_2$, partitions the sample space, and can tell us which part is pertinent.

Often, the precision of the inference about ψ, as provided by such conditionality, varies with values of $\mathbf{T}_2$. In a certain sense, this is similar to the varying precision of samples of different sizes, n.

Examples

1. Suppose a research worker performs an experiment in two stages. Two instruments are available, one of which is very precise, say with variance 10, but also very expensive to operate; the second is cheaper, but gives more variable results, say with a variance of 100. First, one of the two measuring instruments is randomly chosen. Then, one measurement is taken with it. The random variable indicating the instrument selected is ancillary, so that this information should be used and conditional inferences drawn, even though the sample space for the complete experiment contains the possible results from both instruments. (Cox, 1958)

2. Suppose that $Z = Y - \psi$ has a Bernoulli distribution, with Z taking values 1 and -1. Two independent observations of Y are made. A 'natural' estimate (not the m.l.e.) is $\tilde{\psi} = \frac{1}{2}(y_1+y_2)$. Consider, however, $t_2 = y_1 - y_2$. If $t_2 = 0$, we know that $\tilde{\psi} \neq \psi$ and, if $t_2 \neq 0$, that $\tilde{\psi} = \psi$. Thus, t_2 is ancillary. (Hinkley, 1980) □

The first example is an experimental ancillary and the second a mathematical ancillary (Kalbfleisch, 1975). Suppose that, in addition to our usual sample space, described by some model(s), which is the one directly interesting us, there is a second sample space, where the model, P_T, may depend on that of interest, P_Y. An observable random variable, T, on this space is an experimental ancillary if it has the same distribution, P_T, for all choices of $P_Y \in \mathcal{P}_Y$. A mathematical ancillary has the same distribution in the restricted class of all probability distributions, P_ψ.

Thus, an objective of conditioning is to sort possible sets of data into those which are more or less informative. Unfortunately, there is no general means of constructing an ancillary statistic, $\mathbf{T}_2$, and it may not be unique. However, where possible, one useful procedure is to take $(\hat{\psi}, \mathbf{T}_2)$ in one-to-one correspondence with $\mathbf{T}$. If the marginal distribution of $\mathbf{T}_2$ does not depend on ψ, it is ancillary.

Ancillary statistics arise most frequently in the transformation family of models. As we saw in Section 2.7.1, the distribution over orbits does not depend on the parameter.

Example
For the location–scale family of Equation (2.90), the *configuration* (Fisher, 1934)

$$\mathbf{t}_2 = \left(\frac{y_1 - \hat{\mu}}{\hat{\sigma}}, \ldots, \frac{y_n - \hat{\mu}}{\hat{\sigma}}\right)^{\mathrm{T}} \tag{3.35}$$

is ancillary, defining the orbit on which the observations lie. Position on the orbit is defined by the m.l.e., $(\hat{\mu}, \hat{\sigma})$. Thus, any point in the sample space can be represented by the pair, the m.l.e. and the configuration. □

The precision of information obtained from a study should be taken into account in any inferences drawn. This can be stated as the

Ancillarity principle
If an ancillary statistic, $\mathbf{t}_2$, exists, inference on the parameter, $\boldsymbol{\psi}$, should be based on $\mathrm{L}_1(\boldsymbol{\psi}; \mathbf{t}|\mathbf{t}_2)$. □

Because the conditional likelihood is proportional to the unconditional one in Equation (3.34), by the sufficiency principle, this will make no difference for direct likelihood inferences. They always obey this principle. However, the concept of ancillarity will be important, in Chapter 4, for assessing goodness of fit of a model and, in Chapter 6, for attempting to factor the likelihood function to separate nuisance parameters from those of interest.

3.5 Derived functions

A number of functions derived from the likelihood function will be useful, especially to aid us in developing approximate methods for the study of complex likelihood functions which cannot easily be visualized. For continuous distributions, the density approximation of Equation (3.5) is generally used; all such results will depend on the validity of this approximation.

3.5.1 SCORE

The gradient vector of first partial derivatives of the log likelihood function with respect to the parameter vector, with elements

$$\mathrm{U}_{\psi_i}(\boldsymbol{\psi}; \mathbf{y}) \stackrel{\text{def}}{=} \frac{\partial \mathrm{l}(\boldsymbol{\psi}; \mathbf{y})}{\partial \psi_i} \tag{3.36}$$

is called the *score function* or *score vector* (Fisher, 1925, 1935b).

Because, for a given model function, $\mathrm{L}(\cdot)$ is only defined up to a multiplicative constant, under mild regularity conditions, it can be recovered from $\mathbf{U}$. Both the normed likelihood function, R, and the score function,

U, encapsulate all of the 'likelihood information' about the model in the data and only that. Both the mappings $\mathbf{y} \to$ R and $\mathbf{y} \to \mathbf{U}$ are minimal sufficient reductions of the data for a given model.

Example
For a one-dimensional parameter, the relationship between the deviance and the score is

$$\mathrm{D}(\psi) = 2\int_{\psi}^{y} \mathrm{U}(u)du \tag{3.37}$$

□

Score equations The score is important for calculating the maximum likelihood estimates. Thus, the *score equations*, also sometimes called *likelihood, estimating*, or even *normal equations*, are defined to be

$$\mathbf{U}(\hat{\boldsymbol{\psi}}; \mathbf{y}) = \mathbf{0} \tag{3.38}$$

Their solution yields the maximum likelihood estimates.

Examples
1. As we have seen, the likelihood function for the binomial distribution is

$$\mathrm{L}(\pi; y, n) = \binom{n}{y} \pi^y (1-\pi)^{n-y}$$

Then, the score is

$$\mathrm{U}(\pi; y, n) = \frac{y}{\pi} - \frac{n-y}{1-\pi} \tag{3.39}$$

Setting U= 0 and solving for $\hat{\pi}$, we obtain $\hat{\pi} = y/n$.
2. We know that the (approximate) likelihood function for the normal distribution is

$$\mathrm{L}(\mu, \sigma^2; \mathbf{y}) = (2\pi\sigma^2)^{-\frac{n}{2}} \mathrm{e}^{-\frac{\Sigma(y_i-\mu)^2}{2\sigma^2}}$$

Then, the elements of the score vector are

$$\begin{aligned} \mathrm{U}_{\mu}(\mu, \sigma^2; \mathbf{y}) &= \frac{\sum(y_i - \mu)}{\sigma^2} \\ \mathrm{U}_{\sigma^2}(\mu, \sigma^2; \mathbf{y}) &= -\frac{n}{2\sigma^2} + \frac{\sum(y_i-\mu)^2}{2\sigma^4} \end{aligned} \tag{3.40}$$

Setting $\mathbf{U} = \mathbf{0}$ and solving for $(\hat{\mu}, \hat{\sigma}^2)$, we obtain the (approximate) m.l.e.

$$\hat{\mu} = \frac{\sum y_i}{n} \tag{3.41}$$

$$\hat{\sigma}^2 = \frac{1}{n} \sum (y_i - \hat{\mu})^2 \tag{3.42}$$

□

The score equations are most often non-linear and cannot be solved explicitly, as in these examples, to obtain the maximum likelihood estimates. In such cases, some iterative numerical procedure must be used to obtain the solution (see Appendix D).

We shall demonstrate, in Section 5.1.2, that, for fixed, known parameter values, the expected value of the score is zero. Thus, we have both

$$\mathrm{E}[\mathbf{U}(\boldsymbol{\psi}; Y)] = \mathbf{0} \tag{3.43}$$

$$\mathbf{U}(\hat{\boldsymbol{\psi}}; y) = \mathbf{0}$$

3.5.2 OBSERVED AND EXPECTED INFORMATION

As we saw in Section 3.2.3, the amount of information in the data about a parameter is indicated by the width of the likelihood function, with narrower curves containing more information. Often, it is useful to be able to quantify this, something which is difficult to do exactly given the vast variety of forms of curves possible. However, a useful second-order approximation to the amount of information in the data about a parameter in a given model can easily be obtained.

Observed information The Hessian matrix of minus the second partial derivatives of the log likelihood function, with elements

$$\begin{aligned} \mathrm{J}_{\psi_i \psi_j}(\boldsymbol{\psi}; \mathbf{y}) &\stackrel{\text{def}}{=} -\frac{\partial^2 \mathrm{l}(\boldsymbol{\psi}; \mathbf{y})}{\partial \psi_i \partial \psi_j} \\ &= -\frac{\partial \mathrm{U}_{\psi_i}(\boldsymbol{\psi}; \mathbf{y})}{\partial \psi_j} \end{aligned} \tag{3.44}$$

is called the *observed information* (Fisher, 1925). As we have seen, the amount of information generally increases with n, the sample size. Notice that this is not a measure of the amount of relevant information supplied by the data about the data generating mechanism, but only about each parameter in a given model.

Examples

1. For the binomial distribution, the observed information is

$$\mathrm{J}(\pi; y) = \frac{y}{\pi^2} + \frac{n-y}{(1-\pi)^2} \tag{3.45}$$

2. For the normal distribution, the observed information matrix is

$$\mathbf{J}(\mu, \sigma^2; \mathbf{y}) = \begin{bmatrix} \frac{n}{\sigma^2} & \frac{\sum(y_i-\mu)}{\sigma^4} \\ \frac{\sum(y_i-\mu)}{\sigma^4} & -\frac{n}{2\sigma^4} + \frac{\sum(y_i-\mu)^2}{\sigma^6} \end{bmatrix} \tag{3.46}$$

The larger is the variance, the smaller will be the amount of information available about the two parameters for a given sample size, n. As well, their inter-relationship decreases with larger variance ($\mu \neq \hat{\mu}$), as indicated by the off-diagonal elements. □

From this second example, we see that, for known variance, the information about the mean is n/σ^2. The reciprocal square root of this, $\sigma/\sqrt{n}$, gives a measure of the width of the (log) likelihood function in the same units as the parameter. More generally, for each diagonal element of the inverse of an information matrix, $\sqrt{J^{ii}}$ provides an *approximate* (second-order) measure of width in that direction, often called the *standard error* (s.e.), and $1/\sqrt{|\mathbf{J}|}$ provides an approximate global measure of width of a multidimensional (log) likelihood surface. (Both are exact for the normal distribution, which has a quadratic log likelihood function.) As we shall see repeatedly, the latter determinant plays an important role in many approximations in statistical theory.

The statistic given by the value of the observed information at the maximum likelihood estimate is of special interest, and this has also often simply been called the observed information. Because $\hat{\psi}$ is the point of maximum likelihood, $\mathbf{J}(\hat{\psi}; \mathbf{y})$ will generally be positive definite. Its inverse is sometimes called the *dispersion matrix*.

This quantity provides a second-order estimate of the *curvature* of the (log) likelihood function, exact for the normal distribution. As we have seen, the narrower is the curve of the likelihood function, with respect to a parameter, the more information we have about that parameter. In addition, off-diagonal elements indicate the inter-relationship among parameters. In other words, they provide an indication of how much a multidimensional likelihood surface is rotated with respect to the parameter axes.

Example
For the normal distribution, the off-diagonal elements of $\mathbf{J}(\hat{\mu}, \hat{\sigma}^2; y)$ are zero. □

Expected information The expected value of $\mathbf{J}$,

$$\mathbf{I}(\boldsymbol{\psi};\mathbf{y}) \stackrel{\text{def}}{=} \mathrm{E}[\mathbf{J}(\boldsymbol{\psi};\mathbf{y})] \tag{3.47}$$

is called the *Fisher* or *expected information* matrix (Fisher, 1922a). It measures the average information over all possible observations, not just those actually made.

Examples

1. For the binomial distribution, the expected information is

$$\begin{aligned}\mathrm{I}(\pi) &= \frac{n}{\pi} + \frac{n}{1-\pi} \\ &= \frac{n}{\pi(1-\pi)}\end{aligned} \tag{3.48}$$

Less information will be available about π when it is close to 0.5.

2. For the normal distribution, the expected information matrix is

$$\mathbf{I}(\mu,\sigma^2) = \begin{bmatrix} \frac{n}{\sigma^2} & 0 \\ 0 & \frac{n}{2\sigma^4} \end{bmatrix} \tag{3.49}$$

On average, the two parameters are unrelated. □

Generally, standard errors are estimated by substituting the m.l.e. into the expected information. The use of a standard error means that we are approximating the likelihood function by a symmetric normal one:

$$\mathrm{L}_A(\psi) = \mathrm{e}^{-\frac{(\hat{\psi}-\psi)^2}{2s^2}} \tag{3.50}$$

where s is the estimate of the standard error for $\hat{\psi}$. Such approximations generally become increasingly accurate as the sample size increases, so that they are called asymptotic (Chapter 5).

Example

For the binomial distribution, $\hat{\pi} = y/n$, so that the estimated standard error is given by

$$\sqrt{\frac{\hat{\pi}(1-\hat{\pi})}{n}} = \sqrt{\frac{y(n-y)}{n^3}} \tag{3.51}$$

from Equation (3.48). The approximate likelihood function is then

$$\mathrm{L}_A(\pi) = \exp\left[-\frac{n(y-n\pi)^2}{2y(n-y)}\right] \tag{3.52}$$

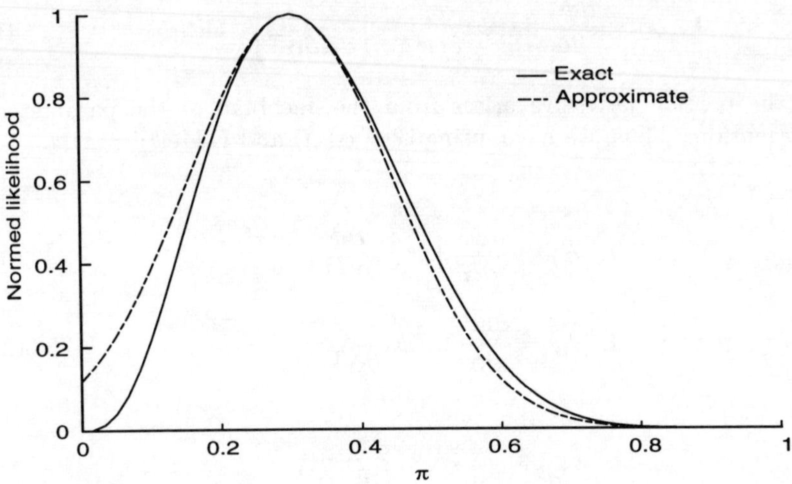

Fig. 3.5. Approximate and exact normed likelihood functions for a binomial experiment.

For $n = 10$, $y = 3$, this is plotted, along with the exact likelihood from Figure 3.2, in Figure 3.5. Notice how the approximate likelihood allows impossible negative values of π to be plausible! □

A number of important relationships hold for the expected information. If we let T_1 and T_2 be statistics derived from $\mathbf{Y}$, then

(1) $0 \leq \mathrm{I}(\psi; t_1) \leq \mathrm{I}(\psi; \mathbf{y})$;
(2) $\mathrm{I}(\psi; t_1, t_2) = \mathrm{I}(\psi; t_1|t_2) + \mathrm{I}(\psi; t_2)$
(3) $\mathrm{I}(\psi; t_1) = \mathrm{I}(\psi; \mathbf{y}) \quad \forall\psi$ iff T_1 is sufficient;
(4) $\mathrm{I}(\psi; t_2) = 0 \quad \forall\psi$ iff T_2 is ancillary.

From these, we see that, if (T_1, T_2) is sufficient, with T_2 ancillary, then $\mathrm{I}(\psi; t_1|t_2) = \mathrm{I}(\psi; \mathbf{y})$. These results follow directly from the factorization theorem.

Notice finally that an inference procedure which uses the expected information will not obey the sufficiency principle, because integration is over the sample space of unobserved values.

Parameter transformations Because the information matrix is only a second-order approximation, it will change when a non-linear transformation is applied to a parameter. Thus, suppose that $\boldsymbol{\alpha} = \boldsymbol{\alpha}(\boldsymbol{\psi})$ is a one-to-one transformation of the parameter $\boldsymbol{\psi}$. Then,

$$\mathbf{U}_\alpha(\boldsymbol{\alpha}) = \frac{\partial \boldsymbol{\psi}^{\mathrm{T}}}{\partial \boldsymbol{\alpha}} \mathbf{U}_\psi(\boldsymbol{\psi}) \tag{3.53}$$

$$\mathbf{J}_\alpha(\boldsymbol{\alpha}) = \frac{\partial \boldsymbol{\psi}^{\mathrm{T}}}{\partial \boldsymbol{\alpha}} \mathbf{J}_\psi(\boldsymbol{\psi}) \frac{\partial \boldsymbol{\psi}}{\partial \boldsymbol{\alpha}^{\mathrm{T}}} - \left[\frac{\partial^2 \boldsymbol{\psi}}{\partial \boldsymbol{\alpha} \partial \boldsymbol{\alpha}^{\mathrm{T}}} \right]^{\mathrm{T}} \mathbf{U}_\psi(\boldsymbol{\psi}) \tag{3.54}$$

where the partial derivative arises from the Jacobian of the parameter transformation. Thus, we have, using $\hat{\boldsymbol{\alpha}} = \boldsymbol{\alpha}(\hat{\boldsymbol{\psi}})$ and $\mathrm{E}[\mathbf{U}(\boldsymbol{\psi})] = \mathbf{U}(\hat{\boldsymbol{\psi}}) = \mathbf{0}$,

$$\mathbf{J}_\alpha(\hat{\boldsymbol{\alpha}}) = \left. \frac{\partial \boldsymbol{\psi}^{\mathrm{T}}}{\partial \boldsymbol{\alpha}} \mathbf{J}_\psi(\hat{\boldsymbol{\psi}}) \frac{\partial \boldsymbol{\psi}}{\partial \boldsymbol{\alpha}^{\mathrm{T}}} \right|_{\boldsymbol{\alpha}=\hat{\boldsymbol{\alpha}}} \tag{3.55}$$

$$\mathbf{I}_\alpha(\boldsymbol{\alpha}) = \frac{\partial \boldsymbol{\psi}^{\mathrm{T}}}{\partial \boldsymbol{\alpha}} \mathbf{I}_\psi(\boldsymbol{\psi}) \frac{\partial \boldsymbol{\psi}}{\partial \boldsymbol{\alpha}^{\mathrm{T}}} \tag{3.56}$$

so that

$$\begin{aligned} \left| \frac{\partial \boldsymbol{\psi}^{\mathrm{T}}}{\partial \boldsymbol{\alpha}} \right|_{\boldsymbol{\alpha}=\hat{\boldsymbol{\alpha}}} &= \sqrt{\frac{|\mathbf{I}_\alpha(\hat{\boldsymbol{\alpha}})|}{|\mathbf{I}_\psi(\hat{\boldsymbol{\psi}})|}} \\ &= \sqrt{\frac{|\mathbf{J}_\alpha(\hat{\boldsymbol{\alpha}})|}{|\mathbf{J}_\psi(\hat{\boldsymbol{\psi}})|}} \end{aligned} \tag{3.57}$$

(If we wished to define measures on Ψ, as we shall in Chapter 8, those given by $\sqrt{|\mathbf{I}(\hat{\boldsymbol{\psi}})|d\hat{\boldsymbol{\psi}}}$ and $\sqrt{|\mathbf{J}(\hat{\boldsymbol{\psi}})|d\hat{\boldsymbol{\psi}}}$ would be parametrization invariant; see Section 3.6.2 below.)

Example
For the normal distribution, suppose that we wish to work with the standard deviation, σ, instead of the variance, σ^2. Then,

$$\begin{aligned} \mathrm{U}(\sigma) &= 2\sigma \left[-\frac{n}{2\sigma^2} + \frac{\sum(y_i - \mu)^2}{2\sigma^4} \right] \\ &= -\frac{n}{\sigma} + \frac{\sum(y_i - \mu)^2}{\sigma^3} \\ \mathrm{J}(\sigma) &= 4\sigma^2 \left[-\frac{n}{2\sigma^4} + \frac{\sum(y_i - \mu)^2}{\sigma^6} \right] - 2 \left[-\frac{n}{2\sigma^2} + \frac{\sum(y_i - \mu)^2}{2\sigma^4} \right] \\ &= -\frac{n}{\sigma^2} + \frac{3\sum(y_i - \mu)^2}{\sigma^4} \end{aligned}$$

Thus, $\mathrm{J}(\hat{\sigma}) = 2n/\hat{\sigma}^2$ while $\mathrm{J}(\hat{\sigma}^2) = n/(2\hat{\sigma}^4)$ so that a parametrization invariant measure would be proportional to $d\hat{\sigma}^2/\hat{\sigma}^2 = d[\log(\hat{\sigma}^2)]$. □

3.5.3 KULLBACK–LEIBLER INFORMATION

As we have seen, all of the information in the observations about a given model is contained in the likelihood function for that assumed model family. The Fisher information is only a second-order approximation. The *Kullback–Leibler information*, closely related to entropy, is defined as the negative log likelihood or one-half the deviance; it contains all of the information in the observations for discriminating between two competing models (Kullback and Leibler, 1951).

The expected information for discriminating between models P_1 and P_2 per observation from the 'true' model, $(\mathcal{Y}, P_1)$, is the Kullback–Leibler (directed) *distance*:

$$\mathrm{K}(P_1, P_2) \stackrel{\text{def}}{=} \int \log\left(\frac{P_1}{P_2}\right) dP_1 \tag{3.58}$$

It is non-negative and zero if $P_1 \equiv P_2$. However, it is an asymmetric measure of 'distance' between two models, because

$$\mathrm{K}(P_1, P_2) \neq \mathrm{K}(P_2, P_1) \tag{3.59}$$

Thus, the Kullback–Leibler *divergence* can be defined as the sum of the 'distances' in the two directions:

$$\mathrm{KLD}(P_1, P_2) = \mathrm{K}(P_1, P_2) + \mathrm{K}(P_2, P_1) \tag{3.60}$$

a generalization of Shannon information

This divergence is related to expected information in the following way:

$$\mathrm{KLD}(P_\psi, P_{\psi+\Delta\psi}) \doteq \sum_i \sum_j \mathrm{I}_{\psi_i\psi_j}(\psi)\Delta\psi_i\Delta\psi_j \tag{3.61}$$

a second-order approximation.

3.5.4 PIVOTAL QUANTITIES

A *pivotal quantity* (Fisher, 1941) is a function of a sufficient statistic and one parameter which has a distribution independent of the parameter. It generally only exists for continuous variables (unless the parameter is discrete) and must be a monotone function of ψ for almost all y. Such quantities are peculiar in that they lie in between statistics and the usual functions of the parameters.

Exact pivotal quantities only exist within the transformation family, and particularly, the location–scale family (Section 2.7.1). Thus, they are primarily of help in the analysis of production and measurement error models. (They will be useful, in Chapter 6, for eliminating nuisance parameters, in Chapter 7, for deriving confidence intervals, and, in Chapter 8, for developing non-informative prior distributions.)

Example
For the location–scale family, given by Equation (2.90),

$$Q_i = \frac{Y_i - \mu}{\sigma} \tag{3.62}$$

is a location–scale pivotal (although it does not fall within Fisher's original definition, involving only one parameter). Further pivotal quantities are

$$Q_1 = \frac{\bar{Y}_\bullet - \mu}{S/\sqrt{n}} \tag{3.63}$$

$$Q_2 = \frac{S}{\sigma\sqrt{n}} \tag{3.64}$$

while

$$Q_{3i} = \frac{Y_i - \bar{Y}_\bullet}{S/\sqrt{n}} \qquad \sum_i Q_{3i} = 0, \sum_i Q_{3i}^2 = n(n-1) \tag{3.65}$$

is ancillary, where $S^2 = \sum(Y_i - \bar{Y}_\bullet)^2/(n-1)$. Notice the relationship to the m.l.e. and the configuration for this family. (Chamberlin and Sprott, 1989) □

The estimate of a pivotal quantity is an ancillary statistic.

A *linear pivotal quantity* is one which, for fixed value of the parameter, is a differentiable one-to-one transformation of the statistic, with Jacobian independent of the parameter, i.e. linear in the parameter of interest.

Example
For n i.i.d. normal random variables, from $\mathrm{N}(\mu, \sigma^2)$, the pivotal quantity, $\mathbf{Q} = (\frac{\bar{Y}_\bullet - \mu}{S/\sqrt{n}}, \log(S/\sqrt{n}) - \log(\sigma))^\mathrm{T}$, is linear for (μ, σ^2). □

Approximate pivotal quantities Exact pivotal quantities will only be available for transformation models, such as the location–scale model. In other cases, an approximate pivotal quantity may be useful. The most common ones are

$$q_1 = \frac{\hat{\psi} - \psi}{\sqrt{\mathrm{J}^{\psi\psi}}} \tag{3.66}$$

$$q_2 = \frac{\partial \mathrm{l}(\psi, \hat{\phi})}{\partial \psi} \sqrt{\mathrm{J}^{\psi\psi}} \tag{3.67}$$

$$q_3 = \text{sign}(\hat{\psi} - \psi)\sqrt{-2\log \mathrm{R}_P(\psi)} \tag{3.68}$$

where $J^{\psi\psi}$ is the appropriate element of $\mathbf{J}^{-1}$, so that the standard error has been used. The derivation of these quantities will be given in Chapter 5. Notice how they arise, respectively, from the m.l.e., the score, and the deviance. In fact, the third is just the directed deviance. q_2 and q_3 are only monotone if the likelihood function has a single maximum.

3.6 More complex models

The normed likelihood function for one parameter, or even two, can easily be plotted and interpreted. However, the study of a high dimensional likelihood or deviance surface can be next to impossible. One can reduce the complexity of the model itself, although this obviously has limits. Thus, one further possibility is to envisage the set of parameters simultaneously, according to some criterion. Another is to attempt to study various elements of the parameter vector in the model function separately. We shall look at all of these possibilities in this section.

3.6.1 MODEL SELECTION

Model selection is concerned with

(1) simplifying a model which is too complex,
(2) choosing among available competing models, and
(3) discovering inadequate aspects of a given set of models.

The third point will be covered in Chapter 4. The first is a special case of the second, both of which we treat in this chapter. Thus, model selection often involves searching for the simplest reasonable model function which will adequately describe the observed data.

As we have seen, normed likelihoods or deviances provide a measure of the distance of each model from the data, of relative goodness of fit. The degrees of freedom provide a corresponding measure of complexity of a set of models; a model with a smaller number of estimated parameters generally smooths the data more. Hence, the normed likelihood must be calibrated with the degrees of freedom to be interpretable. Thus, similar-sized normed likelihoods or deviances, for a series of different models, are not directly comparable if their degrees of freedom differ.

As pointed out above, we shall here look at two different types of model selection:

(1) A complex model function containing many parameters may be under consideration. A simpler nested submodel is to be selected by eliminating some of the parameters.
(2) Several distinct model functions may be in competition.

One model is said to be *nested* in another if the density for the first can be approximated arbitrarily closely by that of the second. Model selection in the first case above is sometimes called model specification. As usual, such a distinction is arbitrary, but useful; additional parameters could be used to index the different functions, for example by embedding them all in a more general model (Section 4.1.1). Both situations require some means of calibrating normed likelihoods for models of different complexity to make them comparable.

Compatibility To understand the problem of comparing model functions of different complexity, let us first look at the clearest case of parameter elimination. This occurs when all parameters are orthogonal to each other, i.e. the likelihood function factors. Thus, suppose that a series of p independent experiments, each yielding n observations, are generated from some distributions, each with one parameter, ψ_j. We wish to see if each parameter could be eliminated, by setting it to some fixed value, say zero. The likelihood function for all parameters being simultaneously zero is the product of the individual functions. Suppose that we require the precision of estimation of any given parameter to be such that the normed likelihood for an individual experiment, j, is at least a for $\psi_j = 0$ to be plausible, i.e. to eliminate that one parameter. In this case of independent experiments, we would like to obtain the same final result whatever sequence we use to eliminate the parameters, one or more at a time, in a series of steps. Then, for all inferences to be *compatible* in this way, the combined likelihood for p experiments need only be greater than a^p for it to be plausible that all parameters are simultaneously zero, i.e. to eliminate p parameters. Compatible inferences maintain the same average precision per parameter no matter how many parameters are present in the model.

We shall soon see that this is still a reasonable procedure even when parameters are inter-related. We shall also see that compatibility helps to ensure that the uncertainty in the model selection process is allowed for in the joint interval estimate for the parameters in the final model. Note that, so far, compatibility does not depend on the form of the distribution used in the model, but only on the likelihood function factoring into a product of terms, each involving only one parameter (Chapter 6).

Parameter elimination Now let us look more closely at how we can interpret a multidimensional likelihood function. For simplicity, suppose that we have a family of models indexed by only two parameters, α and β. We observe data yielding a likelihood function as in Figure 3.6, where three contours of constant likelihood are represented, the wider ones being lower on the surface. The maximum likelihood estimate, $(\hat{\alpha}, \hat{\beta})$, lies at the highest point, the diamond in the centre.

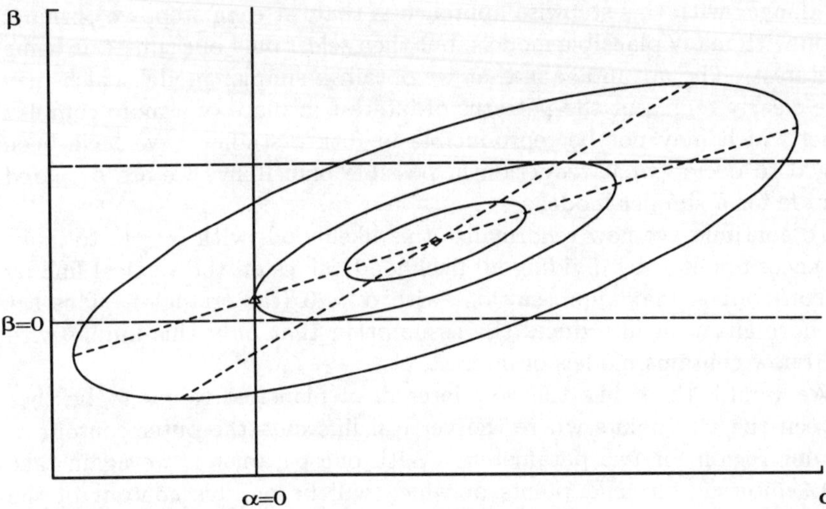

Fig. 3.6. Three contours of constant normed likelihood for a two-parameter model, with submodels for $\alpha = 0$ and $\beta = 0$.

Let us first decide to consider all complete models within the outer contour, say 4%, as being reasonably plausible given the data. We note that the model with $\alpha = \beta = 0$ lies within the region. (In what follows, we shall assume that a zero value makes a parameter disappear from a model, although, in many cases, such as a power, it may be unity or some other value.)

Next, let us, instead, check to see if either parameter individually could be zero. From our previous discussion, we believe that we could use a region of 20% for compatible inferences, because $0.2^2 = 0.04$. If this is the middle contour, we discover that models with $\alpha = 0$, the solid vertical line, and those with $\beta = 0$, the lower dashed horizontal line, lie just on that contour. Thus, both sets of models, or at least one member of each of them, are just at the limit of plausibility. We can decide to eliminate either α or β from our model. If we first eliminate α, the likelihood function for the remaining models of interest now lies on the solid vertical line through $\alpha = 0$ in Figure 3.6.

Notice how, by the decision to eliminate a parameter, we are excluding all models in the old region, here defined by the outer contour, except those in the new subspace intersecting that region, the vertical line. In Figure 3.6, when α is eliminated, all models within the outer contour are excluded, even though they are plausible, except for those on the line segment, $\alpha = 0$, contained within the contour. As well, our new most plausible model, with the constraint $\alpha = 0$, indicated by the triangle on that line, will generally be considerably less plausible than the old most plausible one (the diamond).

The danger with this stepwise approach is that, at each step, we obtain a region with many plausible models, but then select only one subset as being of interest. The advantage is that we obtain a simpler model which may more clearly represent the patterns of interest in place of a more complex model which may not be reproducible in future studies. We have been obliged to decide what reduction in possible plausibility we are prepared to trade for a simpler model.

To continue, we now renormalize this likelihood, with $\alpha = 0$, to make inferences about β by dividing all likelihood values on the vertical line by the constrained maximum, say $\hat{\beta}_\alpha$, with $\alpha = 0$ (the triangle). This act of renormalization introduces the assumption that only this family with $\alpha = 0$ now contains models of interest.

We would, then, like the new interval of plausible values to be that between the two points where the vertical line cuts the outer contour of 4%, our region for two parameters. With one parameter, we again take a 20% interval, the end points of which will lie on this contour of the two-dimensional normed likelihood surface, as required. Notice, however, that if $\alpha = 0$ lies within the 20% contour, the likelihood at $\hat{\beta}_\alpha$ will be greater than that on the 20% contour, so that a 20% interval for β, with α eliminated, will generally be narrower than the 4% contour. By eliminating unnecessary parameters in a model, we can increase the precision of the remaining ones (Altham, 1984).

However, for this likelihood function, the procedure is only approximate (or on average), in contrast to our independent experiments above. This can be seen by moving the $(0, 0)$ point in Figure 3.6 slightly down and to the left. The lines, $\alpha = 0$ and $\beta = 0$, are no longer tangent to the 20% contour, so that neither parameter would be eliminated individually. And yet, the point $(0, 0)$ is still within the 4% contour, so that they can be eliminated jointly. The approximation of this elimination procedure depends primarily on the inter-relatedness (non-orthogonality) of the parameters (Chapter 6) rather than on other aspects of the form of the likelihood function, such as whether or not the log likelihood is quadratic (Chapter 5).

We shall repeatedly see that, when parameters are inter-related (see Section 6.1), i.e. the contours are not parallel to the axes, they cannot be treated individually, except approximately. Here, if α is large (small), then β is likely to be large (small) as well. Exact inferences about one parameter in isolation are not possible; they must be made conditional on the value of the other parameter.

Suppose, now, instead that $\beta = 0$ is the upper horizontal dashed line in Figure 3.6. Then, either α or β can be eliminated individually, but the second parameter will not subsequently be removed. Thus, two distinct models will result, depending on the order of the operations. Indeed, the point for both parameters being zero lies outside the 4% contour, so that they cannot simultaneously be removed. Note also that moving the $(0, 0)$

point a little does not now change the results of the selection procedure. This confirms that we only have compatible inferences approximately or on average.

When parameters are not orthogonal, as indicated by the slope of the contours with respect to the axes, stepwise elimination can yield different results depending on the order followed. We saw above, in Section 3.5.2, that one approximate (second-order) measure of such non-orthogonality is the size of the off-diagonal element of the information matrix. Of course, the likelihood surface will usually have a much more complex shape than that in Figure 3.6, in which case the information matrix only provides a crude indication of width and non-orthogonality.

To summarize, the only way in which successively renormalized likelihoods, as parameters are eliminated, can provide approximately compatible inferences about the set of parameters in a model is if all are calibrated as a^p, where p is the number of estimated parameters and a is a positive constant. If we start with a one-parameter region of 20%, we must have $a = 0.2$ and the comparable two-parameter region will be 4%, and so on for more parameters. The smaller a value chosen for a, the wider will be the interval estimate for one parameter, the less will be the precision, and the less plausible need be the simpler model with parameters eliminated. Thus, the smaller is a, the simpler will be the model chosen. I shall call a the *smoothing factor*, because simpler models smooth the data more.

Non-nested models As we saw in Section 3.2.1, the most complex model which can be fitted to a given set of data is a multinomial distribution, the empirical likelihood. These multinomial probabilities can often be structured in several different ways by the use of various competing model functions, perhaps of different complexity. The latter are just various submanifolds within the space defined by the (saturated) empirical likelihood. The simplest case is that just studied, where successive functions are obtained by adding or removing parameters. However, there is no reason that models need be nested in this way. Comparison of completely different functional forms will often be of interest.

Example
In Section 3.2.1, we compared simple linear regression with two different explanatory variables, as in Equation (3.13), and normal and log normal distributions for responses depending on an explanatory variable, as in Equation (3.15) for fixed and equal numbers of parameters. We are now in a position to extend this to the more general case. □

Thus, our calibration of normed likelihood contours by a^p, according to the complexity of the models, does not require the models to be nested. It is applicable for all model selection procedures on a given set of data. For example, the calibrated deviance can be written

$$\mathrm{D}_c \stackrel{\text{def}}{=} \mathrm{D} - 2p\log(a) \tag{3.69}$$

where p is the number of estimated parameters in the model and a the smoothing factor. Then, a is chosen to obtain the required precision of *one* parameter or the required level of simplicity of the model selected. Equation (3.69) is based on what is sometimes called a *penalized likelihood*, because smaller values of a provide a larger penalty against complex models.

The value, $a = 1/e$, is widely used and is often a reasonable choice; it agrees with the guidelines separating little from positive evidence in Table 3.1. With this value of a, Equation (3.69) is called the Akaike information criterion (AIC). Its reasonableness basically depends on the sample size criterion (Section 3.3.1) having about this value of a. Other choices have also been proposed, including making a a function of the sample size, for example $a = 1/\sqrt{n}$ or $a = 1/\sqrt{\log(n)}$, the former sometimes called the Bayesian information criterion (BIC; see Section 8.6.4). Such choices as these two tend to select rather simple models as n increases, indicating a null model, with only one parameter, as n becomes very large. However, they can be useful to obtain simple models when, for some reason, the sample size is too large for the degree of complexity of model desired.

Interval estimation In Section 3.2.3, we studied interval estimation for a model function with only one unknown parameter. By constructing such a likelihood interval, we discarded models, indexed by the parameter, which the data showed to be unlikely. Model selection is the direct extension of this process to the more complex case where we have multidimensional parameters and often several model functions under consideration. Again, unlikely models are being eliminated. For the different stages of this process to be compatible, we have now seen that interval estimation for a set of parameters must use normed likelihood functions calibrated by a^p where p is the number of parameters in the set (even if $p = 1$). The size of a chosen determines the width of the joint interval estimates. In this way, the uncertainty in model selection is taken into account.

Very often, the model selection process results in no clear cut choice among some of the models (indeed, this is the very idea of requiring an interval estimate!), whether distinct model functions or zero parameter values on the edge of a plausibility region. Then, conclusions should be examined in the light of all of these models – until additional data can be collected.

Further model selection procedures will be described in Chapter 4. Hopefully, the reader need not be reminded that a model selected from a given set of data cannot be critically examined or tested (Section 4.2) from the same data. In a scientific context, a new study has to be performed.

3.6.2 PARAMETRIZATION INVARIANCE

No model has one unique parametrization in the sense that the model is essentially unchanged by any one-to-one transformation of the parameters. The probability of the data will not change. Thus, parameters can be chosen which are most easily interpretable in given circumstances.

Example
For simplicity, the mean is often assumed to vary as a polynomial function of some explanatory variable, for example the quadratic function,

$$\mu_i = \beta_0 + \beta_1 x_i + \beta_2 x_i^2 \tag{3.70}$$

However, the parameters may be more easily interpretable if we write this as

$$\mu_i = \alpha_1 - \alpha_3 (x_i - \alpha_2)^2 \tag{3.71}$$

where α_1 is the maximum of μ_i corresponding to $x_i = \alpha_2$. Unfortunately, this is no longer a linear function of the unknown parameter vector, $\boldsymbol{\alpha}$. On the other hand, for numerical stability in calculations, orthogonal polynomials are preferable:

$$\mu_i = \psi_0 + \psi_1(x_i - \bar{x}_\bullet) + \psi_2[(x_i - \bar{x}_\bullet)^2 - \overline{(x_i - \bar{x}_\bullet)^2}] \tag{3.72}$$

One would hope to obtain the same final results from all three of these formulations. □

Thus, parameters which define a model mathematically, those which are most easily interpretable, and those which are most accurately computable may not coincide. Generally, numerical calculation of the maximum likelihood estimate will be easier if contours of the likelihood surface are close to symmetrical ellipsoids.

An inference procedure is *parametrization invariant* if it yields the same results for all one-to-one transformations of the parameters. Reparametrizing the model and then applying the procedure should yield the same results as first applying the procedure and then translating the results to the new parametrization.

Example
The exponential distribution of Equation (1.14) may be written

$$f(y; \rho) = \rho e^{-\rho y} \tag{3.73}$$

$$= \frac{1}{\mu} e^{-\frac{y}{\mu}}$$

where μ is the mean duration between events and ρ is the risk or intensity of events. Inferences should not depend on whether analysis is performed in terms of the mean, then transformed to intensity, or directly for the intensity. □

Maximum likelihood estimation and inferences from the (normed) likelihood function are parametrization invariant. On the other hand, as we shall see in later chapters, many classical inference procedures and methods of approximation are not, especially those depending on asymptotic normality (Chapter 5; see also Sections 7.4.1, the beginning of 8.3, and 8.5.1).

One commonly used statistic which is not parametrization invariant is the standard error, derived from the information matrix. We saw at the end of Section 3.5.2 that a Jacobian is necessary when the parameters are transformed. This is not taken into account in the standard errors for different parametrizations.

Example
From Equation (3.73), for n independent observations, the estimated standard error for $\hat{\rho}$ in the previous example would be $\hat{\rho}/\sqrt{n}$, while that for $\hat{\mu}$ would be $\hat{\mu}/\sqrt{n} = 1/(\hat{\rho}\sqrt{n})$. □

For small samples, the accuracy of the approximate likelihood in Equation (3.50) depends heavily on the parametrization used (Chapter 5).

3.6.3 NORMED LIKELIHOOD REGIONS

As we saw in Section 3.2.3, the normed likelihood interval for a one-dimensional parameter is easily obtained and interpreted. The same is not true of higher dimensional models; the maximized likelihoods are not necessarily comparable if the maximization has been performed over differing numbers of parameters. From the results in Section 3.6.1, we know that, to obtain compatible inferences, simultaneous regions for the set of parameters in the model can be approximately calibrated as a^p, where p is the number of estimated parameters and a is a smoothing factor.

However, we may often wish to study individually some of the parameters in the model selected. We require appropriate techniques for this which allow for the uncertainty about other parameters in the model which are not currently under study. Thus, the procedures of this section should not be used for model selection, such as determining if a parameter might be zero. They do not adequately take into account the complexity of the models. Their application is rather to interval estimation of a parameter once an appropriate model function has been selected.

Profile likelihood When the parameter vector of a given model is of high dimension, one often wishes to be able to study one parameter component in isolation (Chapter 6). Let us call this component of the parameter of interest, λ, and the other, nuisance, parameters, $\boldsymbol{\phi}$. Then, a (normed) *profile likelihood* for the parameter of interest may be useful. It is defined by

$$\begin{aligned} \mathrm{R}_P(\lambda; \mathbf{y}) &\stackrel{\text{def}}{=} \sup_{\boldsymbol{\psi}|\lambda} \mathrm{R}(\boldsymbol{\psi}; \mathbf{y}) \\ &= \mathrm{R}(\hat{\boldsymbol{\phi}}_\lambda; \mathbf{y}) \end{aligned} \tag{3.74}$$

where $\hat{\boldsymbol{\phi}}_\lambda$ is the m.l.e. of $\boldsymbol{\phi}$ for the given fixed value of λ.

Example
Take n independent observations generated from a normal distribution. If we use the density approximation for continuous probability distributions, the likelihood was given in Equation (3.12), but here is used with common mean, μ.

Suppose that this mean parameter is of interest. For any value of μ, $\hat{\sigma}^2_\mu = \sum(y_i - \mu)^2/n$. Then, the normed profile likelihood is

$$\mathrm{R}_P(\mu) = \left[\frac{\sum(y_i - \hat{\mu})^2}{\sum(y_i - \mu)^2}\right]^{\frac{n}{2}} \tag{3.75}$$

similar to what we already saw in Equation (3.14). Here, this is just a ratio of sums of squared deviations from the mean, where $\sum(y_i - \hat{\mu})^2$ is the smallest such sum for the observed data. □

The profile likelihood function may be plotted in the same way as a one-dimensional normed likelihood.

Example
In Figure 3.6, the two diagonal dashed lines trace the profile likelihoods for α and β. Each of these could be plotted as a curve like those in Figure 3.2. □

However, greater care must be taken because this is the profile of a multidimensional surface. At each point, all other parameters are assumed to be fixed at their m.l.e. so that the width of the likelihood surface in those directions is not taken into account. The imprecision in the estimation of those parameters has not been allowed for. Thus, one may argue that a profile likelihood gives a narrower range of likely values of the parameter of interest than is necessarily appropriate, increasingly so as the number of eliminated nuisance parameters increases.

One solution, perhaps the best when the nuisance parameter is of small dimension, is to plot a series of *cross-sectional* likelihoods for various fixed values of the nuisance parameter.

Modified profile likelihood A normed profile likelihood function does not take into account the uncertainty in the values of the parameters over which it is maximized, if the parameters are interdependent. The more information we have about these parameters, the less uncertain we are. Thus, to calibrate the curve properly, we should divide the profile likelihood by some measure of the amount of information available about the nuisance parameters. Then, a region of models for values of the parameter of interest, λ, will become wider if we are more uncertain about the nuisance parameters, $\boldsymbol{\phi}$.

Example
In Figure 3.6, the uncertainty about β for fixed α is less than that about α for fixed β. This difference is not taken into account by the two profile likelihoods.
□

Notice that this is exactly what we were doing in the conditional density of Equation (2.37)

$$f(\mathbf{t}_2|\mathbf{t}_1;\boldsymbol{\theta}_2) = \frac{f(\mathbf{t}_1,\mathbf{t}_2;\boldsymbol{\theta})}{f(\mathbf{t}_1;\boldsymbol{\theta})} \tag{3.76}$$

The full likelihood for $\boldsymbol{\theta}$ corresponds to the numerator, while all information about the nuisance parameter, $\boldsymbol{\theta}_1$, given the model, is contained in the sufficient statistic, $\mathbf{t}_1$, and hence in the denominator (see Section 6.2.5). In the case of the linear exponential family, the nuisance parameter actually completely disappears from a conditional likelihood based on the left hand side. Unfortunately, this is a special, but useful, case; we here require a more general, if approximate, result.

As we have seen, the information about a parameter is generally indicated by the width of the (normed) likelihood in the direction of that parameter. In Section 3.5.2, we saw that the observed information matrix provides an *approximate* measure of the curvature of the likelihood function (exact for the normal distribution). Thus, an approximate measure of the width of the function in the direction of several parameter components is the reciprocal square root of the determinant of this matrix at their m.l.e. For the nuisance parameters, with fixed λ, this will be $1/\sqrt{|\mathbf{J}_{\phi\phi}(\lambda,\hat{\boldsymbol{\phi}}_\lambda)|}$. However, this is for $\boldsymbol{\phi}_\lambda$ and we are working with $\boldsymbol{\phi}$, so that we apply Equation (3.57) to transform back. The final *modified profile likelihood* is

$$\mathrm{L}_{MP}(\lambda; \mathbf{y}) = \frac{\mathrm{L}_P(\boldsymbol{\lambda}; \mathbf{t})}{\left|\frac{\partial \hat{\boldsymbol{\phi}}_\lambda}{\partial \hat{\boldsymbol{\phi}}}\right| \sqrt{|\mathbf{J}_{\phi\phi}(\lambda, \hat{\boldsymbol{\phi}}_\lambda)|}} \tag{3.77}$$

Notice that the m.l.e. from this function may not be the same as for the profile likelihood.

When calculable, this likelihood is parametrization invariant. In addition, the partial derivative will not involve λ if $\mathbf{J}_{\lambda\phi} = 0$ and it will be one if $\hat{\boldsymbol{\phi}}_\lambda$ does not depend on λ. However, it will be impossible to calculate if the m.l.e. cannot be written in explicit form.

Example
For n independent observations from a normal distribution, $\hat{\mu}_{\sigma^2} = \bar{y}_\bullet$ and the profile likelihood function for σ^2 is

$$\mathrm{L}_P(\sigma^2; \mathbf{y}) = (2\pi\sigma^2)^{-\frac{n}{2}} \mathrm{e}^{-\frac{\Sigma(y_i - \bar{y}_\bullet)^2}{2\sigma^2}} \tag{3.78}$$

For fixed σ^2, the observed information about μ is

$$\mathrm{J}_{\mu\mu}(\sigma^2, \hat{\mu}) = \frac{n}{\sigma^2} \tag{3.79}$$

while the partial derivative for transforming the m.l.e. does not depend on σ^2. Then, the modified profile likelihood is

$$\mathrm{L}_{MP}(\sigma^2; \mathbf{y}) = (2\pi\sigma^2)^{-\frac{n-1}{2}} \mathrm{e}^{-\frac{\Sigma(y_i - \bar{y}_\bullet)^2}{2\sigma^2}} \tag{3.80}$$

which, in this case, not surprisingly, is just the conditional density of Equation (2.38). The normed modified likelihood function will be slightly wider than the corresponding normed profile likelihood function and the m.l.e. will be $\hat{\sigma}^2 = \sum(y_i - \bar{y}_\bullet)^2/(n-1)$ instead of $\hat{\sigma}^2 = \sum(y_i - \bar{y}_\bullet)^2/n$. The division by $n-1$ allows for the uncertainty in the estimation of the mean.

For $n = 10$ observations, suppose that we have $\sum y_i = 50$ and $\sum y_i^2 = 300$. The ordinary and modified likelihood estimates of σ^2 are, respectively, 5 and 5.56. The two deviance plots are given in Figure 3.7. We see that the modified profile likelihood intervals are slightly larger than those for the profile deviance, reflecting the uncertainty about the value of the mean. □

This is another penalized likelihood. In Section 3.6.1, we penalized complex models; here we penalize uncertainty in the nuisance parameters. Intuitively, this approximate correction might be thought to involve multiplying the likelihood function by the standard errors of the nuisance parameters at the value of the parameter of interest, λ. Because we used a normed likelihood, it is rather the ratio of these standard errors to those

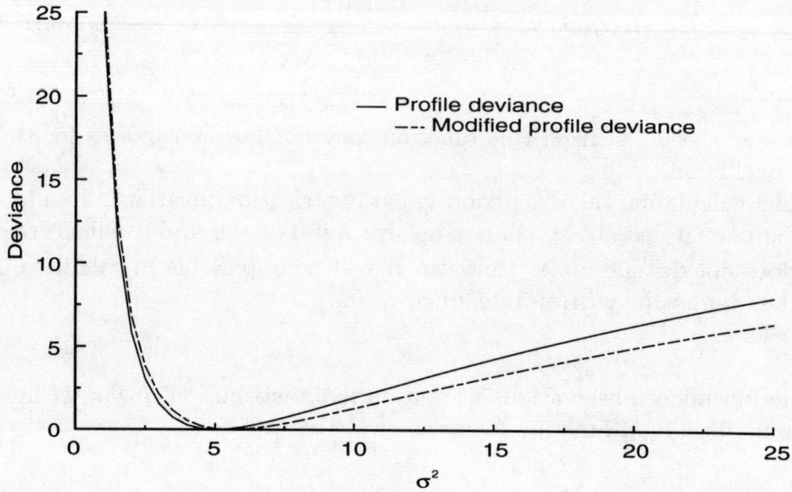

Fig. 3.7. Profile and modified profile deviance plots for the variance of 10 observations from a normal distribution.

at the m.l.e. However, this interpretation can be misleading because the partial derivative also plays an important role in the correction and cannot be ignored (Section 6.3.3). If it is, the resulting function, often called an adjusted profile likelihood, is not parametrization invariant and can yield arbitrary conclusions, depending on the transformation of ϕ used.

If a proper model selection procedure has not been carried out before studying a parameter of interest in this way, so that the nuisance parameters are, in fact, unnecessary, the resulting modified profile likelihood may be narrower than the profile one, but it should be wider than that based on a model with such nuisance parameters eliminated.

We shall study such approximations in more detail in Part III. In particular, we shall see, in Chapter 6, that conditioning on ancillary statistics, obtained when the nuisance parameters have fixed values, makes allowance for the lack of precision of information about them, as in Equation (3.77). Indeed, we already know that the modified profile likelihood for the variance of a normal distribution just given is exact, although we have only used intuitive arguments here. Another example is given below in Section 4.2.2.

3.6.4 PRIOR INFORMATION

In a scientific context, prior theoretical knowledge about some phenomenon under study is expressed in the form of models, while prior empirical information will be contained in the corresponding prior likelihood functions containing the data obtained in previous studies. These may be used to formulate some specific model of special interest to be tested by the new data (Section 4.2). If the evidence is not against this model, then we should combine all of the information that we have about it.

Thus, up until now, we have been primarily concerned with obtaining the information from the current study. But the final (provisional!) conclusions should incorporate all available relevant information. If the circumstances remain constant and the observations are independent, the prior and current likelihood functions can be combined by multiplication. On the other hand, if certain aspects of the situation vary, here over time, this must be modelled (see Section 6.4.1).

Examples
1. A chemical experiment is performed to determine the specific gravity of a compound. If measurement errors can be assumed to follow a normal distribution, the mean gives the quantity desired and the variance expresses the variability of the instrument used. If, subsequently, the experiment is repeated independently in another laboratory, the mean should be the same, but the variance may be different, depending on the type of instrument used and the quality controls applied. But, in either case, the previous information can be incorporated by multiplying the two likelihoods. Because of the increased amount of information, this estimate of the mean is preferable to that only from the current experiment.
2. A new medicine is usually tested a number of times over a period of years, often in different clinics, before being commercialized. Interest may centre on the odds of cure as compared to that under standard treatment. However, the overall rate of cure may vary among clinics, because of the differing types of patients which attend them. Information may or may not be directly available about why this is so.
3. Suppose that, from prior theoretical knowledge, we believe that observations will be adequately represented by a model based on the Poisson distribution with unknown mean, μ. A previous study has reported n_1 observations with mean $\bar{y}_{1\bullet}$. Thus, we can construct the available prior likelihood as

$$\mathrm{L}(\mu; \bar{y}_{1\bullet}) \propto \mathrm{e}^{-n_1\mu}\mu^{n_1\bar{y}_{1\bullet}} \tag{3.81}$$

We now make n_2 new observations and obtain a mean, $\bar{y}_{2\bullet}$, yielding the likelihood

$$\mathrm{L}(\mu; \bar{y}_{2\bullet}) \propto \mathrm{e}^{-n_2\mu}\mu^{n_2\bar{y}_{2\bullet}} \tag{3.82}$$

Combining this with the prior information, we obtain

$$\mathrm{L}(\mu; \bar{y}_{1\bullet}, \bar{y}_{2\bullet}) \propto \mathrm{e}^{-(n_1+n_2)\mu}\mu^{n_1\bar{y}_{1\bullet}+n_2\bar{y}_{2\bullet}} \tag{3.83}$$

from which we can make inferences about μ. Of course, we can also compare this with a model where μ is different in the two studies to check if they are comparable (Chapter 4). □

As in the first example above, it is not always necessary that the model be identical for all studies. If some parameter of interest, such as a treatment effect, is common to the models for a series of studies, then combining them should result in increased precision, i.e. increased knowledge, about that parameter, even if other aspect of the models for some of the studies are different. However, interpretation may often be difficult, because the various models will generally involve conditioning on different factors.

Extreme care must be taken when such prior information is obtained from the scientific literature. Studies with negative results are often not published, so that the prior likelihood based on those in the literature may not be representative, but have an 'optimistic' bias towards positive results (Sterling, 1959; Sterling *et al.*, 1995). As well, if model checking (Chapter 4) is to be performed, as it should, then the raw data, and not just the likelihood function, will be needed from each study.

Independence among successive sets of observations is usually assumed when they are made on different subjects. When previous studies involved the same individuals, the prior information is much stronger. The actual evolution over time can, then, be modelled, as a stochastic process. However, likelihoods for previous observations can no longer simply be multiplied with that for the present ones. Parameters for the dependence links over time must be included in the model.

Example

In the social sciences, a longitudinal study follows a panel of people over time so that each person's changes in, say, some opinion are recorded. The model can allow current opinion to depend on those previously held, thus setting up a stochastic dependence process. □

3.6.5 PREDICTION

When a model, based on a probability distribution, is completely known, it can be used directly to predict future observations. This is very rarely the case in practice. Rather, we generally have a set of more or less plausible models, given observed data, upon which we wish to base such predictions. As is the case with parameter estimation, we shall not just be interested in a point estimate. Rather, a *prediction interval* should be derived, indicating the degree of uncertainty.

When we want to make inferences about a future observation, say y_F, given some data, $\mathbf{y}$, we can treat the former as an unknown quantity, similar to a parameter. However, in contrast to parameters, prediction quantities can eventually be directly observed.

Relative likelihood The simplest approach is, then, to look at the profile likelihood, maximized over the unknown parameters (Fisher, 1959, pp. 128–133)

$$\mathrm{R}_P(y_F; \mathbf{y}) = \sup_{\psi} \frac{f(y_F; \psi) f(\mathbf{y}; \psi)}{f(\hat{y}_F; \hat{\psi}) f(\mathbf{y}; \hat{\psi})} \tag{3.84}$$

where the future observation is independent of those already available. Here, however, we see even more clearly how a profile likelihood gives intervals which are too narrow: the prediction interval for y_F obtained from this likelihood will not be wide enough because we assume that we know the model with the maximum likelihood estimate, $\hat{\psi}$, to be adequate.

Example
For n independent observations from a normal distribution, the profile likelihood for an independent future observation is

$$\mathrm{L}_P(y_F; \mathbf{y}) = (2\pi\hat{\sigma}_F^2)^{-\frac{n+1}{2}} \mathrm{e}^{-\frac{\Sigma(y_i - \hat{\mu}_F)^2 + (y_F - \hat{\mu}_F)^2}{2\hat{\sigma}_F^2}} \tag{3.85}$$

with $\phi = (\mu, \sigma^2)^{\mathrm{T}}$ and the subscript, F, indicating the m.l.e. for arbitrary fixed y_F. The maximum likelihood estimates are $\hat{\mu} = \bar{y}_\bullet$, $\hat{\sigma}^2 = \sum(y_i - \hat{\mu})^2/(n+1)$, and $\hat{y}_F = \hat{\mu}$, so that the predicted value is at the mean of the observations. However, the width of the profile likelihood will be as if we had $n+1$ observations! □

Modified profile likelihood As previously, the modified profile likelihood can be used to take into account the unknown variability, here in μ and σ^2 (Butler, 1989).

Example
For our normal data,

$$\left|\sqrt{\mathbf{J}_{\phi\phi}(y_F, \hat{\mu}, \hat{\sigma}^2)}\right| = \frac{n+1}{\sqrt{2}\hat{\sigma}^3} \tag{3.86}$$

while the partial derivatives do not depend on the parameters. Then, the modified profile likelihood is

$$\mathrm{L}_{MP}(y_F; \mathbf{y}) = (2\pi\hat{\sigma}_F^2)^{-\frac{n-2}{2}} \mathrm{e}^{-\frac{\Sigma(y_i - \hat{\mu}_F)^2 + (y_F - \hat{\mu}_F)^2}{2\hat{\sigma}_F^2}} \tag{3.87}$$

where

$$\begin{aligned} \hat{\mu}_F &= \tfrac{n\bar{y}_\bullet + y_F}{n+1} \\ \hat{\sigma}_F^2 &= \tfrac{\sum(y_i - \bar{y}_\bullet)^2 + (y_F - \bar{y}_\bullet)^2}{n-2} \end{aligned} \tag{3.88}$$

while $\hat{y}_F = \bar{y}_\bullet$ as before. This will give somewhat wider prediction intervals than does the profile likelihood. □

Any prediction will assume that the model used is valid; this is especially critical for extrapolation. It is usually wise to compare the predictions of a number of different reasonable models. Another possibility, if only the prediction, and not the validity of the model, is in question, is to give an average prediction by suitably weighting together those from the individual models (see Section 8.4.1).

3.6.6 PARADOXES

If one reflects on the likelihood principle, as stated in Section 3.1.3, and on the definition of the likelihood function, as given in Section 3.2, one realizes that problems can only arise from two sources:

(1) the set of possible models to describe the data generating mechanism is ill-defined, or
(2) the sample design is inappropriate to obtain information about the data generating mechanism.

All of the possible models which can be considered must yield clearly defined probabilities of the observed data occurring. Otherwise, either the models or the data are inappropriate for the problem at hand.

Nevertheless, the likelihood function has been accused of generating paradoxes. Certain of these involve the maximum likelihood estimate as a point estimate, rather than the likelihood function, and can easily be resolved by examining the latter. Many are based on continuous variables and defining the likelihood function to be the probability density function. These paradoxes are closely related to those arising from conditioning, as in the example of Section B.3.7.

Example
Consider the three-parameter log normal distribution, with density

$$f(y;\mu,\sigma^2,\gamma) = \frac{\mathrm{e}^{-\frac{[\log(y-\gamma)-\mu]^2}{2\sigma^2}}}{(y-\gamma)\sqrt{2\pi\sigma^2}} \qquad y > \gamma \tag{3.89}$$

Suppose that the likelihood function were defined directly in terms of this density, with Y a continuous variable. For a sample of size n, the normed likelihood for γ would approach infinity as γ was allowed to approach the smallest observed value. However, we have seen that a likelihood function, properly defined for the observed data, has a finite maximum given by Equation (3.7). (Hill, 1963; Griffiths, 1980) □

A paradox appears to arise when the sample design is such that there are too few observations for the number of parameters. In such cases, it would often be impossible for the 'true' parameter value, which generated the data, to be the maximum likelihood estimate. This can arise in quite complex and subtle ways, but let us look at a simple example which combines both this and a continuous variable.

Example
For one observation from a normal distribution, the parameters are estimated as $\hat{\mu} = y$ and $\hat{\sigma}^2 = 0$. This is not surprising, because one observation can provide no information about variability. However, with our usual density approximation to the integral, the likelihood function is infinite at this point. Thus, all other parameter values are infinitely less likely than the m.l.e. However, this is no longer possible if the exact integral, with the true unit of measurement, is used. (Barnard, 1965) □

A second inadequacy of the sample design occurs when an inappropriate stopping rule is used for the data generating mechanism under study.

Example
For a series of independent observations from a normal distribution, suppose that the rule is to stop if

$$\bar{y}_\bullet \leq A - \frac{k}{\sqrt{n}} \qquad \text{or} \qquad \bar{y}_\bullet \geq B + \frac{k}{\sqrt{n}} \qquad A \leq B \tag{3.90}$$

If the mean actually lies in the interval $[A, B]$, then the normed likelihood of that value can be made arbitrarily small by the choice of k. This stopping rule prevents the m.l.e. from ever taking its 'true' value if the latter lies in the interval $[A, B]$. □

Finally, the very appropriateness of the likelihood function for choosing among models may be questioned, as in the following example.

Example
Consider the probability distribution,

$$P(Y = \psi) = \frac{1}{3}$$
$$P(Y = 4\psi + i) = \frac{1}{6} \qquad i = 1, 2, 3, 4$$

as shown in Table 3.2. First suppose that this is just one corner of the complete distribution, so that ψ and Y can each have any non-negative integer value. For any observed value, except $y = 0$, only two values of the parameter could possibly have generated it.

Table 3.2. A discrete probability distribution.

ψ	y: 0	1	2	3	4	5	6	7	8	9	10	11	12
0	$\frac{1}{3}$	$\frac{1}{6}$	$\frac{1}{6}$	$\frac{1}{6}$	$\frac{1}{6}$	0	0	0	0	0	0	0	0
1	0	$\frac{1}{3}$	0	0	0	$\frac{1}{6}$	$\frac{1}{6}$	$\frac{1}{6}$	$\frac{1}{6}$	0	0	0	0
2	0	0	$\frac{1}{3}$	0	0	0	0	0	0	$\frac{1}{6}$	$\frac{1}{6}$	$\frac{1}{6}$	$\frac{1}{6}$
3	0	0	0	$\frac{1}{3}$	0	0	0	0	0	0	0	0	0
4	0	0	0	0	$\frac{1}{3}$	0	0	0	0	0	0	0	0

ψ	y: 13	14	15	16	17	18	19	20
0	0	0	0	0	0	0	0	0
1	0	0	0	0	0	0	0	0
2	0	0	0	0	0	0	0	0
3	$\frac{1}{6}$	$\frac{1}{6}$	$\frac{1}{6}$	$\frac{1}{6}$	0	0	0	0
4	0	0	0	0	$\frac{1}{6}$	$\frac{1}{6}$	$\frac{1}{6}$	$\frac{1}{6}$

The likelihood function for one observation is the appropriate column of the table, indicating that the larger of the two possible parameter values is twice as likely as the smaller. However, by looking along a row, except that for $\psi = 0$, we see that, if we choose the smaller ψ for a given y, we shall be correct, on average, two-thirds of the time in repeated trials. Notice, however, that we have placed ourselves in a long run decision-making situation (Part IV), where one parameter value must be chosen, and that, if information is allowed to accumulate through the repeated trials, as soon as two different values of Y are observed, the parameter is known exactly.

Now suppose that ψ can only take the five values shown in the table. The likelihood function will indicate that the correct parameter value is more likely for all observations except $y = 1, 2, 3$, or 4, for which the larger ψ-value is still indicated. In other words, in a decision situation, this choice is only incorrect when $\psi = 0$ and $y \neq 0$. On the other hand, choosing the smaller ψ-value will still be, on average, correct only two-thirds of the time. (Goldstein and Howard, 1991) □

Hopefully, inappropriate models and sample designs will be detected before a 'paradox' in inspecting the likelihood function arises for some observed data!

3.7 Exponential family

We have seen throughout this chapter how inferences are especially simple within the linear exponential family and involve identical procedures for all its members. Many of these results carry over directly to the exponential dispersion family and generalized linear models. In this section, we shall look at a few other such general results. A more technical discussion of a few of them is provided in the appendix at the end of this section.

3.7.1 MAXIMUM LIKELIHOOD ESTIMATES

The log likelihood function for canonical parameters of the linear exponential family,

$$\mathrm{l}(\boldsymbol{\theta}) = \boldsymbol{\theta}^{\mathrm{T}}\mathbf{t} - \kappa(\boldsymbol{\theta}) \tag{3.91}$$

is concave because $\boldsymbol{\theta}^{\mathrm{T}}\mathbf{t}$ is linear and $\kappa(\boldsymbol{\theta})$ is convex. If the sufficient statistic, $\mathbf{t}$, is not on the boundary of its support, a unique maximum exists with finite parameter values.

As we have seen, in the mixed parameter representation, $(\boldsymbol{\mu}_1, \boldsymbol{\theta}_2)$, the components are variation independent. In addition, the m.l.e. of the mean component, $\hat{\boldsymbol{\mu}}_1(\boldsymbol{\theta}_2)$, for fixed value of the canonical component, does not depend on the latter.

Example
The mixed parameter representation for the normal distribution, (μ, σ^2), has $\hat{\mu} = \sum y_i/n$, which does not depend on $\theta_2 = -1/\sigma^2$. The mixed parameter representation, $(\mu/\sigma^2, \mu^2 + \sigma^2)$, has $\widehat{\mu^2 + \sigma^2} = \sum y_i^2/n$, which does not depend on $\theta_1 = \mu/\sigma^2$. □

In the exponential dispersion family, and generalized linear models based on it, location parameters can be estimated independently of the scale parameter.

3.7.2 DEVIANCE

Because the m.l.e. of the mean component, in the mixed parameter representation, does not depend on the canonical (scale or shape) parameter, the latter must be a multiplicative factor in the log likelihood. Thus, an *unscaled deviance*, for the location parameters of an exponential dispersion or generalized linear model, ignoring the scale parameter, is calculable from the observations[1]. This procedure is sometimes called *concentrating out* the scale or shape parameter from the likelihood function.

Example
In a normal distribution, for fixed σ^2, we have the deviance

$$\mathrm{D}(\mu) = -2\{\log \mathrm{L}(\mu, \sigma^2) - \sup_{\mu}[\log \mathrm{L}(\mu, \sigma^2)]\}$$

[1]For users of the statistical package GLIM, what we call the deviance is known there as the scaled deviance, for two-parameter families, while the unscaled deviance is the GLIM deviance.

$$= \frac{\sum(y_i - \mu_i)^2}{\sigma^2} + \frac{n}{2}\log(\sigma^2) - \frac{\sum(y_i - \hat{\mu}_i)^2}{\sigma^2} - \frac{n}{2}\log(\sigma^2)$$
$$= \frac{\sum(y_i - \mu_i)^2}{\sigma^2} \qquad (3.92)$$

because $\hat{\mu}_i = y_i$. The unscaled deviance is then

$$\sigma^2 \mathrm{D}(\mu) = \sum(y_i - \mu_i)^2 \qquad (3.93)$$

The unscaled deviance for the location parameters of a normal distribution is the sum of squares about the mean. Notice how the score equation obtained from Equation (3.92) allows $\hat{\mu}$ to be calculated without involving σ^2. □

3.7.3 ESTIMATION

For the linear exponential family, as we shall see below, the m.l.e.s of the mean value parameters are obtained by equating them to the sufficient statistics, $\hat{\boldsymbol{\mu}} = \mathbf{t}$. Thanks to parametrization invariance, the m.l.e. of the canonical parameters can then be found by transforming that estimate: $\hat{\boldsymbol{\theta}}(\hat{\boldsymbol{\mu}})$. Thus, there is a one-to-one relationship between either of these parameters and the sufficient statistic.

For a generalized linear model (Section 2.5.1) with n observations, known $\mathbf{x}_i$, and canonical link function, we can take the linear structure or linear predictor of Equation (1.9),

$$\xi(\mu_i) = \boldsymbol{\beta}^{\mathrm{T}}\mathbf{x}_i \qquad (3.94)$$
$$= \eta_i$$

where $\mathbf{x}_i^{\mathrm{T}}$ are the rows of the design matrix, $\mathbf{X}$, of explanatory variables, to obtain the canonical regression model. Then, the score has elements

$$U_{\beta_j} = \sum_i \frac{[t(y_i) - \mu_i]x_{ij}}{\mathrm{var}[Y_i]} \frac{\partial \mu_i}{\partial \eta_i} \qquad (3.95)$$

and the observed information is

$$\mathbf{J} = \mathbf{X}^{\mathrm{T}}\mathbf{W}\mathbf{X} \qquad (3.96)$$

where $\mathbf{W}$ is an $n \times n$ diagonal matrix with elements

$$w_{ii} = \frac{1}{\mathrm{var}[Y_i]}\left(\frac{\partial \mu_i}{\partial \eta_i}\right)^2 \qquad (3.97)$$

Thus, the amount of information depends heavily on the design space of the study, defined by $\mathbf{X}$. Generally, the resulting score equations will be non-linear.

Example
For a standard normal linear regression model,

$$\mu_i = \boldsymbol{\beta}^{\mathrm{T}}\mathbf{x}_i \tag{3.98}$$

the score equations are

$$\mathbf{X}^{\mathrm{T}}\mathbf{y} = \mathbf{X}^{\mathrm{T}}\mathbf{X}\boldsymbol{\beta} \tag{3.99}$$

so that the m.l.e. can be obtained explicitly:

$$\hat{\boldsymbol{\beta}} = (\mathbf{X}^{\mathrm{T}}\mathbf{X})^{-1}\mathbf{X}^{\mathrm{T}}\mathbf{y} \tag{3.100}$$

This is also the estimate obtained by 'least squares', a moment estimate.□

3.7.4 SAMPLE SIZE

The procedure for sample size calculation based on the likelihood function, outlined in Section 3.3.1, can easily be generalized to any member of the linear exponential family. The basic reasoning will not be repeated here.

For the (1,1) linear exponential family, the deviance can be written

$$\mathrm{D} = -2\{n\theta(\mu)\hat{\mu} - n\kappa[\theta(\mu)] - n\theta(\hat{\mu})\hat{\mu} + n\kappa[\theta(\hat{\mu})]\} \tag{3.101}$$

where $n\hat{\mu} = n\bar{y}_{\bullet} = \sum y_i$. Suppose that we wish to have just enough observations to be able to distinguish which of the two parameter values, θ_0 and θ_1, is more likely. In the worst case, their likelihoods will be equal and, if the sample size is just large enough, at the borderline of being unlikely. Thus, we can fix the deviance to be equal for these two parameter values, equate them, and solve to obtain the required m.l.e.,

$$\hat{\mu} = \frac{\kappa[\theta(\mu_1)] - \kappa[\theta(\mu_0)]}{\theta(\mu_1) - \theta(\mu_0)} \tag{3.102}$$

Equation (3.101) can also be solved for n at either μ_0 or μ_1:

$$n = \frac{\mathrm{D}}{2\{\hat{\mu}[\theta(\hat{\mu}) - \theta(\mu_0)] + \kappa[\theta(\mu_0)] - \kappa[\theta(\hat{\mu})]\}} \tag{3.103}$$

The value of $\hat{\mu}$ from Equation (3.102) can then simply be substituted into this final equation to obtain the sample size.

Example
Suppose that we are to collect binomial observations, y out of n, and wish to distinguish between the probabilities, π_0 and π_1. The deviances are

$$\mathrm{D}_i = 2n\bar{y}_\bullet \log\left(\frac{\bar{y}_\bullet}{1-\bar{y}_\bullet}\frac{1-\pi_i}{\pi_i}\right) + 2n\log\left(\frac{1-\bar{y}_\bullet}{1-\pi_i}\right) \qquad i=0,1 \quad (3.104)$$

where $\bar{y}_\bullet = y/n$. Equating these expressions for the two models and solving, we obtain

$$\begin{aligned}\hat{\pi} &= \bar{y}_\bullet \\ &= \frac{\log\left(\frac{1-\pi_0}{1-\pi_1}\right)}{\log\left[\frac{(1-\pi_0)\pi_1}{(1-\pi_1)\pi_0}\right]}\end{aligned} \qquad (3.105)$$

the required m.l.e. for π when π_0 and π_1 are equally likely. This is substituted for $\bar{y}_\bullet$ in Equation (3.104) and the required value of n obtained. □

3.7.5 SUFFICIENT AND ANCILLARY STATISTICS

As we saw in Equation (2.7), because the likelihood function of the linear exponential family factors, $\mathbf{T}(\mathbf{Y})$ is the sufficient statistic for $\boldsymbol{\theta}$, and is minimal sufficient if the representation is minimal. However, for continuous distributions, this is only true if measurements are assumed to be made to infinite precision because the density function is used as an approximate likelihood function.

Under standard regularity conditions, it can be shown that the full exponential family contains the only distributions for which the minimal sufficient statistic has the same dimension as the parameter vector. Because the dimension of the minimal sufficient statistic and the parameter vector are identical, there are no ancillary statistics. The m.l.e. is the minimal sufficient statistic.

3.7.6 SCORE AND INFORMATION

Exponential family From Equation (3.91), the score for the linear exponential family is

$$\begin{aligned}\mathrm{U}_{\theta_i}(\boldsymbol{\theta}) &= t_i(\mathbf{y}) - \frac{\partial\kappa(\boldsymbol{\theta})}{\partial\theta_i} \\ &= t_i(\mathbf{y}) - \mathrm{E}_{\theta}[T_i(\mathbf{Y})]\end{aligned} \qquad (3.106)$$

from Equation (2.15), so that the score equations are given by equating the sufficient statistics to their expectations, which are just the mean value parameters:

$$\begin{aligned}\mathbf{t}(\mathbf{y}) &= \mathrm{E}_{\hat{\theta}}[\mathbf{T}(\mathbf{Y})] \\ &= \hat{\boldsymbol{\mu}}\end{aligned} \tag{3.107}$$

The m.l.e. of the mean value parametrization is the moment estimate.

Example
For the binomial distribution,

$$\mathrm{l}(\theta; y, n) = y\theta - n\log(1 + \mathrm{e}^{\theta}) \tag{3.108}$$

and

$$\begin{aligned}\mathrm{U}(\theta) &= y - \frac{n\mathrm{e}^{\theta}}{1 + \mathrm{e}^{\theta}} \\ &= y - n\pi\end{aligned} \tag{3.109}$$

so that $y = n\hat{\pi}$. □

For the canonical representation, the observed information is

$$\begin{aligned}\mathrm{J}_{\theta_i\theta_j}(\boldsymbol{\theta}) &= \frac{\partial^2\kappa(\boldsymbol{\theta})}{\partial\theta_i\partial\theta_j} \\ &= \mathrm{cov}_{\theta}[T_i(\mathbf{Y}), T_j(\mathbf{Y})] \\ &= \mathrm{I}_{\theta_i\theta_j}(\boldsymbol{\theta})\end{aligned} \tag{3.110}$$

so that

$$\begin{aligned}\mathrm{I}_{\theta_i\theta_i}(\boldsymbol{\theta}) &= \mathrm{var}_{\theta}[T_i(\mathbf{Y})] \\ &= \frac{\partial\mu_i}{\partial\theta_i}\end{aligned} \tag{3.111}$$

from Equation (2.15). Thus, the observed information equals the expected information for the canonical representation.

Example
For the binomial distribution,

$$\begin{aligned}\mathrm{J}_{\theta}(\theta) &= \frac{n\mathrm{e}^{\theta}}{(1 + \mathrm{e}^{\theta})^2} \\ &= n\pi(1 - \pi) \\ &= \mathrm{I}_{\theta}(\theta)\end{aligned} \tag{3.112}$$

□

The score for a mean value parameter is

$$\mathrm{U}_{\mu_i}(\boldsymbol{\theta}) = -\mu_i \frac{\partial \theta_i}{\partial \mu_i} + \frac{\partial \theta_i}{\partial \mu_i} t_i(\mathbf{y}) \tag{3.113}$$

and the observed information is

$$\mathrm{J}_{\mu_i \mu_i}(\boldsymbol{\theta}) = \frac{\partial \theta_i}{\partial \mu_i} + \mu_i \frac{\partial^2 \theta_i}{\partial \mu_i^2} - \frac{\partial^2 \theta_i}{\partial \mu_i^2} t_i(\mathbf{y}) \tag{3.114}$$

so that

$$\begin{aligned} \mathrm{I}_{\mu_i \mu_i}(\boldsymbol{\theta}) &= \frac{\partial \theta_i}{\partial \mu_i} \\ &= \frac{1}{\mathrm{I}_{\theta_i \theta_i}(\boldsymbol{\theta})} \end{aligned} \tag{3.115}$$

Thus, the expected information for the mean value parameter is the reciprocal of that for the corresponding canonical parameter, a rather surprising result. However, for this parametrization, the observed and expected information are not equal.

Example
For the binomial distribution,

$$\begin{aligned} \mathrm{I}_{\mu}(\theta) &= \frac{n}{\mu(n-\mu)} \\ &= \frac{1}{n\pi(1-\pi)} \\ &= \frac{\partial \theta}{\partial \mu} \\ &= \frac{1}{\mathrm{I}_{\theta}(\theta)} \end{aligned}$$

□

Exponential dispersion family For the univariate exponential dispersion family of Equation (2.63), the score for the mean parameter is

$$\mathrm{U}_{\xi} = \frac{y - \frac{\partial \kappa^*(\xi)}{\partial \xi}}{\sigma^2} \tag{3.116}$$

so that

$$\mathrm{E}[Y] = \frac{\partial \kappa^*(\xi)}{\partial \xi} \tag{3.117}$$

The observed information is

$$J_{\xi\xi} = \frac{\frac{\partial^2 \kappa^*(\xi)}{\partial \xi^2}}{\sigma^2} \tag{3.118}$$

but, as we shall see in Equation (5.4) below,

$$E[J] = E[U^2]$$

so that

$$\begin{aligned} E\left[\left(Y - \frac{\partial \kappa^*(\xi)}{\partial \xi}\right)^2\right] &= E[(Y-\mu)^2] \\ &= \sigma^2 \frac{\partial^2 \kappa^*(\xi)}{\partial \xi^2} \\ &= \sigma^2 \tau^2 \\ &= \text{var}[Y] \end{aligned} \tag{3.119}$$

As we have already seen in Equation (2.68), the variance is a product of the variance function, τ^2, which is a function of the mean, μ, only, and the dispersion parameter, σ^2.

Examples
1. For the binomial distribution, $E[Y] = n\pi$ and

$$\begin{aligned} E[(Y-\mu)^2] &= n\pi(1-\pi) \\ &= \frac{\mu(n-\mu)}{n} \\ &= \tau^2 \end{aligned}$$

with $\sigma^2 = 1$.
2. For the normal distribution, $E[Y] = \mu$ and $E[(Y-\mu)^2] = \sigma^2$, so that $\sigma^2 = -1/(2\theta_2)$ and $\tau^2 = 1$. For this distribution, the variance does not depend on the mean. □

For an exponential family, the Kullback–Leibler distance, $K(P_{\hat{\theta}}, P_\theta)$, is just the negative normed log likelihood, i.e. the Kullback–Leibler information. In other words, the normed log likelihood is the directed distance between an arbitrary model, indexed by $\boldsymbol{\theta}$, and the m.l.e., taken as the 'true' model. This means that $K(\cdot,\cdot)$ is a convex function of $\boldsymbol{\theta}$. For a curved exponential family, embedded in this family, $\boldsymbol{\theta}(\boldsymbol{\beta})$ is a function of a smaller dimensional parameter, $\boldsymbol{\beta}$. Thus, the maximization of the likelihood function for $\boldsymbol{\beta}$ is equivalent to minimizing the Kullback–Leibler distance between $\hat{\boldsymbol{\theta}}$ and $\boldsymbol{\theta}(\hat{\boldsymbol{\beta}})$.

3.7.7 APPENDIX

Existence of m.l.e. For the canonical exponential family log likelihood function of Equation (3.91), certain conditions are available for the 'existence' of the m.l.e. of the canonical parameter, i.e. when it is finite. (We have already seen that an infinite value can always be transformed to be finite, in which case it will be on the boundary of the new parameter space.) For fixed $\mathbf{t} \in \mathbf{R}^k$, the log likelihood function is a closed concave function on $\mathbf{R}^k$, because $\boldsymbol{\theta}^{\mathrm{T}}\mathbf{t}$ is linear and $\kappa(\boldsymbol{\theta})$ is strictly convex. It is strictly concave on the convex support of the dominating measure, ν. It has a maximum iff $\mathbf{t}$ is in the interior of the support, $\mathcal{T}$, a closed convex set, i.e. $\mathbf{t} \in \mathrm{int}\mathcal{T}$ and, then, the maximum is unique. The m.l.e., $\hat{\boldsymbol{\theta}}$, is the one-to-one mapping on $\mathcal{T}$ and onto $\mathrm{int}\boldsymbol{\theta}$ with inverse $\boldsymbol{\mu} = \mathrm{E}[\mathbf{T}]$.

Conditional distributions For a curved exponential family, an *affine ancillary*, $\mathbf{T}_2$, is such that, for the one-to-one transformation of the minimum sufficient statistic, $\mathbf{T} \rightarrow (\hat{\boldsymbol{\beta}}, \mathbf{T}_2)$, the mapping $\mathbf{T} \rightarrow \mathbf{T}_2$ is affine in $\mathbf{T}$ for $\hat{\boldsymbol{\beta}}$ fixed (Barndorff-Nielsen, 1980). Then, the conditional density is given by

$$f(\hat{\boldsymbol{\beta}}|\mathbf{t}_2;\boldsymbol{\beta}) = c(\mathbf{T}_2)|\mathbf{J}(\hat{\boldsymbol{\beta}})||\mathbf{I}(\hat{\boldsymbol{\beta}})|^{-\frac{1}{2}}|\hat{\boldsymbol{\Sigma}}_{\beta}|^{\frac{1}{2}}a_0(\boldsymbol{\theta})t_0^{\dagger}(\mathbf{y})\mathrm{e}^{\boldsymbol{\theta}\cdot\mathbf{t}(\mathbf{y})} \tag{3.120}$$

where $\mathrm{var}[\mathbf{T}] = \boldsymbol{\Sigma}_{\beta}$ and c is a norming constant. However, $c(\mathbf{T}_2)$, $t_0^{\dagger}(\mathbf{y})$, and occasionally $a_0(\boldsymbol{\theta})$ will be difficult to find. Ways of obtaining accurate approximations to this expression will be discussed in Section 5.7.3.

Example

Suppose that (Y_1, Y_2) are random variables from a bivariate normal distribution with zero mean vector and covariance matrix

$$\boldsymbol{\Sigma} = \begin{pmatrix} \sigma_1^2 & \rho\sigma_1\sigma_2 \\ \rho\sigma_1\sigma_2 & \sigma_2^2 \end{pmatrix}$$

and let $\boldsymbol{\beta} = (\sigma_1^2, \rho\sigma_1\sigma_2, \sigma_2^2)$. For n observations,

$$f(\mathbf{Y};\boldsymbol{\Sigma}) = (2\pi|\boldsymbol{\Sigma}|)^{-\frac{n}{2}}\mathrm{e}^{-\frac{n}{2}\mathrm{tr}(\Sigma^{-1}\hat{\Sigma})} \tag{3.121}$$

so that the normed likelihood is

$$\mathrm{R}(\boldsymbol{\Sigma}) = \left(\frac{|\boldsymbol{\Sigma}|}{|\hat{\boldsymbol{\Sigma}}|}\right)^{-\frac{n}{2}} \mathrm{e}^{-\frac{n}{2}[\mathrm{tr}(\Sigma^{-1}\hat{\Sigma})-1]} \tag{3.122}$$

and the determinant of the observed information

$$|\mathbf{J}(\hat{\boldsymbol{\beta}})| = 4|\hat{\boldsymbol{\Sigma}}|^{-3} \tag{3.123}$$

The exact ancillary is independent of $\hat{\beta}$ so that

$$f(\hat{\mathbf{\Sigma}};\mathbf{\Sigma}) = c|\mathbf{\Sigma}|^{-\frac{n}{2}}|\hat{\mathbf{\Sigma}}|^{\frac{n-3}{2}}\mathrm{e}^{-\frac{n}{2}\mathrm{tr}(\mathbf{\Sigma}^{-1}\hat{\mathbf{\Sigma}})} \tag{3.124}$$

which is called a *Wishart distribution.* In this case, the result is exact. □

Further reading

Blyth (1970) provides a useful discussion of the differences between the scientific inference and decision-making contexts.

The procedure of maximum likelihood estimation dates back to Fisher (1912), although he only first used the term, likelihood, in Fisher (1921); the concept of relative likelihood was also developed at this later time. Fisher also introduced the concepts of degrees of freedom (1922b), sufficiency (1922a), score function (idea in 1925, term in 1935), and information (1925). His use of ancillarity was originally (1925) based on taking successive derivatives of the log likelihood to recapture information lost when the m.l.e. is not sufficient, although he only first used the term in 1934.

Direct likelihood inference originated with Fisher (1922a) and was most clearly expressed in Fisher (1959). Other important works to consult include the writings of George Barnard, especially Barnard *et al.* (1962), Birnbaum (1962), Kalbfleisch and Sprott (1970), Edwards (1972), Lindsey (1973), King (1989), and Frosini (1993). More elementary presentations, from this point of view, can be found in Kalbfleisch (1985a, b) and Lindsey (1995c).

For the central role of the likelihood function in the Bayesian approach, see Berger and Wolpert (1988), and, in the frequentist approach, Barndorff-Nielsen and Cox (1994). The thought provoking collection of essays by Basu in Ghosh (1988) should also be essential reading.

Models for coarsened data can be traced back to Sheppard (1898). The dangers of assuming that variables are measured to infinite precision are clearly documented in Beaton *et al.* (1976) and Dempster and Rubin (1983). Lindsey (1974a, b, 1995a, b) and Lindsey and Mersch (1992) develop statistical models and inference procedures for continuous variables, taking into account the unit of measurement. For a general survey, see Heitjan (1989) and, for an elementary introduction, Lindsey (1995c).

The problem of compatibility of inferences is closely related to that of multiple comparison procedures; for the latter, see, for example, Saville (1990).

Nelder and Wedderburn (1972) introduced the term, deviance. The idea of the directed deviance dates to Lawley (1956) and Sprott (1973). Akaike (1973) introduced the AIC and Schwarz (1978) the BIC for model selection. Barndorff-Nielsen (1983) developed the modified profile likelihood as an approximate correction for extra parameters, but conditioning on information about the nuisance parameters dates back at least to Student's (1908) t test and Fisher's exact test (Fisher, 1958; see Section 4.2.2).

The predictive use of likelihood, introduced by Fisher (1959, pp. 128–133), is discussed by Kalbfleisch and Sprott (1969), Kalbfleisch (1971), Hinkley (1979), Davison (1986), and Butler (1986, 1989).

The elaboration of efficient sample designs dates from the work of Fisher (1960, originally published in 1935) on experimental designs. Wald (1945) and Barnard (1946) introduced sequential designs.

3.8 Exercises

(1) Redo the examples in this chapter, replacing the binomial distribution by the Poisson and the normal distribution by the gamma and inverse Gaussian.

(2) (a) A scientist is checking a river for bacterial pollution. It is not possible to count the number of bacteria in a sample. Only presence or absence of bacteria in each sample can be recorded. Thus, n test tubes, each containing a volume v of river water, are incubated and tested. A negative test indicates that no bacteria are present, while a positive test shows that at least one bacterium was found. Suppose that the probability of z bacteria in a volume v of water is given by a Poisson distribution with mean μv.

i. If y tubes out of n tested give negative results, what is the m.l.e. of μ?

ii. Suppose that 40 test tubes, each containing 10 ml of river water, are incubated, and that 28 give negative results. What is the estimated concentration of bacteria in the river?

iii. Give a reasonable normed likelihood interval for this estimate.

(b) To study the river pollution, as described, it is necessary to specify the volume, v, of water collected in each test tube. If v is too large, all tubes will contain bacteria and give positive reactions; if it is too small, none will. In either case, little information will be provided. One way to foresee this is to collect two or more sets of volumes.

i. Suppose that two sets of 40 test tubes are collected. Those with 10 ml of water gave 28 negative results, while those with 1 ml gave 37 negative results. What is the m.l.e. of μ?

ii. Draw a graph of the deviance for μ and obtain 20% likelihood intervals, based on the two sets of observations taken separately and together.

iii. Suppose that the 40 test tubes with $v = 10$ ml of water had all given positive results. Draw a graph of the deviances.

(Kalbfleisch, 1985b, pp. 7, 15, 20–23)

(3) On each of 200 consecutive working days, 10 items were randomly selected from a production line and tested for imperfections, giving the following results:

Number defective	0	1	2	3	≥ 4	Total
Frequency	133	52	12	3	0	200

Suppose that the number of defective items out of 10 has a binomial distribution with π the probability that an item is defective.

(a) Calculate the m.l.e. of π.

(b) Plot the deviance for these observations.

(c) Calculate the estimated frequencies under the binomial distribution, given the m.l.e. obtained.

(Kalbfleisch, 1985b, pp. 8–9)

(4) Likelihood functions in the exponential family are generally well-behaved concave functions with a unique maximum. Consider instead the Cauchy distribution of Equation (2.94) and suppose that the following eight independent observations are made: $(-4.20, -2.85, -2.30, -1.02, 0.70, 0.98, 2.72, 3.50)$.

(a) Plot the log likelihood function for μ with $\sigma = 0.1$.

(b) What problems might be encountered in trying to find the maximum likelihood estimate numerically?

(c) Compare the maximization techniques of Appendix D in calculating the m.l.e. Try various starting values.

(Brooks and Morgan, 1995)

(5) Certain kinds of electronic components are susceptible to instantaneous failure at any time, but do not deteriorate with age: the probability of failure within a given time period does not depend on the age of the component. Such a lifetime should reasonably be described by the exponential distribution of Equation (1.14), where μ is the expected lifetime of such components.

(a) When 10 components are tested independently, their lifetimes, measured to the nearest day, are

70	11	66	5	20	4	35	40	29	8

i. What values of μ are plausible in light of the data?

ii. Suppose that we are interested in the intensity of events, i.e. the failure rate, $\rho = 1/\mu$, instead of the lifetime. What are plausible values of this parameter?

iii. What is an estimate of the proportion of such components which will last at least 25 days?

(b) Suppose now that the study had only lasted 50 days so that the results were

50+	11	50+	5	20	4	35	40	29	8

The two incomplete observations are censored at 50 days.

i. What is the new likelihood function for these data?

ii. How might you expect the amount of information in the two data sets to compare?

iii. Plot the deviance for the censored data on the same graph as one for the complete data.

(Kalbfleisch, 1985b, pp. 26–29, 34–36, 39)

(6) Two objects, with unknown weights μ_1 and μ_2, are weighed separately and together with the same instrument, to give three measurements, respectively, Y_1, Y_2, and Y_3. Assume that the measurements are independent and normally distributed, with variance equal to one.

(a) What are the likelihood function, score equations, and observed and expected information for this model?

(b) If the observed values are $y_1 = 15.6$, $y_2 = 29.3$, and $y_3 = 45.8$, calculate the m.l.e. of the two means.

(c) Plot the profile deviance for each parameter.

(d) Plot the two-dimensional deviance contours for the two parameters.

(e) Consider the reparametrization as $\psi_1 = \mu_1 + \mu_2$ and $\psi_2 = \mu_1 - \mu_2$. What is the advantage of working with this set of parameters?

(f) Plot the new deviance function in an appropriate way.

(Kalbfleisch, 1985b, pp. 55, 61–64, 66–67, 69)

(7) Deep-groove ball bearings are given endurance tests until failure. For 23 such bearings, the results, in millions of revolutions, are

17.88	28.92	33.00	41.52	42.12	45.60
48.48	51.84	51.96	54.12	55.56	67.80
68.64	68.64	68.88	84.12	93.12	98.64
105.12	105.84	127.92	128.04	173.40	

Such lifetimes may be taken to have the Weibull distribution of Equation (1.4).

(a) What are the score equations for this model?

(b) How may they be solved?

(c) What are the m.l.e.s?

(d) Suppose that we wish to work with the 0.63-quantile, γ, given by $\mu = \gamma^{-\alpha}$. Draw the profile deviances and the contour deviance plots of the two parameters, γ and α.

(Kalbfleisch, 1985b, pp. 56–57, 63, 67)

(8) (a) Suppose that $Y_1, Y_2, \ldots, Y_n$ are independent random variables with a uniform distribution on $[0, \psi], \psi > 0$.

i. Write down the likelihood function.

ii. What are the minimal sufficient statistic and the m.l.e.?
iii. Is there an ancillary statistic?
iv. Is this a member of the exponential family?
(Kalbfleisch, 1985b, p. 281)

(b) Suppose now that the uniform distribution is over $[\psi - 0.5, \psi + 0.5]$.
i. What is the likelihood function?
ii. What are the minimal sufficient statistic and the m.l.e.?
iii. Is there an ancillary statistic?
(Cox and Hinkley, 1974, p. 97)

(9) Two independent random variables, Y_1 and Y_2, have Poisson distributions with means $c\psi$ and $c(1-\psi)$, where c is a known constant.
(a) Write down the likelihood function.
(b) What are the sufficient statistics?
(c) Of what family is this distribution a member?
(d) Now suppose that c is an unknown parameter, say α. How are the answers to the above questions changed?
(Barndorff-Nielsen and Cox, 1994, p. 39)

(10) (a) A machine produces N items per day. The binary random variable, Y_i, is a record of whether a given item is defective (1) or not (0). Once one item is defective, the machine is broken and, although it continues to operate, all subsequent items that day will also be defective. At the end of a representative day, we are interested in the probability, π, of a defective item.
i. What is the likelihood function for such data if observations are recorded sequentially on the production line?
ii. Give the minimal sufficient statistic.
iii. What is the m.l.e.?
iv. For $\sum y_i = 30$, $N = 100$, plot the likelihood function.
v. In a competitor's factory, the machine stops after the first defect. Find an appropriate way to define π and repeat these questions.

(b) Suppose now that production in the first factory cannot be followed throughout the day. Instead, a random sample of $n < N$ items is chosen. Both the sequence number, between 1 and N, and the value of the random variable are recorded for each of these items.
i. What is the likelihood function under these new conditions?
ii. Give the new minimal sufficient statistic.
iii. What can we now say about the m.l.e.?
iv. For $N = 100$, the sample $\{17, 0; 24, 0; 40, 1; 73, 1\}$ is obtained, where the first number in each pair is the order, and the second the defective indicator. What can be said

about the probability of defectives?

v. Plot the likelihood function and compare it to that in the first part of this exercise.

(Basu, 1977, in Ghosh, 1988, pp. 255–256)

(11) For a normal distribution with mean μ and variance σ^2, the parameter of interest is $\lambda = \mu + \sigma^2/2$, the mean of a log normal distribution.
   (a) Write down the likelihood function for λ and σ^2.
   (b) For given λ, what is the m.l.e. of σ^2?
   (c) Write down the profile likelihood function for λ.

   (Barndorff-Nielsen and Cox, 1994, p. 94)

(12) The generalized Pareto distribution has density

$$f(y;\psi) = \frac{(1-\psi_1 y/\psi_2)^{\frac{1}{\psi_1}-1}}{\psi_2} \qquad \psi_2 > 0 \qquad (3.125)$$

   (a) What is the support of Y?
   (b) What distribution is obtained as $\psi_1 \to 0$?
   (c) Write down the likelihood function, with n independent observations, for $\lambda = \psi_2/\psi_1$ and ψ_2.
   (d) Give the m.l.e. for ψ_2 for a fixed value of λ.
   (e) Write down the profile likelihood function for λ.

   (Barndorff-Nielsen and Cox, 1994, pp. 94–95)

(13) Consider independent exponential failure times, with non-informative censoring, from Equation (1.14), as in Exercise (2.16).
   (a) Write down the likelihood function.
   (b) What is the minimal sufficient statistic?
   (c) Does an ancillary statistic exist?

   (Kalbfleisch, 1985b, p. 281)

(14) In a homogeneous Poisson process, events occur with a constant intensity, ρ. Thus, in time t, there will be, on average, $\mu = \rho t$ events.
   (a) Write down the likelihood function for n events in time t.
   (b) In what family is this model
      i. if both N and T are random?
      ii. if the stopping rule is for a fixed time?
      iii. if it is for a fixed number of events?
   (c) What are the sufficient and ancillary statistics in each case?
   (d) Show how this model is related to that for censored exponential failure times in Exercises (3.5) and (3.13) above.
   (e) Relate it also to the models in Exercise (3.2) on volumes of polluted river water.

(15) (a) n Bernoulli trials are to be carried out to estimate the probability of success, π, in such trials. Suppose that $n = 5$ or $n = 100$ depending on the flip of a fair coin.
i. What is the likelihood function?
ii. What is the m.l.e.?
(b) Now suppose that $n = 5$ if the first trial is a success and $n = 100$ if it is a failure.
i. What is the likelihood function?
ii. What is the m.l.e.?

(16) Suppose that Bernoulli trials are being conducted to estimate the Bernoulli probability, π, and that N are planned. For some unforeseen random reason, unconnected with the Bernoulli probability, observation must be discontinued at $n < N$ observations, so that n is a random variable.
(a) What is the likelihood function?
(b) What is the minimal sufficient statistic?
(c) Does this change the way inferences would be made?
(Kalbfleisch, 1985b, p. 292)

(17) Suppose that observations, y_i, $i = 1, 2, 3$, are to be made from a binomial distribution with

$$f(y_i; \pi) = \binom{5}{y_i} \pi^{y_i}(1 - \pi)^{5-y_i}$$

with stopping rule: (1) if $y_1 = 4$ or 5, stop; otherwise observe y_2; (2) if $y_1 + y_2 < 8$, stop; otherwise, observe y_3 and stop.
(a) List the points in the sample space.
(b) Give the corresponding probabilities.
(c) What is the m.l.e. for each point in the sample space?
(d) Is the m.l.e. a sufficient statistic?
(Kempthorne and Folks, 1971, pp. 287–288)

(18) To study the existence of extra-sensory perception (ESP), a number of people are asked to turn up five playing cards from well-shuffled packs, noting beforehand what they thought the cards were going to be, and after what they actually are. Each person is to repeat this five times. We shall assume that everyone records the results honestly.
(a) Develop a suitable model for this study.
(b) The study was conducted by mail, with a considerable number of those contacted not replying. The results showed a global tendency for people to guess correctly much more often than might be expected at random. Develop an explanation for this in terms of stopping rules.

(c) People tend to guess red cards more often than black, and odd numbers more often than even. How can your explanation of the results be checked?

(Fisher in Bennett, 1990, p. 338)

(19) For the Cauchy distribution of Equation (2.94), with $\sigma = 1$, what are

(a) the minimal sufficient statistic;

(b) the m.l.e.;

(c) the observed and expected information?

(Cox and Hinkley, 1974, pp. 25, 114; Kalbfleisch, 1985b, pp. 281–282, 294)

(20) The bivariate gamma hyperbola distribution is

$$f(y_1, y_2; \boldsymbol{\psi}) = \frac{1}{\psi_2^2} \exp\left(-\frac{y_1/\psi_1 + y_2\psi_1}{\psi_2}\right) \tag{3.126}$$

(a) Show that this distribution is a member of the exponential family.

(b) Give the canonical parameters and statistics.

(c) If ψ_2 is known, what is the minimal sufficient statistic?

(d) For unknown ψ_2, give the m.l.e.s.

(e) Derive the score and observed and expected information.

(21) Two random variables, Y_1 and Y_2, have a joint distribution

$$f(y_1, y_2; \psi) = \mathrm{e}^{-y_1\psi - y_2/\psi} \qquad y_1, y_2, \psi > 0 \tag{3.127}$$

a special case of the gamma hyperbola distribution given in Equation (3.126), with $\psi_2 = 1$.

(a) To what family does this distribution belong?

(b) What is the expected information for n observations?

(c) What is the sufficient statistic?

(d) What is the m.l.e.?

(e) Obtain the distribution of this statistic.

(f) What is the expected information derived from this distribution?

(g) Find an ancillary statistic.

(h) Derive the conditional distribution of the m.l.e. given the ancillary.

(i) What is the expected information derived from this distribution?

(j) Does it depend on the sample size, n?

(k) Compare the three values of expected information and draw appropriate conclusions.

(l) Plot the corresponding likelihood functions when we observe $\mathbf{y} = (7, 10, 5, 12, 3)^{\mathrm{T}}$.

(Fisher in Bennett, 1990, pp. 66–70)

(22) Consider a multinomial distribution with

$$\pi_1(\psi) = \frac{1}{6}(1-\psi), \ \pi_2(\psi) = \frac{1}{6}(1+\psi),$$
$$\pi_3(\psi) = \frac{1}{6}(2-\psi), \ \pi_4(\psi) = \frac{1}{6}(2+\psi)$$

(a) What is the minimal sufficient statistic?
(b) Find an ancillary statistic.
(c) Is it unique? If not, is there a preference as to which to use?
(Cox, 1971; Barndorff-Nielsen and Cox, 1994, p. 42)

(23) We have seen that the exponential family contains the only regular distributions with minimal sufficient statistics of the same dimension as the parameter vector. Find a non-regular distribution with such a minimal sufficient statistic.
(Lehmann, 1983, pp. 44–45)

(24) Let X, Y have a bivariate normal distribution with zero means, unit variances, and unknown correlation coefficient, ρ. We make n independent observations of the pairs.
(a) Show that $X_1, \ldots, X_n$ and $Y_1, \ldots, Y_n$ are separately ancillary.
(b) Conditional on each of these ancillary statistics separately, what are the sufficient statistics?
(c) Without the use of ancillaries, what are the sufficient statistics?
(d) After reduction to sufficiency, does an ancillary still exist?
(Lehmann, 1986, pp. 545–546)

(25) A binary signal, Y, is received from one of two sources, labelled by ψ_i. One of two receivers, X, is selected at random with $P(X = 1) = 0.99$, such that

$$\begin{aligned} P(Y=0|X=1;\psi_1) &= P(Y=1|X=1;\psi_2) \\ &= 0.9 \\ P(Y=0|X=2;\psi_1) &= P(Y=1|X=2;\psi_2) \\ &= 0.2 \end{aligned}$$

(a) If $X = 2$ and $Y = 1$ are observed, what is the likelihood ratio for ψ_1 versus ψ_2?
(b) What ancillary statistic have you implicitly used?
(c) What would be the answer if you did not use it?
(Bernardo and Smith, 1994, p. 452)

(26) Inference procedures which depend on an expectation will violate the sufficiency principle.

(a) Give at least one reason why the expectation of simple events should be excluded from this statement. (Hint: review the interpretation of parameters in the exponential family.)
(b) What is the form of the regression model for the link function which is the log mean
  i. for the binomial distribution?
  ii. for the negative binomial distribution?
(c) Find an example of two different sample designs which can yield identical likelihood functions, but which have different expected information.
(d) What would be the effect of replacing the observed by the expected information in the modified profile likelihood?

(27) In the $(1, 1)$ linear exponential family, suppose that we plan to observe two independent samples of equal size, having parameters θ_i. For example, they may pertain to two different treatments for some disease. Find the general formula for the sample size if we wish to be able to detect a difference in mean of δ.

(28) (a) Show that the expected information for the location family does not depend on the location parameter.
(b) Write down the expected information matrix for the location–scale family.
(c) Show that it only depends on the scale parameter.
(Lehmann, 1983, pp. 120–121)

(29) We know that, in most cases, as the sample size increases, more complex models will be required to describe the observations adequately. One way to compensate for this, if simple models are required, would be to use a correspondingly smaller value of the smoothing factor, a, in Equation (3.69). However, in making sample size calculations, from Equation (3.101), we can see that the deviance, in the linear exponential family, is directly proportional to n. Thus, if we want to make inferences for a sample equivalent to one of size, say, m, we could multiply the deviance by m/n, for $m < n$, or, equivalently, modify the normed likelihood to $\mathrm{R}^{m/n}$.
(a) Discuss the merits and inconveniences of such a proposal.
(b) Would it provide valid inferences if $m > n$?
(c) Compare the effects of this approach to that of using a smoothing factor depending on n, such as $a = 1/\sqrt{n}$ or $a = 1/\sqrt{\log(n)}$.

(30) What will be the general form of the modified profile likelihood
(a) for the dispersion parameter in the exponential dispersion family?
(b) for the scale parameter in the location–scale family?
(Barndorff-Nielsen and Cox, 1994, pp. 268–269)

(31) Suppose that the distribution of a random variable is uniform with M discrete possibilities, so that their probabilities are $1/M$. Consider $n \leq M$ observations from this distribution. Then, the normed likelihood of the 'true' distribution will always be $\mathrm{R} \leq (n/M)^n$, indicating that this distribution will always be found to be unlikely. Find the fallacy in this argument.
(Hill, 1975)

(32) We have n Bernoulli observations, with y successes, and wish to predict the number of successes in m subsequent trials.
   (a) Write down the profile relative likelihood.
   (Fisher, 1959, pp. 130–133)
   (b) Derive an appropriate modified profile likelihood.

4
Goodness of fit

Throughout the previous chapters, we have seen various aspects of the model building process. In Chapter 3, we generally assumed that the chosen family of models, $\mathcal{P}$, was suitable, and studied how to rank the relative merits of different members of that family, whether different functional forms or simply different parameter values, in the light of the observed data. We were investigating the validity of prior theoretical knowledge in the light of observed empirical data. Thus, for example, we could claim that sufficient statistics contain all the information about parameters, *given* the model.

Now we shall be concerned more with *scientific discovery*, with the development of new theory. This is a further step in the model selection process which can never be a very formalized procedure. Here, we might say that the only sufficient statistic is the complete set of observations, including all details about how they were collected, because no model is correct. Thus, our approach up until now has provided little general indication of the overall goodness of fit of the (family of) models. We have already seen that any acceptable model should be applicable for previous observations (Section 3.6.4, and also Section 6.4.1). Hopefully, it should also be useful for any future ones, which will provide the ultimate check. However, we generally wish to make as many verifications as possible about the suitability of the model before these become available.

If the family under consideration can be *embedded* in a still larger family, $\mathcal{P} \subset \mathcal{P}^*$, then goodness of fit can be determined, for $\mathcal{P}$, within this larger context, using the inference techniques already described. Obviously, this process of embedding must stop somewhere, the limit being the empirical likelihood, but it can often be sufficient to provide the required measure of global goodness of fit.

A model is supposed to describe the data generating mechanism in the appropriate way of interest. As we have seen, the data collection procedure must be designed so as to obtain relevant information about that mechanism without distorting it. In any given context, several alternative sampling procedures will usually be available. By the sufficiency principle, they may all yield the same likelihood function for a given model function. However, in each case, additional information will usually also be available for checking goodness of fit. For example, for supposedly independent observations, the extra information may involve the order in time in which the data were generated. The amount and type of this information will depend on the sample space defined by the given data collection procedure and, thus, will not be the same for all designs.

However, in many situations, sufficient data may not be available to assess the adequacy of the family of models in terms of likelihood functions. Then, many theoretical assumptions must be made which cannot be directly checked, without collecting more data. For example, we may postulate a normal distribution, but, with limited data, can only check the value of its mean and, perhaps, its variance, but no further details about the form of the curve such as asymmetry. To supplement the more formal inference procedures for such models, relatively *ad hoc* diagnostics of goodness of fit are usually applied. Here, graphics play an important role.

4.1 Likelihood comparisons

4.1.1 EMBEDDING

More complex models than those directly of interest can often be developed. These may attempt to describe the data generating mechanism more fully or may just have an additional (often discrete-valued) parameter indexing a series of possible model functions of interest, that parameter not, itself, being of direct interest. Common examples include adding new explanatory variables or different functions of existing ones, and using more complex probability distributions. Then, the original model of interest will be some special case which can be compared to the complex model through the standard methods which we have studied for the likelihood function. If there is a large difference in likelihood, this indicates that the model of interest is not fitting well.

Examples

1. In Section 3.3.2, we looked at an example of negative binomial trials. Under independence, we saw that the likelihood function is identical to that for binomial trials. However, the former likelihood function, and the underlying model, does not use the relevant necessarily available information on the order in which the results occur. A more complex model can be constructed, where each result depends in some way on what happened previously. If the probability of the observed data is much greater under this more complex model, it indicates that the model of independent trials is unlikely.

2. If we apply the Box–Cox transformation of Equation (2.71) with a normal distribution, we may only be interested in the three models for the values $\alpha = 1, 0, -1$. These index a response having a normal distribution, a log normal distribution, or the reciprocal response having a normal distribution. This is equivalent to comparing the models directly, as in Equation (3.19). □

An advantage of the method of embedding is that, if a suitable complex model is chosen, it can provide an indication of the direction in which lack of fit lies, i.e. of what aspects of the current model describe poorly the patterns in the data. If the additional parameter simply indexes a series of models of interest, then it often may be ignored in the inferences once the model to which it points has been selected. For example, the precision of its estimate will not be of interest. Indeed, this procedure is identical simply to comparing likelihoods of the various models directly, without introducing the discrete parameter. Note, however, that, if prediction intervals (Section 3.6.5) are to be constructed, this uncertainty in model selection should be taken into account.

Saturated models In any situation, the model of interest can be embedded in some maximally complex model which fits the data perfectly. The latter model will have as many parameters as observations, but may not be unique. This is called a *saturated model*. Then, likelihood comparisons can be made with respect to this model. The most general of these, for independent data, is the empirical likelihood of Equation (3.7) (Lindsey, 1974a, b).

Suppose that n independent observations for a response, Y, from some probability distribution, without explanatory variables, are classified into the observed categories of that response, with n_i observations in the category labelled by y_i. A saturated model has a different probability, π_i, for each category, with no functional relationship among them. This is a multinomial distribution with likelihood function

$$\mathrm{L}(\boldsymbol{\pi}) \propto \prod_i \pi_i^{n_i} \tag{4.1}$$

the model upon which empirical likelihood is based. Any distribution, which would be a model function of interest, defines a functional relationship among these probabilities, $\pi_i = P(y_i)$, generally approximated by the density, $\pi_i = f(y_i)\Delta_i$.

Example
A Poisson distribution would have the probabilities related by

$$\pi_i = \frac{\mathrm{e}^{-\mu}\mu^{y_i}}{y_i!} \tag{4.2}$$

This yields a special case of the likelihood in Equation (4.1), which can be compared to the saturated model, where $\hat{\pi}_i = n_i/n_\bullet$, the m.l.e. for the empirical likelihood. □

The resulting deviance function for goodness of fit is

$$\mathrm{D} = 2\sum_i n_i \log\left(\frac{\hat{\pi}_i}{\pi_i}\right) \tag{4.3}$$

This is sometimes approximated by

$$\mathrm{D} = n_\bullet \sum_i \frac{(\pi_i - \hat{\pi}_i)^2}{\pi_i} \tag{4.4}$$

called the Pearson (1900) goodness of fit (see Section 5.5.2).

Another saturated model is commonly used for generalized linear models. Thus, in a linear regression context, a parameter, ψ, usually (a function of) the mean, is allowed to vary with some explanatory variables. This regression model will contain a number of parameters which is less than the number of independent observations, n. When such a model is based on a probability distribution with scalar parameter, or when there are replications for at least some values of the explanatory variables, a special saturated model can be developed, having the same distribution but a different ψ_i for every observation. One potential inconvenience with this approach, as with the empirical likelihood, is that the number of parameters in the saturated model is not necessarily fixed, but may increase with the number of observations. In addition, care must be taken; in certain cases, comparison to such a saturated model may be relatively uninformative about goodness of fit (see also Section 5.4.3).

Examples

1. For Poisson observations with common mean, μ, the saturated model allows a different mean, μ_i, for each observation. Then, the likelihood ratio comparing the two models is

$$\mathrm{LR}(\mu, \boldsymbol{\mu}) = \prod_{i=1}^{n}\left(\frac{\mu}{\mu_i}\right)^{y_i} \mathrm{e}^{-n\mu+\sum \mu_i} \tag{4.5}$$

At the m.l.e., this becomes

$$\mathrm{LR}(\hat{\mu}, \hat{\boldsymbol{\mu}}) = \prod_{i=1}^{n}\left(\frac{\bar{y}_\bullet}{y_i}\right)^{y_i} \tag{4.6}$$

This may easily be extended to the regression case where, for example, $\log(\mu_i) = \beta_0 + \beta_1 x_i$.

2. For n Bernoulli observations, y_i, with probabilities π_i, the deviance for goodness of fit of the model $\pi_i = \pi$, a constant, as compared to the saturated model is given by

$$\begin{aligned} D(\pi) &= 2\sum_{i=1}^{n}\left[y_i \log\left(\frac{y_i}{\pi}\right) + (1-y_i)\log\left(\frac{1-y_i}{1-\pi}\right)\right] \\ &= -2n[\bar{y}_\bullet \log(\pi) + (1-\bar{y}_\bullet)\log(1-\pi)] \end{aligned} \tag{4.7}$$

Because y_i is either zero or one, the saturated model, in fact the maximized empirical likelihood, has $\hat{\pi}_i = y_i$, so that this likelihood is unity no matter what is observed. Thus, the deviance contains only $\bar{y}_\bullet$, the m.l.e. of π, and not the individual observations, and so is, in fact, of little use for measuring goodness of fit. If information is available on the order in which the responses were generated, a more structured model might be developed from that. □

Of course, an alternative saturated model, with a superior fit (unless the model is already multinomial, as in the Bernoulli example), is still the empirical likelihood, which, now, will have a different multinomial distribution for each set of values of the explanatory variables.

The situation is not so simple when the probability distribution has more than one parameter. One will usually be a location parameter and the regression model will generally be constructed in terms of it. In the simplest cases, the other parameter(s) are held constant for all values of the explanatory variables (Section 3.7.2). If the location parameter is allowed to take a different value for each observation when the values of the explanatory variables are all different, there will be more parameters than observations. No information will remain in the data about the other parameter(s) in the probability distribution once the mean parameters are estimated. Although the model is saturated, it is of little use, unless there are replications of observations of the response for at least some values of the explanatory variables.

Example

For a normal distribution (with unknown variance), the regression model usually describes changes in the mean. With a different mean for each independent observation, the variance is estimated as zero. Occasionally, the goodness of fit of a normal regression is judged by the coefficient of multiple determination, one minus the ratio of the variance for the model with the explanatory variables to that without, called R^2. A value of unity indicates a perfect (saturated) fit, but the variance parameter is, then, estimated to be zero in the model. Because such a model is unsatisfactory, closeness of R^2 to one does not provide a reasonable criterion of goodness of fit. □

Thus, extra care must be taken in judging the goodness of fit of regression models based on multiparameter distributions.

4.1.2 ROBUSTNESS

All models are simplified representations of a data generating mechanism and, thus, must be inappropriate to some degree. Some consideration should be given to the sensitivity of the data analysis to this possibility of unsuitable specification. In other words, we would like the important aspects of a likelihood function to be relatively stable or robust under changes of model. If a parameter is not just serving to index models, but takes values of particular interest, such as indicating the size of a treatment effect, then we would like likelihood regions for these parameters to be robust. This is essentially one of the things that we are checking in this chapter, whether through global goodness of fit or diagnostics.

Example
In a two-parameter likelihood function, $L(\psi_1, \psi_2)$, we expect that the profile likelihood for ψ_1 would be more robust than that with a fixed arbitrary value of ψ_2. □

In those cases where point parameter estimates alone are of direct interest, as in many production and measurement error situations, the choice of the distribution may not be of critical importance and the normal distribution is often used. Much of the classical work in robustness (Huber, 1964) concerns such point estimation, usually with symmetric distributions, because of their wide use in such contexts. If extreme (outlying) observations can occasionally be expected, a distribution with flatter tails may be preferable. Often a member of the stable distributions (Section 2.7.2) may be suitable. It is said to be more robust to outliers (Section 4.3.3), in that the parameter estimates will be less influenced by them.

Example
Point estimation of the location of a distribution, using the mean, requires assumptions about the extreme tails of that distribution, where observations will be rare. In contrast, estimation of location by the median is not sensitive to these extreme values. □

When the data generating mechanism is probabilistic, the reason why such extreme values are being produced will generally be of direct interest and an appropriate model should be constructed. Of course, a stable distribution may still be suitable, but we now choose it because of its ability to describe the underlying process, not because it yields more robust estimates. Naturally, if the extreme observations are simply spurious, they should be detected by the diagnostic procedures discussed below and appropriate corrective action taken.

Example
Models with the normal distribution replaced by the Student t (Section C.2) will be more robust to extreme observations, at the cost of one additional parameter, p in Equation (C.10). A small value of p indicates the presence of extreme values, while the distribution approaches normality as $p \to \infty$.□

Thus, robustness involves protection against model mis-specification and resistance of inference conclusions to spurious observations (Sprott, 1978, 1982). As more complex models are constructed, by conditioning on more information, usually by adding more explanatory variables, they will become more valid conditionally, but less robust.

4.1.3 CONTAMINATED OBSERVATIONS

An alternative to choosing a distribution with thicker tails, as in the previous section, is to model the supposedly contaminated observations explicitly, as arising from a distinct process. Thus, if we know that some observations have a different mean than the others, but not which ones, we might choose a finite mixture distribution

$$f(y; \boldsymbol{\mu}, \boldsymbol{\pi}) = \sum_{i=1}^{m} \pi_i f_i(y; \mu_i) \tag{4.8}$$

where $\sum \pi_i = 1$. Fitting such models can be difficult unless a large number of observations is available. Usually, m will be small, only two or three, say. For example, we might have a mixture of acceptable observations together with a few very large and very small ones, each group with a different mean. Then, the model for the data generating mechanism of interest might be that for the central observations. Hopefully, the π_i for these observations will be large and those for the contaminated ones small. Likelihood ratios will indicate if a model with the π_i for the contaminated observations equal to zero is plausible. Thus, this is another example of embedding.

Examples
1. If responses contain some extreme values, one alternative to a stable distribution is a mixture of two normal distributions with the same mean and different variances, say σ^2 and $c\sigma^2$, where c is a known constant. The larger is c, the more will extreme observations be down-weighted in the estimation of the mean.
2. Consider a two-way contingency table, with entries n_{ij}, where a model of independence is of interest. Suppose that this model is found to be implausible and that, say, $n_{11}n_{22} > n_{12}n_{21}$ and $n_{11} > n_{22}$. A model of independence can be set up containing a modified fourth cell (the problem being which one to choose) containing $n_{12}n_{21}/n_{11}$. This is a possible table of uncontaminated observations, while a second table ($m = 2$) contains

three zeros and $n_{22} - n_{12}n_{21}/n_{11}$ in the fourth cell, the contaminating observations. Then, $\hat{\pi}_2$, for the latter, is equal to $(n_{11}n_{22} - n_{12}n_{21})/(n_{11}n_{\bullet\bullet})$. The plausibility of π_2 being zero can then be examined. (Rudas *et al.*, 1994)

3. Workers are contaminated by a toxic chemical. Samples of their cells are taken and the numbers of aberrations counted. Most cells are not affected so that there are too many zero counts for these to be Poisson observations. Some cells have not been contaminated and cannot have events, so that there appear to be too many zeros. A truncated Poisson distribution, Equation (2.41), can be fitted to the non-zero observations and the number of cells in the zero category, which might have had an aberration but did not, estimated by $\exp(-\hat{\mu})$. □

If the observations which are contaminated can be distinguished, with certainty, the problem is simplified. The π_i need no longer be estimated, and the means might, for example, be represented by μ and either $\mu + \delta$ or $\delta\mu$, if all contaminated observations are either larger or smaller than the usual mean.

Example

Suppose that we observe n Poisson random variables, but that some values are extremely large. We decide that the k largest values are contaminated. Then, we can fit Poisson distributions, with mean μ for the $n - k$ smallest observations, and $\delta\mu$, $\delta > 0$, for the k largest ones. □

4.2 Tests of significance

In certain circumstances, it is useful to know if some completely specified model fits sufficiently well in absolute terms. This cannot be done through the likelihood function, which can only be used to *compare* models. Significance tests were specifically developed for this purpose. This will generally occur either as a scientific guess or when a model has been developed by a selection process (Section 3.6.1) on a data set and is to be checked with *new* data. It cannot be used with the data to which model selection was applied. (Statisticians have proposed various sample splitting procedures, but these do not provide *independent* checking of an hypothesized model.)

4.2.1 PURE TESTS

Suppose that a specific, completely defined, model, with $\psi = \psi_0$, *formulated before looking at the data*, is under consideration as a working hypothesis, and that it is the *only* model of interest, at least for the moment. No likelihood *function* is defined, because there is only one model. The sufficiency principle is not applicable. We require some radically different procedure.

Test statistics Suppose next that the probability of the observed value of some *test statistic* can be calculated from the observed data and uses as much of the relevant information in it as is possible. The choice of this test statistic is a mysterious process because there is no simple procedure for optimizing the use of relevant information nor even an objective measure of how much of the information in the data is used (see the discussion at the beginning of Section 1.5). However, two suitable statistics are the m.l.e. and the 'normed likelihood function' for the fixed model; these and others will be studied in Chapters 5 and 7.

If the observed value of this statistic is a rare or improbable event, according to some criterion, given the completely specified model, as compared to other possible values of that statistic (which were not observed), then the hypothesis might be taken to be questionable, and, in extreme cases, rejected. As with all inferences, this conclusion is only provisional, because future observations may reverse it, indicating that the observation was indeed a rare coincidence.

On the other hand, when the hypothesis is not rejected by such a test, the research worker might conclude that

- the unreasonableness of the working hypothesis does not seem sufficient to warrant modification, or
- deviation from the hypothesis is indicated, confirming previous suspicions, but more data are required to confirm this.

Thus, the calculated probability does not relate directly to the real world, but to a model and to the investigator's attitude to it.

Probabilities, as opposed to likelihoods, are here being used for inference, for the first time, but only negatively. They may allow us to pinpoint one particular model as unreasonable in light of the data without leading to a path for improvement. The reasoning in such a testing procedure is that *any* observed data set always has rather small probability, even if it is the most probable one for a given model. This is especially so as the number of observations increases. Thus, this probability has no absolute meaning, and some other criterion must be found.

Relative probability One reasonable possibility to judge rareness is to compare the *relative probability* of the observed data under the model with ψ_0 to that of the most probable similar data under that model:

$$\frac{P(y;\psi_0)}{\sup_u P(u;\psi_0)} \tag{4.9}$$

Notice the similarity to a normed likelihood. However, here we compare probabilities of two data sets, one not observed, for one model, instead of the probabilities of the observed data set under two models. It is a probability comparison, not a likelihood one.

Table 4.1. Two possible distributions for a five-valued variable.

y	$f(y,\psi_1)$	$f(y,\psi_2)$
0	0.80	0.75
1	0.15	0.15
2	0.05	0.05
3	0.00	0.04
4	0.00	0.01

Example
Consider a random variable, Y, which can take only five values and which follows a probability distribution, $f(y;\psi)$. The distributions for two possible parameter values are given in Table 4.1. Suppose that we observe $y = 2$. Both models have the same likelihood. If we ignore ψ_2 and take ψ_1 as the only model of interest, the relative probability of the data is 0.0625. On the other hand, if we took only ψ_2, the relative probability would be 0.0667. This latter model indicates that the data are slightly less rare. □

A problem with this method is that it depends heavily on the probability of one unobserved data point. (It may also not be used because asymptotic results are difficult to obtain!)

P-values The more usual procedure is, instead, to look at the total probability of the set of all possible similar observations, under the model, which are at least as rare as that observed. This cumulated probability, given the hypothesis, is called the *P-value*; it is also known as the *tail probability*, because it is usually obtained from the extremities of the distribution. This procedure yields a *pure* or *Fisherian test of significance.* The particularity of this method is that it depends on the probabilities of unobserved data points in the tail of the distribution, about which information is difficult to obtain, because they are rare; hence it is very sensitive to model choice.

The appeal of P-values is that they allow one to make an absolute statement about one model, in the light of the data, without comparing it to any other. Such a statement is not possible from the likelihood function, which compares different models. Obviously, if the hypothesized model is rejected, another will have to be found to replace it, so that our assumption of only one model available can be, at best, momentary. Of course, this is a basis for scientific discovery: the research worker may be prompted by the test to look for, and to interpret, unexpected patterns in the data.

A danger is that, if the model to test is specified after inspecting the data, one can be chosen, even unconsciously, which makes the observations rare.

Table 4.2. Exact distribution of the m.l.e. of a binomial distribution for $n = 10$.

y	$\hat{\pi}$	$p(\hat{\pi}; 0.5)$	$p(\hat{\pi}; 0.7)$
0	0.0	0.0010	0.0000
1	0.1	0.0098	0.0001
2	0.2	0.0439	0.0014
3	0.3	0.1172	0.0090
4	0.4	0.2051	0.0368
5	0.5	0.2461	0.1029
6	0.6	0.2051	0.2001
7	0.7	0.1172	0.2668
8	0.8	0.0439	0.2335
9	0.9	0.0098	0.1211
10	1.0	0.0010	0.0282

Ordinate test One way in which the P-value can be defined is by

$$\int_A f(\hat{\psi}; \psi_0) d\hat{\psi} \tag{4.10}$$

where the set $A = \{\hat{\psi} : f(\hat{\psi}; \psi_0) \leq f(\hat{\psi}_o; \psi_0)\}$ with $\hat{\psi}_o$ the observed value of the m.l.e. If we check, in this way, to see if the observed value of the m.l.e. is improbable, we have an *ordinate test.* For continuous random variables, the probabilities depend critically on the measure for the random variable, which should be the unit of measurement.

Example
For a binomial distribution with fixed value $n = 10$, we observe $y = 1$ so that $\hat{\pi}_o = 0.1$. Suppose that our hypothesis is that $\pi_0 = 0.5$. The distribution of the m.l.e. under any fixed hypothesis can be calculated exactly, as shown in Table 4.2 for two values of π. The P-value probability of an m.l.e. at least as improbable as that observed is the sum of all values

$$\begin{aligned}\binom{10}{y} 0.5^y 0.5^{10-y} &\leq \binom{10}{1} 0.5^{10} \\ &= 0.0098\end{aligned}$$

so that the P-value is 0.0216, a relatively rare occurrence. □

In pure significance tests, we are only concerned with what has happened, as compared to what might have happened. No long run frequency arguments, concerning what might happen in the future, are involved (Chapter 7). However, because we compare what was observed to

Table 4.3. Three sets of binomial observations.

n	y	$\hat{\pi}$
20	15	0.750
200	115	0.575
2000	1046	0.523

phenomena which were not observed, or, perhaps, do not even use all of the information in the data, this approach to inference could be said to violate the sufficiency principle, if it were applicable. It will not necessarily give results compatible with those from the likelihood function, and, in fact, will only do so if ψ is a location parameter (or a transformed scale parameter, as in Section 2.4.2) in an ordinate test (Seidenfeld, 1979, pp. 128–129).

Examples

1. Consider again Table 4.1, with the observation $y = 2$. If ψ_1 is the only model of interest, the P-value is 0.05 while, if it is ψ_2, the P-value is 0.10, yielding the same conclusion (in this case) as the relative probability, although both models are equally supported by the likelihood. Here, ψ is not a location parameter. (Cox, 1958)

2. Consider the three sets of binomial observations given in Table 4.3. For each set, the P-value is about 0.021 under the model, $\pi_0 = 0.5$. However, in the larger data sets, the m.l.e. of the binomial probability is much closer to π_0. Thus, as n increases, an m.l.e. ever closer to the hypothesized value becomes rare. P-values act in a way similar to a likelihood function, in that increasing the number of observations increases the precision. (See the discussion by Morris in Berger and Sellke, 1987.) □

Fixed significance levels Often, what has been done is to choose a fixed probability level, α, such as 0.05, as indicating rarity. This may be explained by historical reasons: only tables giving a few probability levels were available. Now, computer software provides exact values which should be used.

Thus, instead of inspecting and reporting the observed P-value, a statement was simply made as to whether or not the result was significant, i.e. reached this 0.05 value. In effect, the full data set had been replaced by a single binary observation, z say, indicating whether or not the actual observations were in the rare set, say $\mathcal{Y}_\alpha$, of possible observations. For some test statistic, T, this set might, for example, be

$$\mathcal{Y}_\alpha = \{\mathbf{Y} : T(\mathbf{Y}) \leq t(\mathbf{y})\} \tag{4.11}$$

Then, $z = 1$ if $\mathbf{y} \in \mathcal{Y}_\alpha$ and zero otherwise. $T(\mathbf{Y})$ would be the test statistic such that

$$P(T(\mathbf{Y}) \leq t(\mathbf{y}_o); \psi_0) = P(z = 1; \psi_0) \qquad (4.12)$$
$$= \alpha$$

so that here small values of $T(\mathbf{Y})$ are rare. Note that, in many cases, $\mathcal{Y}_\alpha$ will not be uniquely defined; it might be based, for example, on the m.l.e., with $t(y) = \hat{\psi}$, or the log likelihood function, $t(y) = l(\psi; y)$. (If the m.l.e. is the sufficient statistic, these will be the same.)

This procedure of fixing the significance level can be seen to be equivalent to using a new reduced likelihood function,

$$\mathrm{L}(\psi) = P(\mathcal{Y}_\alpha; \psi) \qquad (4.13)$$

where $P(\mathcal{Y}_\alpha; \psi_0) = \alpha$. Thus, a drawback of significance tests with a fixed probability level was that they involved a drastic loss of information, resulting from using only the reduced likelihood function. They provide no distinction between observations on the border of significance and those very rare under the model tested. Instead of fixing the probability level, α, and only indicating whether or not it has been attained, the actual P-value should be reported.

4.2.2 EXACT TESTS

In fact, a more frequent, applied situation for such testing than specifying a complete model will occur when ψ_0 is a unique feature of the model for the data generating mechanism, of special prior interest, thus warranting temporarily ignoring all other possibilities.

Example

In a regression model specified in advance of collecting the data (not obtained by some stepwise model selection procedure from the data!),

$$g(\mu_i) = \beta_0 + \beta_1 x_{1i} + \beta_2 x_{2i}$$

the value of, say, $\beta_1 = 0$ is of special interest, because it means that the explanatory variable, x_{1i}, does not influence the mean response. Nevertheless, if the scientist really believed this to be true *a priori*, x_{1i} would not be included in the model in the first place. □

As in this example, only some of the parameter values may be specified, as the feature to be tested, others remaining as nuisance parameters.

As we already saw above, in certain simple cases, it is possible to construct exact distributions for test statistics for the size of sample observed. In such circumstances, the improbability of the observed statistic can also be calculated. For the normal distribution, the well-known, but generally mis-used, tests based on the Chi-squared, Student t, and F distributions (Section 5.6, Chapter 7, and Appendix C) are exact. Tables of tail probabilities are widely available and computer statistical packages all provide them.

Example
For observations from a normal distribution with known variance, σ^2, we can test the hypothesis that the mean is μ_0 with the statistic

$$T(\mathbf{Y}) = \frac{\bar{Y}_\bullet - \mu_0}{\sigma/\sqrt{n}} \tag{4.14}$$

which is a pivotal quantity with an N(0,1) distribution. If the variance is unknown, but is not specified by the hypothesis, the statistic

$$T(\mathbf{Y}) = \frac{\bar{Y}_\bullet - \mu_0}{S/\sqrt{n}} \tag{4.15}$$

another pivotal quantity, where $S^2 = \sum(Y_i - \bar{Y}_\bullet)^2/(n-1)$, may be used. This has an exact Student t distribution from which an exact P-value can be obtained. (Student, 1908) □

Again in this example, the task is complicated by the presence of a nuisance parameter not specified by the hypothesis. We have relied on a conditioning argument, hinted at in Sections 2.2.3 and 3.6.3 above and to be developed further in Chapter 6 below. Another famous illustration is *Fisher's exact test* for the 2×2 contingency table (Fisher, 1958, pp. 96–97; Yates, 1934).

Examples
1. Consider logistic regression for independent binary random variables, $Y_1, \ldots, Y_n$:

$$\log\left(\frac{\pi_i}{1-\pi_i}\right) = \beta_0 + \beta_1 x_i \tag{4.16}$$

for known x_i. Because this is a canonical regression model, the minimal sufficient statistic for $\boldsymbol{\beta}$ is $\mathbf{T} = (\sum Y_i, \sum x_i Y_i)^{\mathrm{T}}$.

Our parameter of interest is β_1. We wish to test if it could be zero, i.e. for no effect of x_i. To avoid having to specify a fixed value for β_0, we would like to have a (conditional) distribution only containing the parameter of interest. From Section 2.2.3, we know that this is possible within the linear exponential family. The sufficient statistic for the nuisance parameter, β_0, is $T_0 = \sum Y_i$ and the marginal density of $\mathbf{T}$ is

$$f(t_0, t_1; \beta_0, \beta_1) = \frac{c(t_0, t_1)e^{\beta_0 t_0 + \beta_1 t_1}}{\prod(1 + e^{\beta_0 + \beta_1 x_i})} \tag{4.17}$$

where $c(t_0, t_1)$ is the number of distinct subsets of size t_0 (the number of successes) which can be formed for the observations and which sum to t_1. The marginal density for t_0 is, then,

$$f(t_0; \beta_0, \beta_1) = \frac{\sum_u c(t_0, u)e^{\beta_0 t_0 + \beta_1 u}}{\prod(1 + e^{\beta_0 + \beta_1 x_i})} \tag{4.18}$$

It follows immediately by division that

$$f(t_1|t_0; \beta_1) = \frac{c(t_0, t_1)e^{\beta_1 t_1}}{\sum_u c(t_0, u)e^{\beta_1 u}} \tag{4.19}$$

is the required conditional distribution on which inference for β_1 alone can be based. This can be used in the same way as a modified profile likelihood for β_1 (Sections 3.6.3 and 6.2.5; see Lindsey, 1975).

However, here we are interested in P-values which can be calculated from Equation (4.19), although this will be time consuming if n is large.

2. In the preceding example, suppose that the known x_i only take two values. Let us sort the data so that the first n_1 observations have $x_i = 1$ and the remaining $n_2 = n - n_1$, $x_i = 0$. Then

$$c(t_0, t_1) = \binom{n_1}{t_1}\binom{n_2}{t_0 - t_1} \tag{4.20}$$

with

$$\sum_u c(t_0, u) = \binom{n}{t_0} \tag{4.21}$$

so that

$$f(t_1|t_0; \beta_1) = \frac{\binom{n_1}{t_1}\binom{n_2}{t_0 - t_1}e^{\beta_1 t_1}}{\sum_{u=\sup(t_0 - n_2, 0)}^{\inf(t_0, n_1)} \binom{n_1}{u}\binom{n_2}{t_0 - u}e^{\beta_1 u}} \tag{4.22}$$

defines a conditional distribution which can be used as a modified profile likelihood.

For the special case where the model of interest is $\beta_1 = 0$, we have the P-value given by

$$\frac{\sum_{u=t_1}^{\inf(t_0, n_1)} \binom{n_1}{u}\binom{n_2}{t_0 - u}}{\binom{n}{t_0}} \tag{4.23}$$

i.e. from a *hypergeometric distribution.* This is Fisher's exact test for no difference, or independence, between two samples (coded by x_i) in a 2×2 contingency table.

Suppose, now, that we have the following table:

	$y=0$	$y=1$
$x=0$	0	4
$x=1$	3	2

The P-value is 0.167, so that, for the model with $\beta_1 = 0$, i.e. no difference (independence) between the two samples, the results in this table are not a rare event under this model.

Now consider a second table:

	$y=0$	$y=1$
$x=0$	19	4
$x=1$	14	12

The P-value for this table is 0.039, indicating that this would be a rather improbable result for the model with no difference between the two samples.□

In these examples, the likelihood function of Equation (4.17) has been factored into two parts, given by Equations (4.18) and (4.19). The latter can been used as a type of modified profile likelihood and for the test because it only contains β_1. However, it does not use all of the information in the data about β_1 because Equation (4.18) is generally not a completely flat likelihood in this parameter. In allowing for the uncertainty about the nuisance parameter, β_0, we have apparently lost information about the parameter of interest, β_1. The same thing occurred in the modified profile likelihood of Equation (3.77) for σ^2 which could also be used for a test about that parameter: the Chi-squared distribution mentioned above.

4.2.3 SUFFICIENT AND ANCILLARY STATISTICS

As we have seen, in most situations, more information relevant to the data generating mechanism is available in the data than that required to write down the likelihood function for a model of prior interest. Thus, for example, whenever the sufficient statistic, $\mathbf{t}$, for the parameters in a model is of smaller dimension than the vector of responses, $\mathbf{y}$, additional information is available for checking the goodness of fit of that model. From the factorization theorem (Section 3.4.1), we have

$$f(\mathbf{y}; \boldsymbol{\psi}) = f(\mathbf{t}; \boldsymbol{\psi}) f(\mathbf{y}|\mathbf{t})$$

In Chapter 3, we have used the first factor as the likelihood function for $\boldsymbol{\psi}$, assuming the adequacy of the model family. The second factor, $f(\mathbf{y}|\mathbf{t})$, is not used directly in the likelihood function for the model of prior interest. It can yield no direct information about plausible values of $\boldsymbol{\psi}$, given the model function, but does contain information on goodness of fit which it may be possible to use.

Example
Consider a vector of n observations, $\mathbf{y}$, which we assume to be independent, from a Poisson distribution with mean μ. The minimal sufficient statistic for the mean is the sum of the observations, $t = y_\bullet = \sum y_i$, which has a Poisson distribution with mean $n\mu$. Then, the conditional distribution is

$$\begin{aligned} f(\mathbf{y}|t) &= \frac{\prod_i \frac{e^{-\mu}\mu^{y_i}}{y_i!}}{\frac{e^{-n\mu}(n\mu)^{y_\bullet}}{y_\bullet!}} \\ &= \binom{y_\bullet}{y_1 \cdots y_n}\left(\frac{1}{n}\right)^{y_\bullet} \qquad (4.24) \\ &= \binom{y_\bullet}{y_1 \cdots y_n}\left(\frac{\bar{y}_\bullet}{y_\bullet}\right)^{y_\bullet} \end{aligned}$$

a uniform distribution, which is a special case of the multinomial distribution. Note the similarity to Equation (4.6). □

If $\mathbf{y}$ is somehow extreme in the equivalence class of observations defined by $\mathbf{t}$, it can throw doubt on the adequacy of the model family. However, the distribution of $\mathbf{Y}$, given $\mathbf{t}$, will often be complex. Then, a simpler distribution, not depending on ψ, may be that for an ancillary statistic, which can be used as a test statistic.

Example
When $n = 2$, Equation (4.24) can be examined relatively easily, but for larger samples, this becomes difficult or impossible. Then, some special aspect of $\mathbf{y}$ may be examined.

The variance may not be close to the Poisson mean, in which case the ancillary statistic, $\sum(y_i - \bar{y}_\bullet)^2/n$, can be studied, usually as a ratio to $\bar{y}_\bullet$ in a test statistic called the coefficient of dispersion. Its value may indicate the presence of over- or underdispersion (Section 1.2). Or the number of zero values may be of prior concern; their number can be compared to the number of empty cells in the uniform distribution. □

Exact tests can also be developed for checking global goodness of fit of distributions, using such conditional arguments.

Example
Goodness of fit of n observations hypothesized to be from a Poisson distribution with common mean can be obtained from Equation (4.24). The P-value is the sum over all possible observation sets, $\mathbf{y}$, of size n with

$\sum y_i = t$, the observed value, which have conditional probability less than or equal to that observed:

$$\binom{t}{y_1 \cdots y_n} \left(\frac{1}{n}\right)^t$$

□

For other distributions, the development of such tests can be complex, and perhaps for this reason, they have been little used.

4.3 Diagnostics

Likelihood comparisons, using embedding and saturated models, and tests of significance usually only provide global indications of lack of fit of a model of interest to the data at hand. If the current model is found to be wanting, a saturated model would not usually be an acceptable replacement. Instead, unexpected ways in which the data depart from the model must be studied in order to discover means of improving or replacing the old model of interest. Thus, this is a supplementary procedure in the model selection process. For example, if the old model assumes independence among observations, the new model might require some specific form of dependence.

If one is aware of the type of departure, embedding will be the path to follow. The procedures outlined in this section are more exploratory. Basically, we shall decompose global measures of goodness of fit into the individual terms arising from each independent observation. This may allow us to discover which aspects of the models of interest are unsatisfactory, leading to development of more complex or functionally different ones, a means of scientific discovery.

Such diagnostics procedures continue in the line of our approach of studying ancillary statistics. However, here we look at individual values, corresponding to each separate observation. If we think of pivotal quantities, for a location–scale family, the configuration is ancillary. In other families, the three approximate pivotal quantities of Equations (3.66) to (3.68) can be decomposed in a similar way into the maximum likelihood estimate and an approximate configuration. As we shall see, the latter yields three forms of residuals used in diagnostics.

4.3.1 DEPARTURES FROM MODELS

Two types of departures from a model may be distinguished:

(1) the observations on one or a small number of individuals may not be well represented by the model, or

(2) the whole data set may show an unsuspected systematic departure from the model.

Table 4.4. Four invented sets of data, from Anscombe (1973).

X_1	Y_1	X_2	Y_2	X_3	Y_3	X_4	Y_4
10	8.04	10	7.46	8	6.58	10	9.14
8	6.95	8	6.77	8	5.76	8	8.14
13	7.58	13	12.74	8	7.71	13	8.74
9	8.81	9	7.11	8	8.84	9	8.77
11	8.33	11	7.81	8	8.47	11	9.26
14	9.96	14	8.84	8	7.04	14	8.10
6	7.24	6	6.08	8	5.25	6	6.13
4	4.26	4	5.39	19	12.50	4	3.10
12	10.84	12	8.15	8	5.56	12	9.13
7	4.82	7	6.42	8	7.91	7	7.26
5	5.68	5	5.73	8	6.89	5	4.74

The study of such departures is known as *diagnostics*. In the following presentation, we shall primarily be concerned with general regression situations, such as generalized linear models, but many of the ideas extend rather naturally to other contexts.

In order to understand the types of problems which can appear, it will be useful to refer to several simple but extreme data sets, which illustrate the ideas.

Example
Consider the four well-known sets of data given in Table 4.4. If we assume a normal distribution, with simple linear regression, all four data sets have exactly the same fitted model:

$$\mu_i = 3.0 + 0.5x_i$$

with the same variance estimate. Thus, if we blindly fitted such a linear regression model, without checking the data, we would be led to the same conclusions for all four data sets. However, if we plot this regression line, along with each set of observed values, as in Figures 4.1 to 4.4, we see that the four data sets are not at all similar. (Anscombe, 1973) □

In realistic cases, departures from a model will be much more subtle than those in these extreme data sets, which are obvious from the plots. As well, if several explanatory variables are present in a model, it will not even be easy to represent the complete model graphically. Thus, these sets will only serve to highlight how diagnostics can aid us in interpreting data in the light of a model, with little attempt at realism.

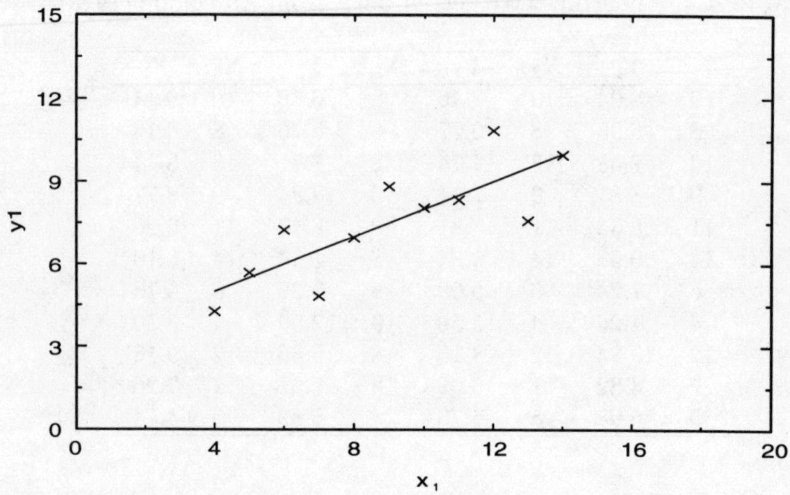

Fig. 4.1. Plot of the first data set from Table 4.4, with the fitted linear regression line.

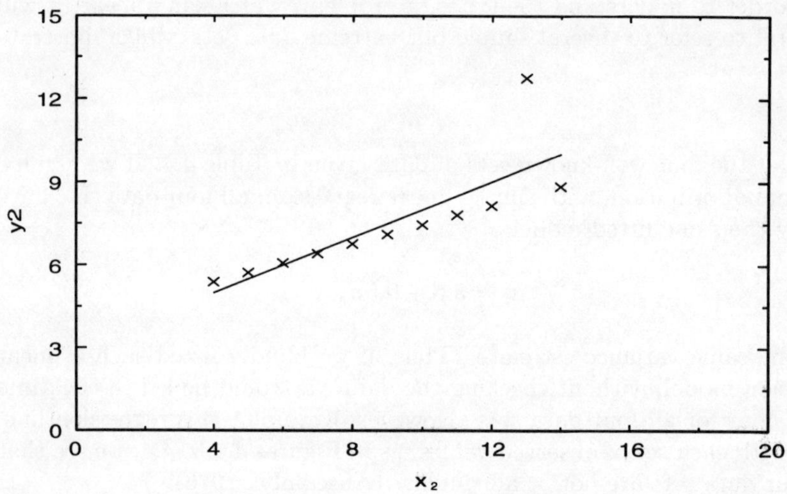

Fig. 4.2. Plot of the second data set from Table 4.4, with the fitted linear regression line.

4.3.2 RESIDUALS

The study of departures from a model has classically primarily been concerned with the estimated expected values of the responses, $\hat{\mu}_i$, under some model and with their differences from the observed responses, called *resid-*

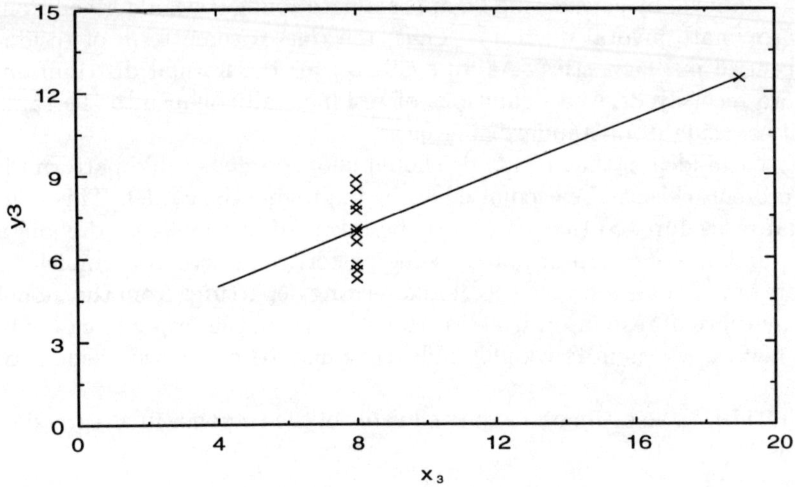

Fig. 4.3. Plot of the third data set from Table 4.4, with the fitted linear regression line.

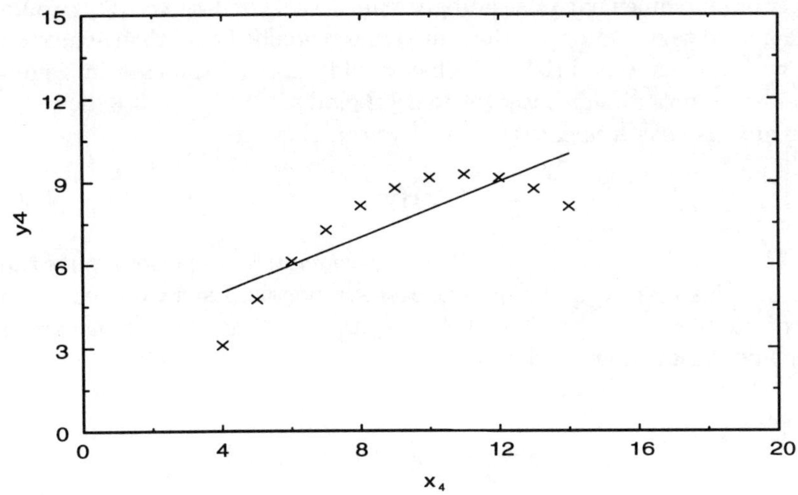

Fig. 4.4. Plot of the fourth data set from Table 4.4, with the fitted linear regression line.

uals. This approach was originally developed for normal linear regression models. It is, more generally, most useful in transformation models, such as the location–scale family (Section 2.7.1). However, as we shall soon see, the concept of residual has subsequently acquired several more general definitions, which are of much wider application. These can arise, in the way suggested above, by considering an approximate configuration derived from an approximate pivotal quantity. Thus, the role of some form of residuals is central in diagnostics. As we shall see, for the normal distribution, the more recently derived definitions of residuals all collapse to the same quantities, aside from standardizing factors.

The basic idea is that residuals should show no detectable pattern. If one is present, this can be accounted for by improving the model. This is an informal procedure, so that care must be taken: if one looks hard enough at any random collection of points, some pattern is always discernible!

Plots of residuals are very useful in detecting departures from the model for many types of response variables. However, when the response can only take a few values, as in Bernoulli trials, they may be of limited usefulness.

Hat matrix The estimated expected value of the response in a regression model is

$$\begin{aligned} \mathrm{E}_{\hat{\psi}}[Y_i] &= \hat{\mu}_i \\ &= \hat{Y}_i \end{aligned} \tag{4.25}$$

which is often, somewhat misleadingly, called the *fitted value.* This could mistakenly be taken to imply that all values 'should' be at their expected value, with no random variation. That would never be the case in a non-deterministic, probability-based, *statistical* model.

For models with a linear structure, we can show that

$$\hat{\mathbf{Y}} = \mathbf{H}\mathbf{y} \tag{4.26}$$

Then, $\mathbf{H}$, with elements h_{ij}, is called the *hat matrix*, because it puts the hat on y. This matrix is idempotent and symmetric; its trace is equal to p, the dimension of the parameter vector, and its elements cannot exceed one for linear regression models.

Examples

1. For normal linear regression, from Equation (3.100),

$$\begin{aligned} \hat{\mathbf{Y}} &= \mathbf{X}\hat{\boldsymbol{\beta}} \\ &= \mathbf{X}(\mathbf{X}^{\mathrm{T}}\mathbf{X})^{-1}\mathbf{X}^{\mathrm{T}}\mathbf{y} \end{aligned} \tag{4.27}$$

so that

$$\mathbf{H} = \mathbf{X}(\mathbf{X}^{\mathrm{T}}\mathbf{X})^{-1}\mathbf{X}^{\mathrm{T}} \tag{4.28}$$

The variance of $\hat{Y}_i$ is $h_{ii}\sigma^2$ and is thus not constant.

2. For a generalized linear model, $\mathbf{X}$ in the above hat matrix is replaced by $\mathbf{W}^{\frac{1}{2}}\mathbf{X}$, where the weights, from Equation (3.97), are

$$\mathbf{W} = \text{diag}[\mathrm{J}_{\eta_i\eta_i}(\eta)] \tag{4.29}$$

and η is the linear structure. The hat matrix,

$$\mathbf{H} = \mathbf{W}^{\frac{1}{2}}\mathbf{X}(\mathbf{X}^{\mathrm{T}}\mathbf{W}\mathbf{X})^{-1}\mathbf{X}^{\mathrm{T}}\mathbf{W}^{\frac{1}{2}} \tag{4.30}$$

now generally depends on the parameter values through $\mathbf{W}$. □

The reciprocal of the diagonal element, $1/h_{ii}$, called the *effective replication*, can be roughly interpreted as the number of observations providing information about $\hat{Y}_i$. If this value is close to unity, the observation is isolated or self-estimating. Because of the continuity of an unsaturated model, neighbouring points generally provide additional information in predicting the response at a given point in the design space through smoothing.

Deviance residuals The deviance for the goodness of fit of a model can be measured using the likelihoods comparing the model of interest embedded in a saturated model, as in Section 4.1. We can decompose this as *deviance residuals* (Pregibon, 1981). These may indicate which individual observations contribute most to this global measure of lack of fit. They are usually standardized by dividing by $\sqrt{1-h_{ii}}$ and are defined as the (corrected) contribution of the ith observation to the directed deviance of Equation (3.27):

$$\varepsilon_i^D = \frac{\text{sign}(\tilde{\eta}_i - \hat{\eta}_i)\sqrt{2\mathrm{l}(\tilde{\eta}_i; y_i) - 2\mathrm{l}(\hat{\eta}_i; y_i)}}{\sqrt{1-h_{ii}}} \tag{4.31}$$

where $\tilde{\eta}_i$ is the value of the linear structure, η, which maximizes the unconstrained likelihood for the data, the saturated model. For exponential dispersion and generalized linear models, the deviance can be calculated for fixed scale parameter, and then an estimate of that parameter supplied.

Examples

1. For the normal distribution, the deviance residuals are

$$\varepsilon_i^D = \frac{y_i - \hat{\mu}_i}{\sqrt{\hat{\sigma}^2(1-h_{ii})}} \tag{4.32}$$

from Equation (3.92).

2. For the logistic regression, the deviance residuals are

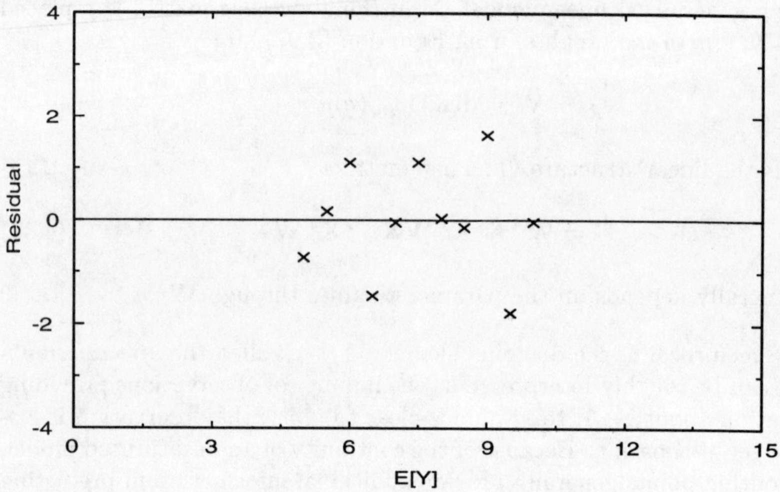

Fig. 4.5. Plot of the standardized deviance residuals for the first data set from Table 4.4.

$$\varepsilon_i^D = \frac{\pm\sqrt{2y_i \log\left(\frac{y_i}{n_i\hat{\pi}_i}\right) + 2(n_i - y_i)\log\left(\frac{n_i - y_i}{n_i - n_i\hat{\pi}_i}\right)}}{\sqrt{1-h_{ii}}} \tag{4.33}$$

where the sign is chosen to be the same as that of $y_i - n_i\hat{\pi}_i$.

3. Let us study the residuals from our four data sets as compared to the estimated expected values, $\hat{\mu}_i = 3.0 + 0.5x_i$. For this simple linear normal regression, the deviance residuals for these data are

$$\varepsilon_i^D = \frac{y_i - (3.0 + 0.5x_i)}{\hat{\sigma}\sqrt{1-h_{ii}}}$$

These are plotted in Figures 4.5 to 4.8. We only find the desired constant variability in the first plot. The three other graphs illustrate extreme forms of systematic variation in residuals which may be found when an incorrect model has been fitted. Of course, for these simple examples, the deviations from the model were already evident in the plots of the data and fitted values above. □

Deviance residuals are of wider application than only to regression models. For example, they can be fruitfully inspected for probability distributions with constant location parameter and for stochastic processes (Lawson, 1993).

Effective design matrix The $n \times p$ matrix, with elements

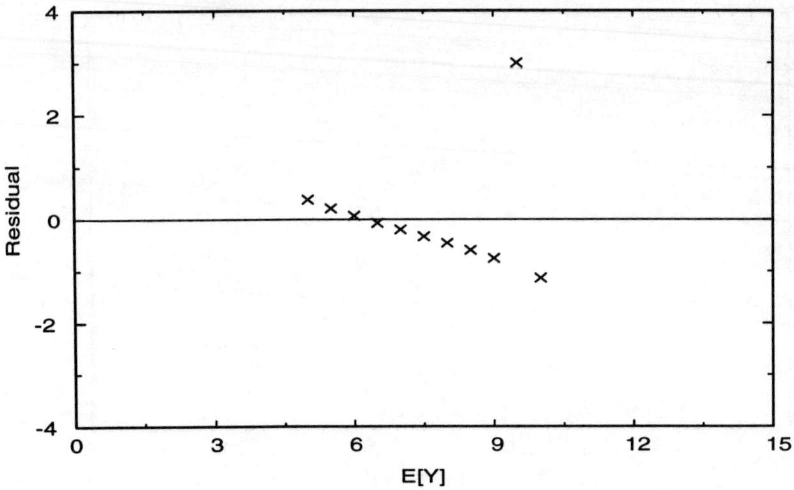

Fig. 4.6. Plot of the standardized deviance residuals for the second data set from Table 4.4.

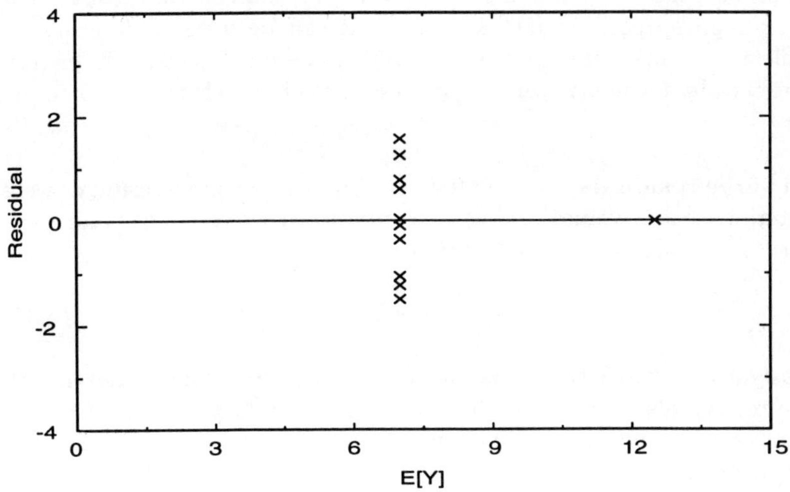

Fig. 4.7. Plot of the standardized deviance residuals for the third data set from Table 4.4.

$$A_{ij} = -\frac{\partial \varepsilon_i^D}{\partial \psi_i} \tag{4.34}$$

is generally a function of the unknown parameters, except in normal linear regression, where it is proportional to $\mathbf{X}$. Its value at the maximum

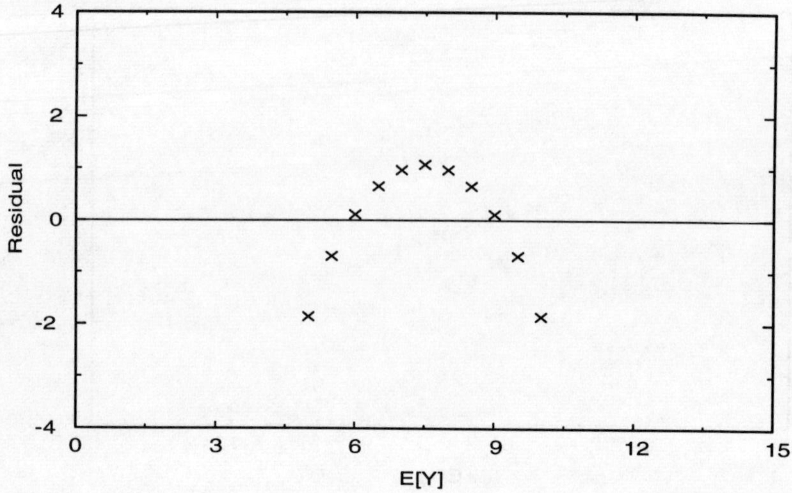

Fig. 4.8. Plot of the standardized deviance residuals for the fourth data set from Table 4.4.

likelihood estimate is called the effective design matrix, indicating what type of design space is in fact available. It can be used in determining how efficiently values of explanatory variables were assigned in an experiment, especially for non-linear and/or non-normal models (Ross, 1990, pp. 69–70).

Fitted value residuals As pointed out above, the original residuals were based on the fitted values. Thus, the *raw fitted value residual* for each response is its difference from its fitted value:

$$\varepsilon_i^R = y_i - \hat{Y}_i \tag{4.35}$$

Obviously, this construction makes most sense for a location(–scale) model. A more reasonable residual for the exponential family might involve comparing observed and fitted values on the canonical scale rather than the mean value scale as here.

In normal linear regression models, and Gauss linear models based on other members of the location–scale family, the variance of the response variable is assumed constant for all values of the explanatory variable. One might expect this to correspond to the residuals having about the same variability across the graph. However, even for such a regression, the variance of the raw residuals is not constant, as an explanatory variable changes, but is larger for responses near the mean of that variable. Thus, it is useful to standardize the raw fitted value residuals by dividing them by their standard error to obtain a standardized (internally) *studentized* (fitted value) *residual*:

$$\varepsilon_i^F = \frac{y_i - \hat{\mu}_i}{\sqrt{(1 - h_{ii})\widehat{\text{var}[Y_i]}}} \tag{4.36}$$

For discrete data models, such as the binomial and Poisson, this is also sometimes called the *standardized Pearson residual* because the square, $(y_i - \hat{Y}_i)^2/\widehat{\text{var}[Y_i]}$, is the contribution of the ith observation to the generalized Pearson (score) statistic of Equation (4.4). An externally studentized residual involves calculating $\widehat{\text{var}[Y_i]}$ differently for each i, using all observations except the ith, thus providing an estimate of precision independent of the observation. There is a monotone relationship between internally and externally studentized residuals, so that both are not required (Ellenberg, 1976).

In most models, other than those based on the normal distribution, the variance of the response variable cannot be assumed to be constant. For example, in generalized linear models, it is a function of the mean. For such models, the standard error varies with the mean. This is properly taken into account by Equation (4.36).

Examples

1. For a logistic regression model, the studentized residuals are

$$\varepsilon_i^F = \frac{y_i - n_i\hat{\pi}_i}{\sqrt{n_i\hat{\pi}_i(1 - \hat{\pi}_i)(1 - h_{ii})}} \tag{4.37}$$

2. For the normal regression model, from Equations (4.27), (4.28), and (4.35), the variance of the raw residuals is given by

$$\text{var}[\boldsymbol{\varepsilon}^R] = \sigma^2(\mathbf{I}_n - \mathbf{H}) \tag{4.38}$$

where $\mathbf{I}_n$ is the identity matrix. Thus, the studentized residuals are

$$\varepsilon_i^F = \frac{y_i - \hat{\mu}_i}{\sqrt{\hat{\sigma}^2(1 - h_{ii})}} \tag{4.39}$$

They are the (standardized) proportional contributions of the raw residuals to the estimated standard deviation and are identical to the deviance residuals, given above in Equation (4.32) for this model.

3. The fitted value residuals from our four data sets were plotted above as deviance residuals. □

Except for the normal distribution, these residuals will generally have a skewed distribution. A wide variety of transformations have been suggested, but these will not be pursued here because of the limited general usefulness of such residuals.

Score residuals Maximum likelihood estimates for a model are obtained by solving the score equations. For independent responses, these involve a sum of terms, one for each observation, set equal to zero. Thus, we can inspect the individual terms of the sum, called *score residuals*, to see which are furthest from zero. As above, these will be standardized by correcting for their standard deviation:

$$\varepsilon_i^S = \frac{\mathrm{U}_i}{\sqrt{(1-h_{ii})w_{ii}}} \tag{4.40}$$

where U_i is the term in the score for the ith response and w_{ii} is the ith diagonal element of the weight matrix in Equation (4.29).

In a generalized linear model, if the score for the linear structure, η, is used, the score residuals are identical to the studentized fitted value residuals given above. However, as with deviance residuals, these residuals are also of wider application than only to regression models.

Likelihood residuals Another possibility is to compare the likelihood for a fitted model for the complete set of observations to that when each observation, in turn, is omitted. This requires substantial calculation; it may be approximated by what have been called *likelihood residuals*,

$$\varepsilon_i^L = \mathrm{sign}(\tilde{\eta}_i - \hat{\eta}_i)\sqrt{h_{ii}(\varepsilon_i^S)^2 + (1-h_{ii})(\varepsilon_i^D)^2} \tag{4.41}$$

a weighted average of deviance and score residuals (Williams, 1987).

The sum of squares of any of these residuals provides a global measure of goodness of fit, which may be more or less useful, as discussed above in Section 4.1.

Residual plots Residuals can be plotted against a variety of statistics and other indices, each providing different information about departures from the model. It may be helpful to calculate them not only for the m.l.e. of the parameters but also for more extreme values at the edges of their plausible likelihood intervals.

In an *index plot*, the residuals are shown against the corresponding observation number. Ordering in this way may make identification of departures from the model easier.

Example

For the first data set of Table 4.4, the index plot of fitted value residuals is given in Figure 4.9. □

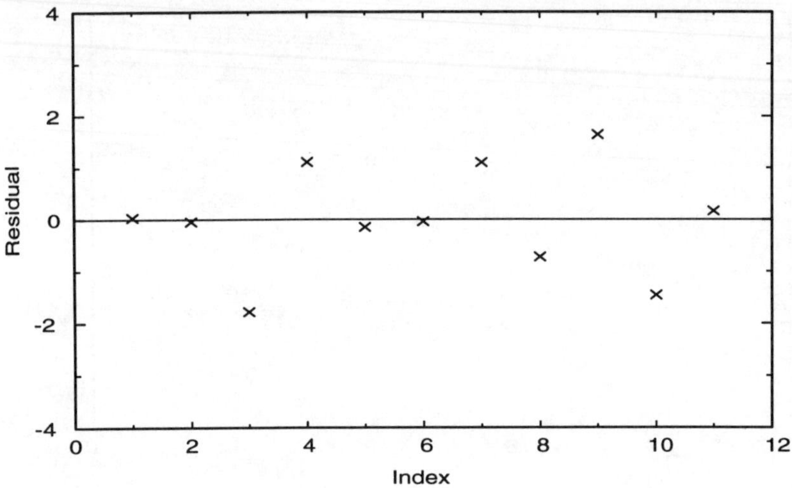

Fig. 4.9. Index plot of standardized fitted value residuals for the first data set of Table 4.4.

If the order has intrinsic meaning, for example as the order of collection of the data in time, the plot may indicate systematic variability in this sense. In this case, it may also be helpful to plot the residuals against their *lagged* values at one or more previous time points to check for dependence.

Residuals can be plotted against the estimated means or estimated linear structure, as was done in the graphs above. They may also be plotted against each of the available explanatory variables, including those not in the model under consideration.

Example
For the first data set above, the plot of fitted value residuals against x_i is shown in Figure 4.10. □

Finally, a *normal probability* or *Q–Q* (quantile) *plot* shows the standardized residuals, arranged in ascending order, against an approximation to their expected values, which is given by a standard normal distribution, $\Phi^{-1}[(i - \frac{3}{8})/(n + \frac{1}{4})]$. If the model fits well, this should yield a straight line at 45°. If the distribution of residuals is too skewed, the line will not pass through the origin, and will be J shaped. If it is too short or too long tailed, the line will be curved, as an S or inverted S respectively.

Example
For the first data set above, the normal probability plot of fitted value residuals is given in Figure 4.11. As might be expected, it follows the 45° line reasonably well. □

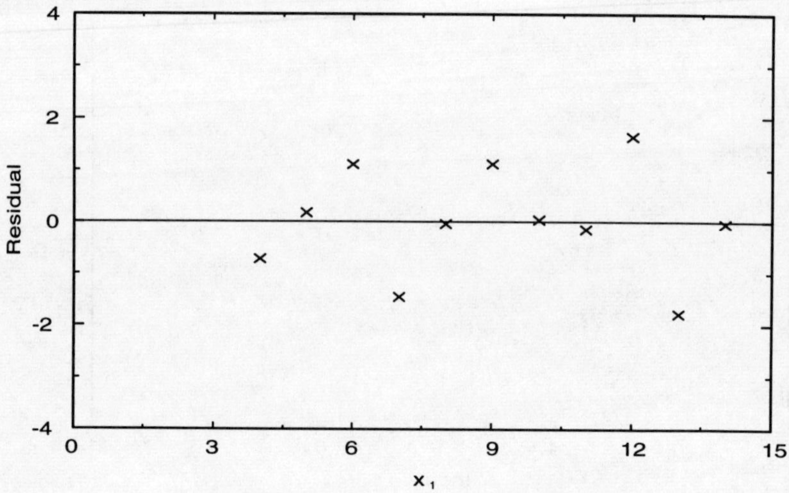

Fig. 4.10. Plot of standardized fitted value residuals against x_i for the first data set of Table 4.4.

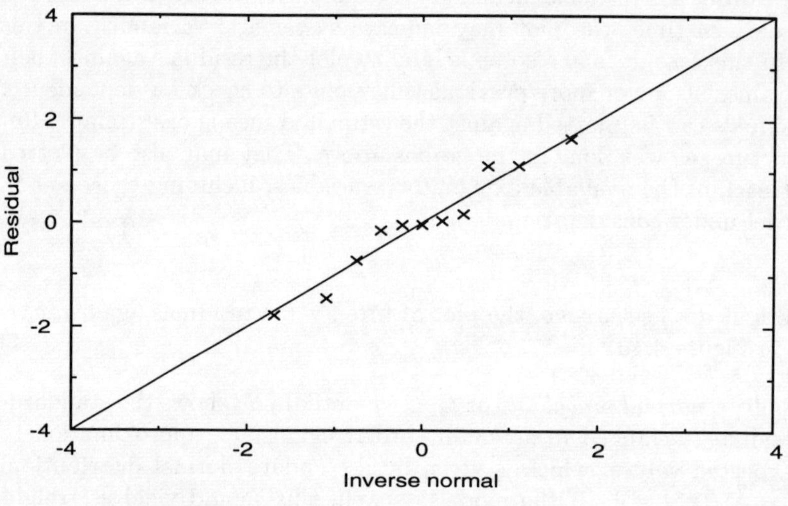

Fig. 4.11. Normal probability plot of standardized fitted value residuals for the first data set of Table 4.4.

Another similar graph is the *half-normal plot* which uses the absolute values of the residuals against $\Phi^{-1}[(i + n - \frac{1}{8})/(2n + \frac{1}{2})]$.

All of these graphs can be inspected, both to look for patterns in the residuals and to detect certain extreme values.

4.3.3 ISOLATED DEPARTURES

When only a very few observations do not fit the model, several possibilities may be considered:

- there may be some error in choosing certain members of the population sampled;
- the population may not be homogeneous for the factors considered in the model;
- there may be some error in recording the results on the part of people doing the recording or transcribing them;
- the individuals giving the responses may have erred, for example when respondents do not understand a question in a sample survey;
- some rare, but possible, occurrence may have been observed;
- the model may not be properly specified to account for completely acceptable observations, thus pointing to unforeseen aspects of the phenomenon under study.

If there is no error, one will eventually have to decide if the departure is important enough to modify the model to take it into account or if it is simply assumed to be a rare event. This will be discussed in the next section.

Outliers Any individual observation which departs in some way from the pattern of the main body of the data is called an *outlier*. Notice, however, that such extreme observations can only be determined in relation to some model.

Examples

1. In the four extreme data sets, the second and third, as seen in Figures 4.6 and 4.7, each contain an outlier. The second, apparently more related to the response, might be a recording error, while the third, linked to the values of the explanatory variables, would seem to point to an error of design.

2. The data set, $\{240, 194, 215, 194, 450, 240, 215, 215\}$, appears to contain the outlier 450. However, by what model have we judged this? A uniform or normal distribution? In fact, these are weekly salaries of a random sample of employees in a firm, where the large figure is for a manager and the others for junior staff, a proportion reflecting the pay structure of the company. (McPherson, 1989) □

Outliers may be due to extreme values of the random variable (the response) or of one or more of the explanatory variables. The possibility that the ith observation is an outlier can be studied by fitting the model without that observation. A large increase in normed likelihood will indicate that it may be an outlier. If one wishes to check all observations in this way, the approximation using the likelihood residuals of Equation (4.41) can considerably reduce the number of calculations required.

In most situations, it is rarely, if ever, wise simply to eliminate an outlier, unless it is known to be an error. In such case, it should, if possible, be corrected! Eliminating one outlier and refitting the model will quite often result in a second outlier appearing, and so on. It is usually preferable either to find out why the model cannot easily accommodate the observation or to accept it as a rare value.

Note, however, that the definition of an observation may not always be clear in this context. If Bernoulli trials are aggregated as frequencies, for example in a contingency table, should one trial or one whole frequency be omitted from the table?

Influence and leverage An *influential* observation is one which, if it is changed by a small amount or omitted, will modify substantially the parameter estimates of the model. It is an observation which may have undue impact on conclusions from the model. However, it may not be an outlier, in the sense that it may not be too far from the main body of data. It may have a small residual.

Example
Of our two data sets with outliers, the third has a very influential point, which completely determines the slope of the model, but has zero residual. □

Leverage is one indication of how much influence an observation has on the model estimation, in fact on prediction by the fitted values. A general definition of the leverage of the jth observation on the ith fitted value is the magnitude of the derivative of the ith fitted value with respect to the jth response value. In a (generalized) linear model, this is given by the hat matrix, $\mathbf{H}$. Thus, a measure of the leverage effect of the observation, i, in the determination of $\hat{\mu}_i$ is the diagonal element of the hat matrix, h_{ii}, with $0 \leq h_{ii} \leq 1$ (in non-linear regression, this constraint does not hold). This is the reciprocal of the effective replication. It is a measure of the distance of that observation from the remaining ones, in the space of these variables. Note, however, from Equations (4.29) and (4.30), that, except for linear normal models, this depends both on the explanatory variables and on the parameter estimates. The trace of $\mathbf{H}$ is equal to p, the dimension of the vector of regression parameters, $\boldsymbol{\beta}$, so that the average leverage is p/n. Values much larger than, say, twice this should be examined.

Example
The fitted value residuals are plotted against the leverage for the second and third data sets in Figures 4.12 and 4.13. The average leverage is 2/11. In the first plot, we see one large residual, but no exceptional leverage values,

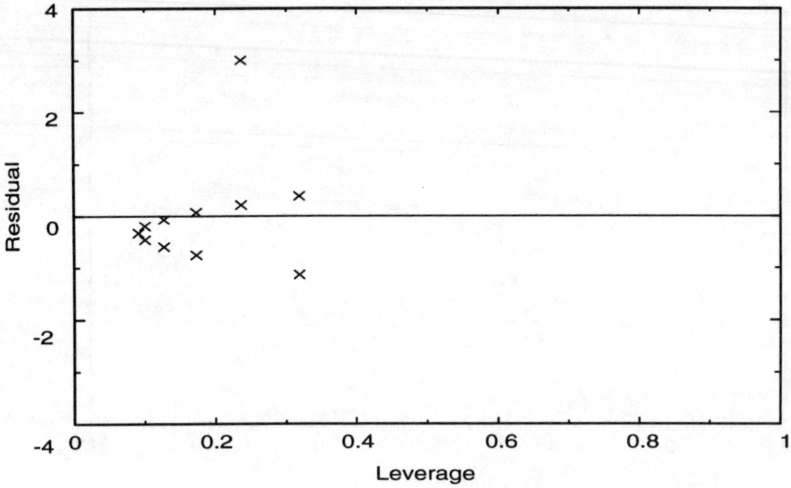

Fig. 4.12. Leverage plot for the second data set of Table 4.4.

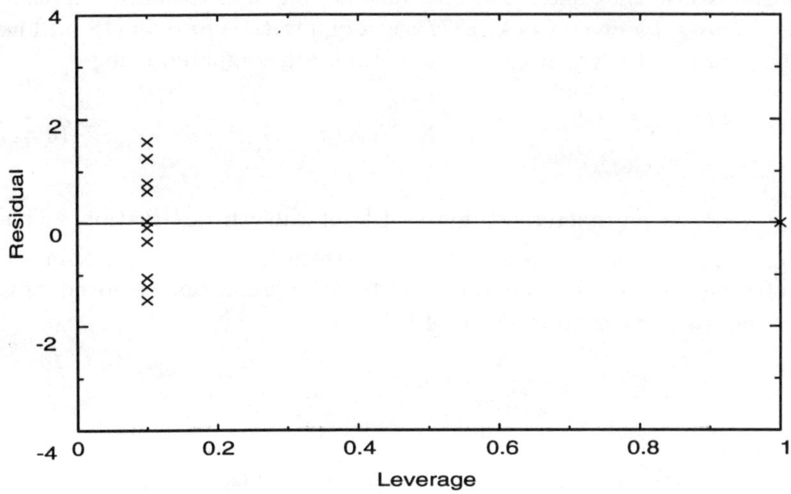

Fig. 4.13. Leverage plot for the third data set of Table 4.4.

and, in the second, the high leverage of one point, whereas there are no extreme residuals. □

Cook's distances To examine how each observation affects the complete set of parameter estimates, the model can be fitted with and without each observation and the likelihoods compared. This involves fitting $n+1$ mod-

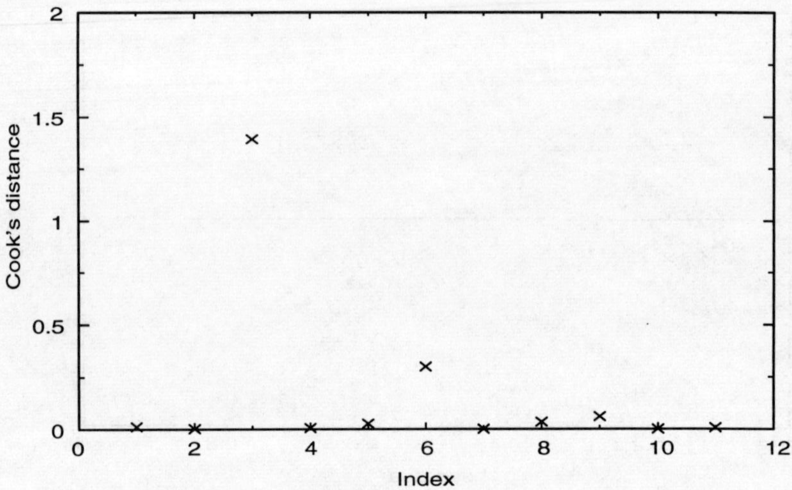

Fig. 4.14. Plot of Cook's distances for the second data set of Table 4.4.

els. We saw above that likelihood residuals are one approximation to this. Another, *Cook's distance* (Cook, 1977), is a quadratic approximation. The estimates, with and without each observation, are compared using

$$\mathrm{C}_i = \frac{1}{p}(\hat{\boldsymbol{\beta}} - \hat{\boldsymbol{\beta}}_{(i)})^{\mathrm{T}}\mathbf{X}^{\mathrm{T}}\mathbf{W}\mathbf{X}(\hat{\boldsymbol{\beta}} - \hat{\boldsymbol{\beta}}_{(i)}) \tag{4.42}$$

where $\hat{\boldsymbol{\beta}}_{(i)}$ is the parameter estimate without the ith observation. This statistic measures the squared distance between $\hat{\boldsymbol{\beta}}$ and $\hat{\boldsymbol{\beta}}_{(i)}$. Again, to avoid refitting the model n times, one with each observation removed, this distance can be further approximated by

$$\mathrm{C}_i \doteq \frac{h_{ii}(\varepsilon_i^F)^2}{p(1 - h_{ii})} \tag{4.43}$$

It is a combination of leverage and residuals and is most usefully presented as a plot against index values.

Example
Cook's distances are given as index plots for the second and third data sets in Figures 4.14 and 4.15. In the first plot, we see one large distance, for the outlier. On the other hand, this measure does not detect the point of high leverage in the third data set which has zero residual. □

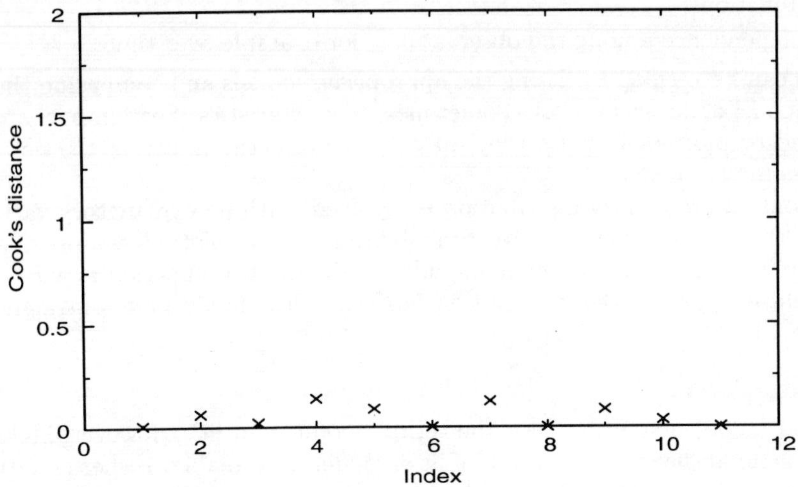

Fig. 4.15. Plot of Cook's distances for the third data set of Table 4.4.

4.3.4 SYSTEMATIC DEPARTURES

Systematic departures from a model can often be detected from the residual plots already described. Certain patterns may appear when the residuals are plotted against some other statistic or against explanatory variables.

Example
As seen in the residual plot of Figure 4.8, the fourth extreme data set shows systematic departure. The relationship is clearly curved, perhaps quadratic, and not linear. □

Recall our general regression model of Equation (1.7),

$$g(\mu_i) = h(\mathbf{x}_i, \boldsymbol{\beta})$$

where μ_i is the mean of the random variable following some probability distribution. Mis-specification of such a model may come about in a number of ways, some of which include

- an incorrect probability distribution,
- an incorrect specification of the way in which the mean changes with the explanatory variables,
  - ∗ the systematic component, $h(\cdot)$, may be mis-specified,
  - ∗ the link function, $g(\cdot)$, may not be appropriate,
- missing explanatory variables,
- incorrect functions of the explanatory variables in the model, including missing interactions among them, or not enough such different

functions,

- dependence among the observations, for example over time.

These can be verified by fitting the appropriate models and comparing the likelihoods, as in Section 4.1. Sometimes, score statistics (Section 5.5) are checked or plotted, because they do not involve actually fitting the new, more complex model.

If only a probability distribution is involved, with no explanatory variables, the empirical cumulative probabilities can be plotted against the probabilities from the estimated cumulative distribution function in a *P–P* (percentile) *plot.* If the distribution fits well, this should give a straight line at 45°.

Further reading

Pearson (1900) first introduced the term, 'goodness of fit'. However, little useful general theory on goodness of fit of models is available. Fisher (1950) introduced the (negative) log likelihood as a measure of goodness of fit. The importance of ancillary information derives from the work of Cox (1958) and Fisher (1959). The ideas on embedding come from Lindsey (1974a, b). On mixture models for contaminated observations, see Barnard (1954) and Box and Tiao (1968).

Significance testing, as presented in this chapter, is Fisherian; see Fisher (1958, 1959), who first used the term, significance level; P-value is more recent. The exact test was introduced simultaneously by Fisher (1934 edition of 1958) and Yates (1934; see also 1984).

Box (1953) first pointed out the need to consider robustness, which was subsequently developed by Huber (1964). Sprott (1978, 1982) has shown the importance of models in robustness.

On the other hand, there does exist an abundant literature on diagnostics. However, until recently, most of it centred on normal theory linear regression models and this has heavily influenced its extension to other contexts. In addition, a mentality of looking for observations to exclude, rather than for an improved model, underlies much of it. The reader may like to consult Barnett and Lewis (1984) or Cook and Weisberg (1982). The essential early reference on non-normal diagnostics is Pregibon (1981); see also Pierce and Schafer (1986) and Davison and Tsai (1992).

One serious handicap of most goodness of fit procedures is that they work with a point estimate of the model, as if it were the only one being retained as plausible, without any measure of precision. Bayesian procedures (Chapter 8) generalize those in this chapter by averaging over possible estimates (Gelman *et al.*, 1995, pp. 167–175).

4.4 Exercises

(1) (a) For n independent observations from a normal distribution with

unit variance, derive the conditional distribution of the observations, given the sufficient statistic for the mean.

(b) How might this distribution be used for goodness of fit?

(2) In five binary trials, the first four give zero and the last a one. Our model to test is that we have Bernoulli trials with $\pi_0 = 0.5$.

(a) What is the P-value if the number of trials was fixed at five?

(b) If trials continued until a one was obtained?

(c) Why?

(Cox and Hinkley, 1974, p. 31)

(3) (a) In a classical example, Fisher (1960, p. 11) describes an experiment to determine if a lady can detect whether milk or tea was put first into a cup. She will taste eight cups, four being randomly assigned to each order of mixture, all of which information she knows before beginning. After tasting all eight, she will be asked to assign four to each order.

i. What is the likelihood function?

ii. Would it change if the cups had not been randomized, but the lady did not know the order beforehand?

iii. What are possible advantages of having randomized in this experiment?

iv. Construct an exact significance test for the hypothesis that the lady cannot distinguish between the two orders.

v. What is the maximum number of mistakes that the lady can make so that the hypothesis is still rejected with a P-value of at least 0.05?

(b) Suppose now that the experiment were performed with complete randomization, so that each cup could have the mixture in either order and each order is not fixed to have four cups.

i. What is the likelihood function now?

ii. Construct an exact significance test for the hypothesis that the lady cannot distinguish between the two orders.

iii. What is the maximum number of mistakes that the lady can now make so that the hypothesis is rejected with a P-value of at least 0.05?

(4) The following 140 observations, counts of events, are made for which a Poisson distribution is under consideration:

Count	Frequency
0	124
1	12
2	2
3	2

(a) What is the value of the sufficient statistic for the mean of this distribution?

(b) List the probabilities of all possible samples of size 140 with this value.

(c) What is the exact probability of all samples at least as rare as that observed?

(d) Calculate the deviance for these observations.

(e) What is the exact probability of all samples with deviance at least as large as that observed?

(f) Calculate the Pearson goodness of fit statistic.

(g) What is the exact probability of all samples with this statistic at least as large as that observed?

(h) Calculate the coefficient of dispersion, i.e. the ratio of the estimated variance to the estimated mean.

(i) What is the exact probability of all samples with this statistic at least as large as that observed?

(j) Discuss how these different results can be interpreted.

(Fisher, 1950)

(5) Fisher (1960, p. 30) gives the heights (in) of young maize (*Zea mays*) plants as recorded by Darwin, where the last column provides differences between pairs of cross-fertilized and self-fertilized plants grown in the same pot:

Crossed	Self-fertilized	Difference ($\frac{1}{8}$ in)
$23\frac{1}{2}$	$17\frac{3}{8}$	49
12	$20\frac{3}{8}$	−67
21	20	8
22	20	16
$19\frac{1}{8}$	$18\frac{3}{8}$	6
$21\frac{1}{2}$	$18\frac{5}{8}$	23
$22\frac{1}{8}$	$18\frac{5}{8}$	28
$20\frac{3}{8}$	$15\frac{1}{4}$	41
$18\frac{1}{4}$	$16\frac{1}{2}$	14
$21\frac{5}{8}$	18	29
$23\frac{1}{4}$	$16\frac{1}{4}$	56
21	18	24
$22\frac{1}{8}$	$12\frac{3}{4}$	75
23	$15\frac{1}{2}$	60
12	18	−48

(a) What might be an appropriate model to describe the differences, d_i, between pairs, if we are interested in determining the plausibility of crossed plants being taller?

(b) Some observations appear to be outliers. If this is taken into account, do your results change?

(c) Consider now a model for the differences based on the distribution,

$$f(d_i; \mu, \sigma, \alpha) \propto \left[1 + \left(\frac{d_i - \mu}{\sigma}\right) \frac{1}{\alpha}\right]^{-\frac{\alpha+1}{2}} \tag{4.44}$$

Note that this gives, as a limiting case when $\alpha \to \infty$, a normal distribution. Does this model fit better than your previous model?

(d) To what family is it related?

(e) How can you explain your results, taking into account interaction among plants grown in the same pot?

(Sprott, 1978)

(6) Fisher (1958, p. 121) gives the effect on 10 patients of two supposedly soporific drugs in producing sleep:

Patient	Drug A	Drug B	Difference (hours)
1	+0.7	+1.9	+1.2
2	−1.6	+0.8	+2.4
3	−0.2	+1.1	+1.3
4	−1.2	+0.1	+1.3
5	−0.1	−0.1	0.0
6	+3.4	+4.4	+1.0
7	+3.7	+5.5	+1.8
8	+0.8	+1.6	+0.8
9	0.0	+4.6	+4.6
10	+2.0	+3.4	+1.4

Repeat the steps of the preceding question with these data in order to determine if one of the drugs is superior for producing sleep.

(Sprott, 1982)

(7) For multiple linear regression models based on the normal distribution, the coefficient of multiple determination,

$$R^2 = \frac{\sum(y_i - \bar{y}_\bullet)^2 - \sum(y_i - \hat{\mu}_i)^2}{\sum(y_i - \bar{y}_\bullet)^2} \tag{4.45}$$

mentioned in an example in Section 4.1.1, is often used as an indication of goodness of fit, especially by non-statisticians. It is said to give the proportion of the variance of Y explained by the explanatory variables.

(a) Discuss what this might mean.

(b) Investigate what, in fact, this coefficient measures, using your knowledge of comparison of models.

(Anderson-Sprecher, 1994)

(8) (a) For the normal distribution, show that the standardized fitted value residuals, ε_i^F, are ancillary.

(b) What is their distribution?

(c) Given that $\sum \varepsilon_i^F = 0$ and $\sum(\varepsilon_i^F)^2$ is a constant, what statistics could be studied to detect global departures from normality? (See also Exercise (5.13).)

(Barndorff-Nielsen and Cox, 1994, p. 48)

(9) Consider again a linear regression model based on the normal distribution.

(a) What is the maximum observable value of the internally studentized residuals, $(y_i - \hat{\mu}_i)/\sqrt{\widehat{\text{var}[Y_i]}}$ for n observations, when the linear regression contains p parameters?

(b) What is the maximum observable value of ε_i^F?

(c) Suitably standardized residuals are often assumed to follow a standard normal distribution when looking for extreme values. Discuss the effect of these results on such an approach.

(Gray and Woodall, 1994)

(10) T_4 cells/mm^3 were counted in blood samples from 20 patients in remission from Hodgkin's disease and 20 patients in remission from disseminated malignancies, as shown in the following table (Altman, 1991, p. 199):

Hodgkin's Disease		Non-Hodgkin's Disease	
396	288	375	675
568	1004	375	700
1212	431	752	440
171	795	208	771
554	1621	151	688
1104	1378	116	426
257	902	736	410
435	958	192	979
295	1283	315	377
397	2415	1252	503

We are interested in whether the counts differ, on average, between the two types of diseases.

(a) Fit models with and without two different means based on the following distributions:

i. normal,

ii. log normal,

iii. Poisson,
iv. negative binomial.

(b) Compare the conclusions as to the possibility of a difference in means.

(c) Compare residual plots for the models with difference in means.

(d) Which model fits best? Justify your conclusions.

Part III

Approximations

5
Asymptotics

If exact analytic results for the likelihood function of some model are difficult to obtain or are complex to calculate, it is often useful to have available approximations. Indeed, this was precisely what we did for the likelihood function of continuous distributions in Section 3.2.1, for the approximate likelihood based on the standard error in Section 3.5.2, for model selection in Section 3.6.1, and for profile likelihoods in Section 3.6.3. Numerical approximations are very important in statistics. Some of the common techniques, with which the reader should be familiar, are outlined in Appendix D. In this chapter, we shall be concerned with deriving some of the more commonly used analytic approximations.

Often n independent observations are made, in a random sample, assuming the same parametric statistical model. In such situations, it may be interesting to study what would happen if we could imagine observing a much larger sample or perhaps the whole theoretically infinite population. Many statisticians go further and claim that maximum likelihood estimation is premised on the viewpoint according to which the data collection process can accrue information in a way that can lead, in the limit, to all parameters in a given model becoming known. As we saw in Chapter 1, this, in fact, makes little sense from a modelling perspective, except perhaps when production and measurement error models are involved. However, it is the major assumption underlying the original derivation of many of the classical results to be exposed in this chapter.

In Chapter 2, we saw how parameters of models from the exponential family can be interpreted in terms of expectations of observable quantities. Here, we can interpret parameters in terms of limiting sequences of observable quantities. Thus, the general idea is that the amount of information about the parameter should increase with the sample size. This need not generally be the case, especially if observations are not independent, and hence must be checked in specific circumstances. Mathematical convergence of sequences of statistics to some fixed value, as n grows large, can be demonstrated. It is their empirical interpretation, as some true quantity, which is questionable. Instead, in the modelling context, we shall be interested in how such methods can provide usable small sample approximations to complex likelihood functions.

Hence, one common way of deriving the required approximations is by letting the amount of information, usually measured by n, become arbitrarily large. When this occurs, many likelihood functions tend towards the shape of that for a normal distribution, i.e. their logarithm is a quadratic function; we saw this in the example of the likelihood for the binomial distribution in Section 3.2.3 and applied it in using the standard error in Section 3.5.2. This is known as a *large sample asymptotic approximation.* Such approximations can be obtained by means of series expansions, usually in $1/\sqrt{n}$, where higher order terms are ignored. When observations are not independent, as, for example, in many stochastic process models, fairly strong assumptions sometimes have to be made although such asymptotic approximations can frequently be obtained using martingales (Section B.4).

Unfortunately, a large sample size is of little aid in making inferences if variability is high or unstable in the data, as in most cases where stochastic models are involved. Even in more common situations, large sample asymptotics may not work, for example, if the number of observations actually made cannot be controlled. This can occur with sequential stopping rules (Section 3.3.2) and with missing data mechanisms such as drop-outs (Section 6.4.2) where it may not be conceptually possible for the number of observations recorded to become very large. Then, *small dispersion* and *high precision asymptotics* for small samples, as described below, may be of more practical importance than large sample asymptotics.

In this chapter, we shall be concerned with certain special statistics, the m.l.e., the deviance, and the score, and their distributions, and with how they can be used to obtain approximations to complex likelihood functions. Up until now, the latter two have not been viewed in this light, i.e. simply as statistics, functions of the random variable.

Because asymptotic results often involve the central limit theorem, certain special distributions related to the normal distribution are frequently required. These are reviewed in Appendix C.

5.1 Preliminaries

5.1.1 REGULARITY

The development of asymptotic approximations generally involves the manipulation of integrals and derivatives. For these operations to be possible, certain conditions are usually necessary.

The log likelihood function for a (one-dimensional) parameter is said to be *regular* if, in an open neighbourhood of the 'true' parameter value, it can be approximated by a concave quadratic function, i.e. the log likelihood for a normal distribution, whose maximum point converges to that 'true' value as the sample size increases. These conditions can break down in several ways:

- the 'true' value may lie on the boundary of the parameter space so

that a quadratic function around it is impossible;
- the likelihood function for a parameter may not be defined when some other parameter takes a particular value;
- limiting values of a parameter may lead to the likelihood function taking a different functional form.

The generalization to multidimensional likelihood surfaces is direct.

Examples
1. The canonical logit link function

$$\log\left(\frac{\mu}{n-\mu}\right) = \log\left(\frac{\pi}{1-\pi}\right) \tag{5.1}$$

for the binomial distribution, where $\mu = n\pi$, does not have a finite value if $\pi = 0$ or 1, these being boundary values of the parameter space. Recall Figure 3.1.
2. In the exponential growth model of Equation (1.8),

$$\mu_t = \alpha e^{\beta t}$$

at $\alpha = 0$, the likelihood function for β is undefined.
3. The Box–Cox transformation is often defined as in Equation (2.71)

$$\frac{y^\alpha - 1}{\alpha}$$

As $\alpha \to 0$, this becomes $\log(y)$. However, the Jacobian, $y^{\alpha-1}$, does have a unique representation. □

In Sections 2.1.5 and 2.8.4, we saw the regularity conditions for the exponential family. More generally, regularity will be required for asymptotic approximations in several ways:
- series expansions can be made, locally, to the order required;
- expectations can be differentiated, under the integral, with respect to parameters;
- distinct parameter values represent distinct distributions.

In summary, regularity conditions are important for asymptotic *approximations* to likelihood functions. This is because such approximations usually involve a normal distribution centred at the 'true' value. Then, for example, if the latter is on a boundary, a normal distribution about that value is impossible. On the other hand, these conditions are not relevant to exact comparisons of (properly defined) likelihoods, as in Chapters 3 and 4.

5.1.2 MEAN AND VARIANCE OF THE SCORE STATISTIC

For the demonstrations of various asymptotic results to follow in subsequent sections, we shall require two results about the score, considered as a statistic with the parameters fixed at their 'true' values.

The expected score is zero.
proof: By definition, we have $\int_{\mathcal{Y}} f(y;\boldsymbol{\psi})dy = 1$, for fixed $\boldsymbol{\psi}$, so that

$$\int_{\mathcal{Y}} \frac{\partial f(y;\boldsymbol{\psi})}{\partial \boldsymbol{\psi}} dy = \mathbf{0}$$

if $f(\cdot)$ is sufficiently regular for the integral and derivative to be interchanged. Now,

$$\begin{aligned}\mathbf{U} &= \frac{\partial \log(f)}{\partial \boldsymbol{\psi}} \\ &= \frac{1}{f}\frac{\partial f}{\partial \boldsymbol{\psi}}\end{aligned}$$

so that the expectation,

$$\begin{aligned}\int_{\mathcal{Y}} \mathbf{U} f dy &= \int_{\mathcal{Y}} \frac{\partial f}{\partial \boldsymbol{\psi}} dy \\ &= \mathbf{0}\end{aligned}$$

as required. □

Example
Consider a discrete uniform density with

$$f(Y = i; \psi) = \frac{1}{\psi} \quad i = 1, \ldots, \psi \tag{5.2}$$

where ψ is the unknown number of categories. Here, the support of Y depends on the unknown parameter. We have

$$\frac{d}{d\psi} \sum_{i=1}^{\psi} \frac{1}{\psi} = 0 \tag{5.3}$$

Thus, the sum and derivative cannot be interchanged, so that the above proof cannot be used to find the expected score for this distribution. □

Score or estimating equations (Section 3.5.1) are said to be unbiased if their expectation is zero: $\mathrm{E}[\mathbf{U}] = 0$.

The variance of the score statistic is the expected information.

proof: We have

$$\int_{\mathcal{Y}} \frac{\partial^2 f(y;\boldsymbol{\psi})}{\partial\boldsymbol{\psi}\partial\boldsymbol{\psi}^{\mathrm{T}}} dy = \mathbf{0}$$

for $\boldsymbol{\psi}$ fixed, provided the appropriate regularity conditions hold. Now,

$$\begin{aligned}\mathbf{J} &= -\frac{\partial \mathbf{U}}{\partial \boldsymbol{\psi}} \\ &= -\frac{\partial}{\partial \boldsymbol{\psi}}\left[\frac{1}{f}\frac{\partial f}{\partial \boldsymbol{\psi}}\right] \\ &= \frac{1}{f^2}\left[\frac{\partial f}{\partial \boldsymbol{\psi}}\right]\left[\frac{\partial f}{\partial \boldsymbol{\psi}}\right]^{\mathrm{T}} - \frac{1}{f}\frac{\partial^2 f}{\partial\boldsymbol{\psi}\partial\boldsymbol{\psi}^{\mathrm{T}}}\end{aligned}$$

so that

$$\frac{\partial^2 f}{\partial\boldsymbol{\psi}\partial\boldsymbol{\psi}^{\mathrm{T}}} = \mathbf{U}\mathbf{U}^{\mathrm{T}} f - \mathbf{J} f$$

Integrating, we obtain

$$\int_{\mathcal{Y}} \mathbf{U}\mathbf{U}^{\mathrm{T}} f dy - \int_{\mathcal{Y}} \mathbf{J} f dy = \mathbf{0}$$

or

$$\mathrm{E}[\mathbf{U}\mathbf{U}^{\mathrm{T}}] = \mathrm{E}[\mathbf{J}] \tag{5.4}$$

but, because $\mathrm{E}[\mathbf{U}] = \mathbf{0}$, this is identical to

$$\mathrm{var}[\mathbf{U}] = \mathbf{I} \tag{5.5}$$

as required. □

Because $\mathbf{J} = -\frac{\partial \mathbf{U}}{\partial \boldsymbol{\psi}}$, this result also implies that

$$\begin{aligned}\mathrm{E}[\mathbf{U}\mathbf{U}^{\mathrm{T}}] &= \mathrm{E}\left[-\frac{\partial \mathbf{U}}{\partial \boldsymbol{\psi}}\right] \\ &= \mathrm{var}[\mathbf{U}]\end{aligned} \tag{5.6}$$

Example

The Laplace distribution is defined by

$$f(y;\psi) = \frac{1}{2}\mathrm{e}^{-|y-\psi|} \tag{5.7}$$

Then,

$$U_\psi(\psi; y) = \text{sign}(y - \psi) \tag{5.8}$$

and the second derivative is zero, except at $y = \psi$, where there is a singularity. However, the variance of the score statistic can be obtained from the left hand side of Equation (5.6) and is one for all ψ. (Barndorff-Nielsen and Cox, 1994, p. 26) □

5.2 Small sample asymptotics

In this section, we shall outline the general approximation techniques available. Although many involve letting n become large, we are only interested in those which provide reasonable results for small n; hence the title of this section.

Example
Suppose that some statistic, T_n, calculated from the data, asymptotically provides a reasonable estimate of a parameter of interest. Then, the sequence which is identically equal to zero for $n < 20\,000$ and elsewhere equal to T_n is asymptotically equally reasonable but of no use for small samples. □

Such methods are especially important for complex problems. However, in this chapter, we shall usually give relatively simple examples of applications, where such approximations may not really be necessary, but comparison can be made with exact likelihoods. In this way, we can ascertain how good the approximations are. A brief outline of some of the more complex techniques for approximating an arbitrary probability distribution, and its likelihood function, often by a normal distribution, is provided in an appendix to this chapter; even that is rather complex. The reader is advised to consult the references provided at the end of the chapter for more details.

5.2.1 LARGE SAMPLE ASYMPTOTICS

A large sample approximation involves studying what happens to some function of the given fixed model when the number of observations is allowed to become large, although it is generally only useful if the approximation is still reasonable for small samples. For simplicity, we generally assume, in this chapter, that the sample size is fixed by the design, and is not random, so that it can, conceptually, be allowed to increase. Most often, the function will be the log likelihood or the score, because they are additive for (conditionally) independent observations. In the simplest cases, a Taylor series expansion is applied to the log likelihood and higher order terms ignored. Because this involves derivatives, the score and information will appear in the resulting terms. Then, the law of large numbers and the central limit theorem can be applied, again, because of the additive nature of the functions. For these reasons, the mean and the variance of the score statistic will often be required, and have been given above.

When such simple techniques are used, the resulting approximate distribution will generally be normal (if it is to first order), or some distribution, such as the Chi-squared, derived from it (if it is to second order). Necessary details about these are provided in Appendix C. Such approximations will not be parametrization invariant. Indeed, as we shall see, their validity for reasonable-sized n is generally highly dependent on the parametrization and is questionable for many.

Often, with a bit more effort, better approximations can be obtained directly for a probability distribution. These are outlined in the appendix to this chapter. The saddlepoint approximation is a particularly important method. The one application with which we shall be primarily concerned, in this chapter and the next, is the p^* formula of Section 5.3.5.

5.2.2 SMALL DISPERSION ASYMPTOTICS

One alternative to studying the validity of approximations as n becomes large is to let the dispersion parameter, when the model contains one, become small.

For a proper dispersion family,

$$f(y;\mu,\sigma^2) = \mathrm{e}^{t(y,\mu)/\sigma^2 - a^\ddagger(\sigma^2) - b(y)} \tag{5.9}$$

from Equations (2.56) and (2.57), let us take y_m to be the point of global maximum of $t(y,\mu)$. Define

$$\mathrm{K}(\mu) = -\left.\frac{\partial^2 t(y,\mu)}{\partial y^2}\right|_{y=y_m} > 0 \tag{5.10}$$

Then, by the Laplace approximation to an integral (Section 5.7.1), we have

$$\begin{aligned} \mathrm{e}^{a^\ddagger(\sigma^2)} &= \int_{\mathcal{Y}} \mathrm{e}^{t(y,\mu)/\sigma^2 - b(y)} dy \\ &\doteq \sqrt{\frac{2\pi\sigma^2}{\mathrm{K}(\mu)}} \mathrm{e}^{t(y_m,\mu)/\sigma^2 - b(y_m)} \end{aligned} \tag{5.11}$$

so that

$$f(y;\mu,\sigma^2) \doteq \sqrt{\frac{\mathrm{K}(\mu)}{2\pi\sigma^2}} \mathrm{e}^{[t(y,\mu)-t(y_m,\mu)]/\sigma^2 - b(y) + b(y_m)} \tag{5.12}$$

Applying a second-order Taylor series expansion to $t(y,\mu)$ near y_m, we find that

$$\frac{y-y_m}{\sigma} \sim \mathrm{N}[0, \mathrm{K}^{-1}(\mu)] \quad \text{for } \sigma^2 \to 0 \tag{5.13}$$

for any proper dispersion family. Notice that, from Equation (5.11), neither $t(y_m,\mu)$ nor $\sqrt{\mathrm{K}(\mu)}\exp[b(y_m)]$ depends on μ (Jørgensen, 1987b).

5.2.3 HIGH PRECISION ASYMPTOTICS

The density approximation to the likelihood function which we have been using for continuous distributions carries the assumption that the precision of measurement is sufficiently great so that Equation (3.5) is close to Equation (3.4). Because of the complexity of the latter for most models, the appropriateness of the approximation can usually only be studied by series expansions. Most often, even the exact maximum likelihood estimate, and not just the shape of the likelihood function, will depend on the unit of measurement, Δ.

Example

For the exponential distribution, the exact likelihood has an analytic form, obtained by integrating Equation (3.4):

$$\mathrm{L}(\mu;\mathbf{y}) = \left[\exp\left(\frac{\Delta}{2\mu}\right) - \exp\left(-\frac{\Delta}{2\mu}\right)\right]^n \exp\left(-\frac{\sum_{i=1}^n y_i}{\mu}\right) \tag{5.14}$$

From this, the exact m.l.e. can also be obtained in closed form:

$$\begin{aligned}\hat{\mu}_e &= \frac{\Delta}{\log\left(\frac{2\bar{y}+\Delta}{2\bar{y}-\Delta}\right)} \\ &= \bar{y}\left[\frac{z}{\log\left(\frac{2+z}{2-z}\right)}\right]\end{aligned} \tag{5.15}$$

where $z = \Delta/\bar{y}$. □

In contrast to this example, for most continuous distributions, including even the normal distribution, the required integral in Equation (3.4) cannot be written in closed analytic form.

High precision asymptotics, thus, let $\Delta \to 0$, i.e. measurement precision becomes infinitely great (Fisher, 1922a). Interest will centre on how small Δ must be for Equation (3.5) to be reasonable. Notice, in the above example, that the correction, with respect to the approximate m.l.e., $\bar{y}$, does not depend on n. This is typical of exact maximum likelihood estimates. The usual approximate ones are not improved by increasing the sample size, if the unit of measurement remains the same.

Example
For the exponential distribution, with fixed $\bar{y}$, $\hat{\mu}_e$ is a monotone decreasing function of Δ, independent of n, with $\lim_{\Delta \to 0} \hat{\mu}_e = \bar{y}$. □

5.2.4 CRITERIA FOR APPROXIMATIONS

Various criteria for comparing an approximate likelihood to the exact one may be considered. Let us denote the approximation by the subscript A. Note, in general, that $\mathrm{l}(\psi_1) - \mathrm{l}(\psi_2) = \mathrm{O}_p(n)$. An approximation is

(1) *globally pointwise equivalent* if

$$[\mathrm{l}_A(\psi_1) - \mathrm{l}_A(\psi_2)] - [\mathrm{l}(\psi_1) - \mathrm{l}(\psi_2)] = \mathrm{o}_p(n) \tag{5.16}$$

for all fixed values of ψ_1 and ψ_2;

(2) *support interval equivalent* if the (normed) likelihood interval, (ψ_l, ψ_u), with $\psi_u - \psi_l = \mathrm{O}_p(n^{-1/2})$, approximated by (ψ_{lA}, ψ_{uA}), has

$$|\psi_{lA} - \psi_l| + |\psi_{uA} - \psi_u| = \mathrm{o}_p(n^{-1/2}) \tag{5.17}$$

(3) *maximum likelihood equivalent* if

$$\hat{\psi}_A - \hat{\psi} = \mathrm{o}_p(n^{-1/2}) \tag{5.18}$$

i.e. the difference between the maximum likelihood estimates tends to be negligible relative to their error;

(4) *information equivalent* if

$$\mathrm{J}_A(\hat{\psi}_A) - \mathrm{J}(\hat{\psi}) = \mathrm{o}_p(n) \tag{5.19}$$

i.e. the exact and approximate likelihood functions have approximately the same (second-order) curvature at their respective maxima;

(5) *locally pointwise equivalent* if, for the sequences $\psi_{1n}, \psi_{2n} = \hat{\psi}_A + \mathrm{O}_p(n^{-1/2})$,

$$[\mathrm{l}_A(\psi_{1n}) - \mathrm{l}_A(\psi_{2n})] - [\mathrm{l}(\psi_{1n}) - \mathrm{l}(\psi_{2n})] = \mathrm{o}_p(1) \tag{5.20}$$

i.e. the likelihood ratio close to $\hat{\psi}_A$ is first-order accurate.

The last four modes of equivalence are closely related and are all essentially different from the first. Thus, we may contrast support interval and globally pointwise equivalence as being the essential modes (Tsou and Royall, 1995).

In summary, when applying asymptotic approximations, we should note several points (Cox, 1988).

- Although having a precise mathematical meaning, series expansions where $n \to \infty$ are only a convenient statistical fiction.
- One looks for a good approximation for the particular n in an observed data set, so that those approximations working best for small n are to be preferred.
- In certain models, one or more of several indices, such as the number of observations at each point in time, or the number of time points, might be allowed to grow. The choice will usually depend on the particular aspect of the likelihood function to be approximated.
- If possible, bounds on error should be obtained, if not analytically, then perhaps numerically, for example by simulation.
- Nevertheless, the commonly used asymptotic results for n large, for continuous distributions, use densities and are most often even more approximate, or invalid, if the precision cannot be assumed to be infinite.

We are now ready to look at some of the approximations to likelihood functions which have been developed. In fact, often it is easier to work with the deviance, because of its additive nature.

5.3 Maximum likelihood estimate statistic

In the linear exponential family, the m.l.e. is the sufficient statistic for the parameter. In other cases, it is simply a point estimate. In either case, a first requirement is that an approximate likelihood function also have its maximum at this point. Thus, the m.l.e. can be a useful starting point for developing approximations. The major advantage is that an n-dimensional problem may possibly be reduced to a p-dimensional one, i.e. from the size of the sample to that of the parameter vector. If the m.l.e. is not sufficient, information will be lost. The loss will also depend on the closeness of any approximation used.

Throughout the remainder of this chapter, the notation, $\hat{\boldsymbol{\psi}}$, will be used, as usual, for the observed m.l.e., while an upper case symbol, $\hat{\boldsymbol{\Psi}}$, will refer to the corresponding statistic or random variable.

5.3.1 EXACT DISTRIBUTION

The distribution of the m.l.e. is a marginal distribution. It can be obtained by an appropriate one-to-one transformation of the random variable and integration over the remaining dimensions of the new statistics. In some cases, this can be done exactly, as the following examples show. From the factorization theorem (Section 3.4.1), we may expect that, at least if sufficient statistics are involved, an exact distribution of the m.l.e. will yield the exact likelihood function.

Examples

1. The binomial density,

$$f(y;\pi,n) = \binom{n}{y}\pi^y(1-\pi)^{n-y}$$

has $\hat{\pi} = y/n$. The random variable, $\hat{\Pi}$, can only take $n+1$ distinct values, $0, \frac{1}{n}, \ldots, 1$. Integration is not necessary, because y and $\hat{\pi}$ have the same dimension, so that

$$f(\hat{\pi};\pi,n) = \binom{n}{n\hat{\pi}}\pi^{n\hat{\pi}}(1-\pi)^{n(1-\hat{\pi})} \tag{5.21}$$

The distributions for two possible values of π were given in Table 4.2. Note that, as might be expected, in each case, it has a maximum at $\hat{\pi}$.

2. The normal density,

$$f(\mathbf{y};\mu,\sigma^2) = (2\pi\sigma^2)^{-\frac{n}{2}}\mathrm{e}^{-\frac{\Sigma(y_i-\mu)^2}{2\sigma^2}}$$

has $\hat{\mu} = \sum y_i/n$ and $\hat{\sigma}^2 = \sum(y_i - \hat{\mu})^2/n$. We shall take the transformation, $\mathbf{y} \rightarrow (\hat{\mu}, \sqrt{n}\hat{\sigma}, a_1, \ldots, a_{n-2})^\mathrm{T}$, where $a_i = (y_i - \hat{\mu})/\hat{\sigma}$ are ancillary statistics. The determinant of the Jacobian is

$$\frac{n(\sqrt{n}\hat{\sigma})^{n-2}}{|a_n - a_{n-1}|}$$

where the denominator is a function of the first $n-2$ values of a_i. Then,

$$\begin{aligned} f(\hat{\mu},\hat{\sigma}^2;\mu,\sigma^2) &= \int\cdots\int \frac{n(\sqrt{n}\hat{\sigma})^{n-2}\mathrm{e}^{-\frac{n}{2\sigma^2}[\hat{\sigma}^2+(\hat{\mu}-\mu)^2]}}{|a_n-a_{n-1}|(2\pi\sigma^2)^{\frac{n}{2}}}da_1\ldots da_{n-2} \\ &= c\frac{(\sqrt{n}\hat{\sigma})^{n-2}}{\sigma^n}\mathrm{e}^{-\frac{n}{2\sigma^2}[\hat{\sigma}^2+(\hat{\mu}-\mu)^2]} \\ &= \left[c_1\frac{1}{\sigma}\mathrm{e}^{-\frac{n}{2\sigma^2}(\hat{\mu}-\mu)^2}\right]\left[c_2\frac{(\sqrt{n}\hat{\sigma})^{n-2}}{\sigma^{n-1}}\mathrm{e}^{-\frac{n\hat{\sigma}^2}{2\sigma^2}}\right] \end{aligned} \tag{5.22}$$

where c, c_1, and c_2 are normalizing constants. Thus, $\hat{M}$ and $\hat{\Sigma}^2$ have independent distributions.

From the first factor, we can see that $\hat{M}$ has a normal distribution, $\mathrm{N}(\mu, \sigma^2/n)$, so that $c_1 = \sqrt{n/2\pi}$. Because the normal distribution is a member of the exponential dispersion family, we already knew this from the convolution property (Section 2.5.1).

In the second factor, let us transform our statistic, $\sqrt{n}\hat{\sigma}$, to $\hat{\chi}^2 = n\hat{\sigma}^2/\sigma^2$, with Jacobian $\sigma^2/2\sqrt{n}\hat{\sigma}$. We, then, have

$$
\begin{aligned}
f_2(\hat{\sigma}^2;\sigma^2) &= \frac{c_2}{2}\frac{(\sqrt{n}\hat{\sigma})^{n-3}}{\sigma^{n-3}}\mathrm{e}^{-\frac{n\hat{\sigma}^2}{2\sigma^2}} \\
&= \frac{c_2}{2}(\hat{\chi}^2)^{\frac{n-3}{2}}\mathrm{e}^{-\frac{\hat{\chi}^2}{2}}
\end{aligned}
\tag{5.23}
$$

which is a Chi-squared distribution with $n-1$ degrees of freedom and

$$
c_2 = \frac{1}{\Gamma(\frac{n-1}{2})2^{\frac{n-3}{2}}}
$$

□

From these examples, we see, as was expected, that the likelihood derived from the exact distribution of the m.l.e., if it is a sufficient statistic, is just the usual one. In addition, we can note that the fact that the distribution for the m.l.e. of two parameter elements factors does not mean that we obtain a separation of the parameter elements themselves. For the normal distribution, the distribution of $\hat{\mu}$ involves both μ and σ^2.

5.3.2 CONSISTENCY

At the beginning of this section, we suggested that an important requirement for an approximate likelihood was to have its maximum at the exact m.l.e. However, in Section 5.2.3, we discovered that this is not generally true for the density approximation to the likelihood function of continuous distributions in Equation (3.5). It is time to investigate this further.

A statistic, $\mathbf{T}_n$, for n independent observations, is said to be a *weakly consistent* estimate of a parameter, ψ, if it converges in probability to the 'true' value, ψ, and *strongly consistent* if it converges with probability one. Consistency is often said to be a minimum requirement for a point estimate. However, these conditions suppose that a model actually can be correct.

If $f(y;\psi)$ is continuous with respect to $\psi \in \Psi$, the density is bounded, Ψ is finite dimensional and compact, and the sample design is such that n can, conceptually, tend to infinity, then the m.l.e. of ψ, calculated from that density, is consistent (given infinite measurement precision). As well, the conditional m.l.e.s of Chapter 6 below are consistent (Andersen, 1970).

Example

Suppose that pairs of measurements of n different objects are made using the same instrument. Each such pair will have a different location parameter, but the same dispersion parameter. This latter parameter is of particular interest. However, the parameter space is not of finite dimension because the number of location parameters increases with n, and the ordinary m.l.e. is not consistent. □

Two more detailed examples of the estimation of the precision of measuring instruments in the presence of incidental parameters will be given in Chapter 6. A consistent estimate of the parameter of interest can usually be obtained by using a conditional likelihood or a random effects model, both studied in that chapter.

Notice that the non-consistency of the m.l.e. in such examples does not follow from the conditions given above; they are sufficient but not necessary. Maximum likelihood estimates can be consistent without these conditions being fulfilled. For an example, see Section 6.5.1.

Most such consistency results for continuous distributions assume infinite precision of measurement, i.e. $\Delta \to 0$. Generally, the m.l.e. from Equation (3.5), which is invariably used, will not be consistent for $\Delta > 0$ (McNeil, 1966).

Examples

1. From Equation (5.15), $\hat{\mu} = \bar{y}$ in the exponential distribution will be increasingly too large as Δ increases.

2. The usual m.l.e. of the variance of a normal distribution, $\sum(y_i - \bar{y})^2/n$ is too large if $\Delta > 0$. An approximate correction, introduced by Sheppard (1898), involves subtracting $\Delta^2/12$. □

As we have seen, the error in Equation (3.5) does not depend on n. Thus, it becomes relatively more important as n increases, because the approximate likelihood function becomes narrower about the wrong value, eventually excluding the 'true' value. Hence, most of the classical large sample asymptotic results in this chapter do not involve consistent estimates for continuous variables if $\Delta > 0$, which empirically always will be the case. Maximum likelihood estimation, based on the usual density approximation of Equation (3.5), cannot be justified by the claim that, in the limit, as information accumulates, parameters take their 'true' values.

Fisher consistency An estimate is *Fisher consistent* if, when calculated for the whole population, not necessarily infinite, it is equal to the 'true' parameter value. In other words, a Fisher consistent statistic is a function of the observed frequencies which takes the exact parameter value when these frequencies are replaced by their expectations (Fisher, 1922a, 1959, p. 146).

Examples

1. Suppose that the observable categories have probabilities π_i, with $\sum \pi_i = 1$, and observed frequencies n_i. Then, the linear function, $T =$

$\sum c_i n_i / n_\bullet$, will be a consistent estimate of the function $\psi = \sum c_i \pi_i$. (Fisher, 1959, p. 144)

2. For the (1,1) linear exponential family, a statistic $T(Y)$ is Fisher consistent for a parameter ψ iff

$$T[\mu(\psi)] = \psi \tag{5.24}$$

□

As Fisher makes clear, this definition must take into account the unit of measurement. Notice also that it applies in cases where the asymptotic consistency is difficult to apply because the sample size cannot tend to infinity, as with some sequential designs (Section 3.3.2) and missing data mechanisms (Section 6.4.2).

5.3.3 NORMAL APPROXIMATION AND EFFICIENCY

We can now derive our first approximate likelihood function based on a normal distribution. Under mild regularity conditions, for n independent observations, we saw in Section 5.1.2 that the mean of the score statistic is zero,

$$\mathrm{E}[\mathbf{U}(\boldsymbol{\psi})] = \mathbf{0} \tag{5.25}$$

and its variance is the expected information,

$$\begin{aligned} \mathrm{E}[\mathbf{U}\mathbf{U}^{\mathrm{T}}] &= \mathrm{E}\left[-\frac{\partial \mathbf{U}}{\partial \boldsymbol{\psi}}\right] \\ &= \mathbf{I} > \mathbf{0} \end{aligned} \tag{5.26}$$

Using these, we can show that, for independent observations,

The m.l.e. has an asymptotic normal distribution.

proof: We can expand the score statistic in a Taylor series about the 'true' value, $\boldsymbol{\psi}$:

$$\mathbf{U}(\hat{\boldsymbol{\Psi}}) = \mathbf{U}(\boldsymbol{\psi}) + \frac{\partial \mathbf{U}(\boldsymbol{\psi})}{\partial \boldsymbol{\psi}^{\mathrm{T}}} \frac{\hat{\boldsymbol{\Psi}} - \boldsymbol{\psi}}{1!} + \cdots \tag{5.27}$$

The left hand side is zero, from Equation (3.38), so that

$$\mathbf{U}(\boldsymbol{\psi}) \doteq \mathbf{J}(\boldsymbol{\psi})(\hat{\boldsymbol{\Psi}} - \boldsymbol{\psi}) \tag{5.28}$$

ignoring higher order terms. By the law of large numbers, because $\mathbf{J}(\boldsymbol{\psi})$ is a sum,

$$\lim_{n \to \infty} \mathbf{J}(\boldsymbol{\psi}) = \mathrm{E}[\mathbf{J}(\boldsymbol{\psi})]$$

$$= \mathbf{I}(\boldsymbol{\psi}) \text{ almost surely} \tag{5.29}$$

so that

$$(\hat{\boldsymbol{\Psi}} - \boldsymbol{\psi}) \doteq \mathbf{I}^{-1}(\boldsymbol{\psi})\mathbf{U}(\boldsymbol{\psi}) \tag{5.30}$$

(This approximation will fail if the log likelihood is appreciably far from being quadratic.)

The mean and variance of the right hand side are $\mathbf{0}$ and $\mathbf{I}^{-1}(\boldsymbol{\psi})$. Then, because $\mathbf{U}(\boldsymbol{\psi})$ is a sum, by the central limit theorem, asymptotically

$$\hat{\boldsymbol{\Psi}} \sim \mathrm{N}[\boldsymbol{\psi}, \mathbf{I}^{-1}(\boldsymbol{\psi})] \tag{5.31}$$

□

For dependent observations, martingale convergence (Section B.6.3) can be used to obtain similar results.

This result demonstrates why Equation (3.66) is an approximate pivotal quantity. A similar proof can be used to demonstrate asymptotic normality of the conditional m.l.e. of Chapter 6; see Andersen (1970).

Approximate deviance An approximation to the likelihood, or to the deviance, obtained by assuming that the distribution under consideration is reasonably approximated by a normal distribution, can be derived from Equation (5.31):

$$\mathrm{D}_{N_1}(\boldsymbol{\psi}) \stackrel{\text{def}}{=} (\hat{\boldsymbol{\psi}} - \boldsymbol{\psi})^{\mathrm{T}}\mathbf{I}(\boldsymbol{\psi})(\hat{\boldsymbol{\psi}} - \boldsymbol{\psi}) \tag{5.32}$$

the deviance for a normal distribution. This generally yields an approximation which is support interval equivalent to the exact deviance.

Example

The (density approximation) likelihood for n i.i.d. observations from the exponential distribution is based on

$$f(\mathbf{y};\mu) = \mu^{-n}\mathrm{e}^{-\frac{\Sigma y_i}{\mu}} \qquad \mu > 0 \tag{5.33}$$

with $\hat{\mu} = \sum y_i/n$ and the corresponding deviance is

$$\mathrm{D}(\mu) = 2n\left[\frac{\hat{\mu}}{\mu} - \log\left(\frac{\hat{\mu}}{\mu}\right) - 1\right] \tag{5.34}$$

The expected information is

$$\mathrm{I}(\mu) = \frac{n}{\mu^2} \tag{5.35}$$

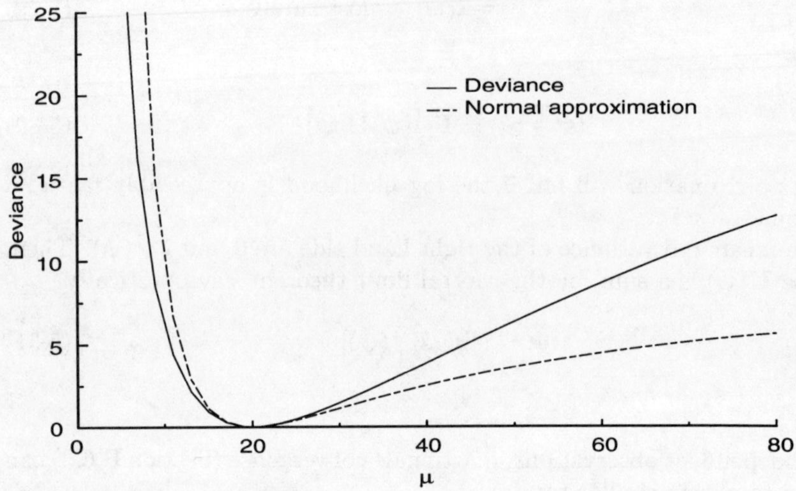

Fig. 5.1. The 'exact' deviance and the normal approximation for the mean of an exponential distribution.

so that a normal approximation is

$$D_{N_1}(\mu) = \frac{n(\hat{\mu} - \mu)^2}{\mu^2} \tag{5.36}$$

These 'exact' and approximate deviances are plotted for $n = 10$ and $\hat{\mu} = 20$ in Figure 5.1. The approximation is not too unreasonable, considering that $n = 10$. However, the exact 10% likelihood interval is $(11, 43)$, approximated by $(12, 62)$. □

Normal approximations to the deviance are not parametrization invariant (Section 3.6.2). Thus, in the exponential family, an approximate likelihood, derived in this way for the canonical parameter, will not give the same results as one derived for the mean value parameter.

Example
Suppose that, in the exponential distribution, we work with the canonical parameter, $\theta = -1/\mu$, instead of the mean value parameter. Then,

$$f(\mathbf{y}; \theta) = (-\theta)^n e^{\theta \Sigma y_i} \qquad \theta < 0 \tag{5.37}$$

$\hat{\theta} = -n / \sum y_i$, and the deviance is

$$D(\theta) = 2n \left[\frac{\theta}{\hat{\theta}} - \log\left(\frac{\theta}{\hat{\theta}}\right) - 1 \right] \tag{5.38}$$

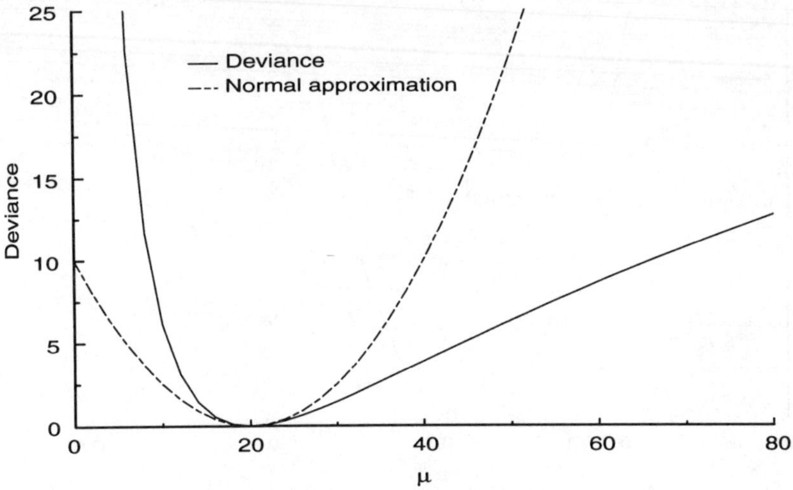

Fig. 5.2. The 'exact' deviance and the normal approximation, based on the canonical parameter, but plotted in terms of the mean, of an exponential distribution.

The expected information is now

$$\mathrm{I}(\theta) = \frac{n}{\theta^2} \tag{5.39}$$

so that the normal approximation is

$$\begin{aligned} \mathrm{D}_{N_1}(\theta) &= \frac{n(\hat{\theta} - \theta)^2}{\theta^2} \\ &= \frac{n(\hat{\mu} - \mu)^2}{\hat{\mu}^2} \end{aligned} \tag{5.40}$$

If we plot this approximate deviance, along with the 'exact' one, for the same data, we obtain the results in Figure 5.2. With this parametrization, the approximation is much worse. Here, the approximate 10% likelihood interval is $(7, 33)$, as compared to the exact $(11, 43)$. Thus, this normal approximation here provides a more reasonable approximation for the deviance of μ than of θ. □

The normal approximations to the deviance are asymptotically equivalent. However, as we can see, for small n, the accuracy of the result depends much more on the parametrization used for the model than on the size of n (Sprott, 1973, 1975a). In Section 2.4.2, we looked at parameter transformations for the exponential family. The value, $\delta = 1/3$ in Equation (2.54), close to quadratic in the parameter, gives the best approximation

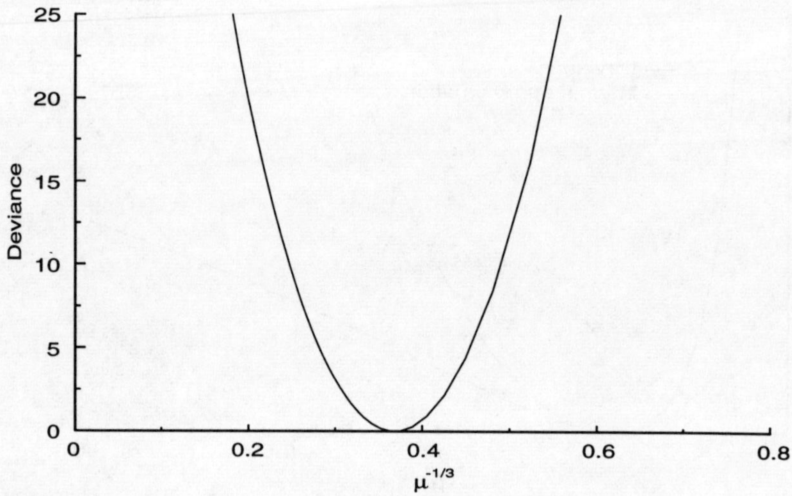

Fig. 5.3. The 'exact' deviance for the transformed mean of an exponential distribution.

to a normal likelihood.

Example
For the exponential distribution, the transformation is $(-\theta)^{\frac{1}{3}} = \mu^{-\frac{1}{3}}$. If we replot the 'exact' deviance for our example, transforming the scale in this way, we have the graph of Figure 5.3, which is much closer to the symmetric normal form. □

We shall study one way of checking the appropriateness of a parametrization for normal approximation in Section 6.3.4.

Because, ψ is typically unknown, any consistent estimate of $\mathbf{I}(\psi)$, such as $\mathbf{I}(\hat{\psi})$ or $\mathbf{J}(\hat{\psi})$, can be used without affecting the limiting distribution. However, in small samples, it is generally preferable to use the observed information, $\mathbf{J}(\hat{\psi})$, especially if $\mathbf{J}$ can have large fluctuations around its expected value. Generally, $\lim_{n\to\infty} \mathbf{J}(\hat{\psi}) = \mathbf{I}(\psi)$, but this will often not be true for stochastic processes; see Barndorff-Nielsen and Cox (1994, pp. 306–314). $\mathbf{J}(\hat{\psi})$ provides a more relevant measure of precision of the m.l.e. for the observed data than $\mathbf{I}(\hat{\psi})$, which is an average over the sample space. The observed information, in effect, provides conditioning on the ancillary statistics. This is generally true in all that follows (Efron and Hinkley, 1978; Hinkley, 1980).

Independence, sufficiency, and efficiency If the off-diagonal elements of the information matrix are zero (called information orthogonality; see Section 6.1.3), the m.l.e.s are *asymptotically independent.* Because the m.l.e. has an asymptotic normal distribution, it is *asymptotically sufficient* (Wald, 1943).

Equation (5.31) implies (see Section C.1) that, asymptotically,

$$(\hat{\boldsymbol{\Psi}} - \boldsymbol{\psi})^{\mathrm{T}} \mathbf{I}(\boldsymbol{\psi})(\hat{\boldsymbol{\Psi}} - \boldsymbol{\psi}) \sim \chi^2_p \tag{5.41}$$

where again p is the dimension of $\boldsymbol{\psi}$. An estimate of $\boldsymbol{\psi}$ with these characteristics is said to be *asymptotically efficient* (Fisher, 1922a), where, as we shall see in Chapter 7, $\mathbf{I}^{-1}(\boldsymbol{\psi})$ is known as the *Cramér–Rao lower bound* for $\mathrm{var}[\hat{\boldsymbol{\Psi}}]$. However, superefficient estimates, with variance smaller than this lower bound for some values of the estimate, can exist (Cox and Hinkley, 1974, pp. 304–305; Lehmann, 1983, pp. 406–408).

5.3.4 STANDARD ERRORS

Asymptotic standard error As we saw in Section 3.5.2, the square root of the diagonal of $\mathbf{I}^{-1}(\hat{\boldsymbol{\psi}})$ is known as the (asymptotic) standard error of $\hat{\boldsymbol{\psi}}$. It can be thought of as an approximate 'standard deviation' for the estimate. In many cases where it is used, the assumption is implicitly made that the variance remains constant for different values of the parameter, $\boldsymbol{\psi}$. As we saw in Chapter 2, this is generally not true for distributions other than the normal.

Examples

1. For the binomial distribution,

$$\mathrm{I}(\pi) = \frac{n}{\pi(1-\pi)} \tag{5.42}$$

so that the standard error of $\hat{\pi}$ is $\sqrt{\pi(1-\pi)/n}$.

2. For the normal distribution,

$$\mathbf{I}(\mu, \sigma^2) = \begin{bmatrix} \frac{n}{\sigma^2} & 0 \\ 0 & \frac{n}{2\sigma^4} \end{bmatrix} \tag{5.43}$$

so that $(\hat{\mu}, \hat{\sigma}^2)$ are asymptotically uncorrelated, with standard errors given by

$$\left(\frac{\sigma}{\sqrt{n}}, \sqrt{\frac{2}{n}}\sigma^2\right) \tag{5.44}$$

□

Standard errors may provide a reasonable measure of precision of parameter estimates if the likelihood function is close to multivariate normal with covariances given by the expected information matrix. For results to be reasonable, it is not always sufficient that the profile likelihood for each parameter element be close to normal because of the inversion of the information matrix. In many cases, standard errors can be extremely unreliable and often misleading in the same way as the approximate likelihood function in Figure 5.2. Because this is a quadratic approximation to the log likelihood function, the larger is the distance between $\boldsymbol{\psi}$ and $\hat{\boldsymbol{\psi}}$, the larger will be the discrepancy between this statistic and the exact deviance.

For a single parameter, if $\mathrm{I}(\hat{\psi})$ is used in Equation (5.32), as we did in Equation (3.50), we have a normal distribution centred on the m.l.e. with variance equal to the square of the standard error.

Example
Equation (5.40) is equivalent to using $\mathrm{I}(\hat{\mu})$ in place of $\mathrm{I}(\mu)$, i.e. the standard error of $\hat{\mu}$, in Equation (5.36). On the other hand, the result in Equation (5.36) is what we would obtain in Equation (5.40) if we used the standard error for $\hat{\theta}$. □

Wald statistic From the standard error, we can derive the *Wald statistic*,

$$(\hat{\boldsymbol{\Psi}} - \boldsymbol{\psi})^{\mathrm{T}} \left(\sqrt{\mathrm{diag}[\mathrm{I}^{-1}(\hat{\boldsymbol{\Psi}})]}\right)^{-1} \tag{5.45}$$

which, asymptotically, has a standard multivariate normal distribution, $\mathrm{N}(\mathbf{0}, \mathbf{I}_p)$, or its square a Chi-squared distribution (Wald, 1943).

Because the Wald statistic is not parametrization invariant, it can be replaced by an invariant form, based on what is called the *sample space derivative* of the log likelihood (Barndorff-Nielsen, 1990; Pierce and Peters, 1992),

$$-\frac{1}{2}\frac{d\mathrm{D}(\psi)}{d\hat{\psi}}\sqrt{\mathrm{J}(\hat{\psi})} \tag{5.46}$$

For a canonical parameter of the exponential family, the two are identical.

Example
For the exponential distribution, with canonical parameter θ,

$$-\frac{1}{2}\frac{d\mathrm{D}(\theta)}{d\hat{\theta}} = \frac{n\theta}{\hat{\theta}^2} - \frac{n}{\hat{\theta}} \tag{5.47}$$

and $\mathrm{J}(\hat{\theta}) = n/\hat{\theta}^2$ so that we obtain $\sqrt{n}(\theta - \hat{\theta})/\hat{\theta}$, the usual Wald statistic for this parameter, while, for the mean value parameter μ,

$$-\frac{1}{2}\frac{d\mathrm{D}(\mu)}{d\hat{\mu}} = \frac{n}{\mu} - \frac{n}{\hat{\mu}} \tag{5.48}$$

and $\mathrm{J}(\hat{\mu}) = n/\hat{\mu}^2$ so that we have $\sqrt{n}(1/\mu - 1/\hat{\mu})\hat{\mu}$, which is the same as the above, but different from the usual Wald statistic for this parameter. This corresponds to the approximate likelihood of Equation (5.36), an example of which was plotted in Figure 5.1. □

Delta method Occasionally, it is useful to know how such asymptotic results are affected by some transformation, $h(\psi)$, of a parameter, and of the corresponding estimate, $\hat{\psi}$. If we expand the transformation as a Taylor series about the 'true' parameter value,

$$h(\hat{\psi}) = h(\psi) + (\hat{\psi} - \psi)\frac{\partial h(\psi)}{\partial \psi} + \frac{1}{2}(\hat{\psi} - \psi)^2\frac{\partial^2 h(\psi)}{\partial \psi^2} + \cdots \tag{5.49}$$

we see that, for $\hat{\psi} - \psi$ small,

$$\mathrm{E}[h(\hat{\psi})] \doteq h(\psi) + \frac{1}{2}\frac{\partial^2 h(\psi)}{\partial \psi^2}\mathrm{var}[\hat{\psi}] \tag{5.50}$$

(5.51)

to second order, and

$$\mathrm{var}[h(\hat{\psi})] \doteq \left[\frac{\partial h(\psi)}{\partial \psi}\right]^2 \mathrm{var}[\hat{\psi}] \tag{5.52}$$

to first order. This is called the *delta method.* It provides a simple approximate means of obtaining the standard error of a reparametrization without calculating the new information matrix. But it also demonstrates how such approximate likelihood functions are not invariant under reparametrization.

Dimension of models Suppose, now, that $\boldsymbol{\psi}$ has dimension p, but that it is a function of $\boldsymbol{\lambda}$, of dimension $r < p$, which indexes the 'true' model that generated the observations. Then,

$$\mathrm{var}[h(\boldsymbol{\psi}(\hat{\boldsymbol{\lambda}}))] \leq \mathrm{var}[h(\hat{\boldsymbol{\psi}})] \tag{5.53}$$

for large n, where h is, again, an arbitrary function (Altham, 1984). This implies that parameters of interest will be more precisely estimated, for example, will have smaller standard errors, in a model with fewer rather than more parameters, as we saw in Section 3.6.3 for normed likelihood regions.

Example
Suppose that the 'true' model is

$$\xi_i = \beta_0 + \beta_1 x_{1i} \tag{5.54}$$

some canonical generalized linear model, where β_1 is the parameter of interest. If we fit, instead, the model

$$\xi_i = \beta_0 + \beta_1 x_{1i} + \beta_2 x_{2i} \tag{5.55}$$

where the additional explanatory variable, x_{2i}, is not needed, the precision of $\hat{\beta}_1$ will be reduced. □

5.3.5 MODIFIED PROFILE LIKELIHOOD APPROXIMATION

Often, it is necessary to have an even better approximation to the distribution of the m.l.e. In the appendix to this chapter (Section 5.7.3), we show how, for the one-parameter exponential family, the conditional distribution of a statistic, given another statistic, can be obtained as a saddlepoint approximation. But this can be used more generally as an asymptotic approximation to the conditional distribution of the m.l.e., given the ancillary:

$$p^*(\hat{\psi}|\mathbf{t}_2;\psi) = c(\mathbf{t}_2,\psi)\sqrt{|\mathbf{J}(\hat{\psi})|}\mathrm{R}(\psi)[1+\mathrm{O}(n^{-\frac{3}{2}})] \tag{5.56}$$

where $\mathrm{R}(\psi)$ is the normed likelihood. This equation, called the p^* formula, has the following properties:

(1) the norming constant, c, is parametrization invariant;
(2) p^* is parametrization invariant;
(3) it is invariant under one-to-one transformations of the data;
(4) it is exact for the location–scale family;
(5) for the linear exponential family, where no conditioning is involved, it is the (renormalized) *saddlepoint approximation*, which is highly accurate and often exact.

As we shall see, similar modifications of the likelihood function appear quite often in statistical theory (for example, Sections 6.3 and 8.3.3).

Example
For the exponential distribution, the m.l.e. is minimal sufficient, so that there is no ancillary, and

$$\mathrm{J}(\mu) = -\frac{n}{\mu^2} + \frac{2n\hat{\mu}}{\mu^3} \tag{5.57}$$

so that

$$J(\hat{\mu}) = \frac{n}{\hat{\mu}^2} \tag{5.58}$$

Then, we have

$$\begin{aligned} p^*(\hat{\mu};\mu) &= c_1(\mu)\frac{\sqrt{n}}{\hat{\mu}}\left(\frac{\hat{\mu}}{\mu}\right)^n e^{n(1-\frac{\hat{\mu}}{\mu})} \\ &= c(\mu)\frac{\hat{\mu}^{n-1}}{\mu^n}e^{-\frac{n\hat{\mu}}{\mu}} \end{aligned} \tag{5.59}$$

which is a gamma distribution with second parameter, n, the exact result for the distribution of the m.l.e. (assuming infinite measurement precision). As expected, the deviance is the usual one given above in Equation (5.33).□

5.4 Deviance statistic

A second approach to obtaining an approximate likelihood function is to work with the distribution of the likelihood itself, or, more easily, with that of the deviance. Obviously, if we know the deviance, an approximation is not necessary, so that these distributional results will be primarily useful for obtaining approximate significance tests (see also Section 5.6 and Chapter 7).

5.4.1 EXACT DISTRIBUTION

The deviance, $D(\psi)$, is used to compare the support for different possible models, given the observations. Thus, it is a function of the observations, as well as of the (unknown) parameters. For a fixed value of the parameters, the deviance is a statistic, and, thus, has a distribution. In certain simple cases, this can be calculated exactly.

Examples

1. For the binomial model, only $n+1$ different values of the deviance are possible, corresponding to the possible numbers of successes, y. Their distribution is given by the binomial probabilities for a fixed value of the parameter, π_0.

For $n = 10$, the distributions are given for two parameter values in Table 5.1. (These are the same probabilities as in Table 4.2.) Notice that *large* values of the deviance statistic are rare. For the binomial model, the set of possible values of the deviance statistic changes with the fixed value of the parameter, π_0, whereas the possible values of the m.l.e. did not. The probability of a deviance statistic greater than a given constant depends on the fixed value of π_0. Thus, $P(D > 4; \pi = 0.5) = 0.0216$, whereas $P(D > 4; \pi = 0.7) = 0.0387$.

Table 5.1. Exact distribution of the deviance for a binomial distribution for $n = 10$.

y	$D(0.5; \hat{\pi})$	$p(D; 0.5)$	$D(0.7; \hat{\pi})$	$p(D; 0.7)$
0	13.8629	0.0010	24.0795	0.0000
1	7.3613	0.0098	15.8832	0.0001
2	3.8549	0.0439	10.6822	0.0014
3	1.6457	0.1172	6.7784	0.0090
4	0.4027	0.2051	3.8408	0.0368
5	0.0000	0.2461	1.7435	0.1029
6	0.4027	0.2051	0.4516	0.2001
7	1.6457	0.1172	0.0000	0.2668
8	3.8549	0.0439	0.5146	0.2335
9	7.3613	0.0098	2.3264	0.1211
10	13.8629	0.0010	7.1335	0.0282

2. For n i.i.d. observations from a normal distribution with σ^2 known, the deviance is

$$D(\mu) = \frac{n(\hat{\mu} - \mu)^2}{\sigma^2} \tag{5.60}$$

Suppose that the 'true' mean is μ_0 so that $\hat{M} \sim N(\mu_0, \sigma^2/n)$. In standardized form, we have $z(\hat{\mu}) = \sqrt{n}(\hat{\mu} - \mu_0)/\sigma$ where $Z \sim N(0,1)$, so that

$$D(\mu_0) = z^2(\hat{\mu}) \tag{5.61}$$

which, thus, has a Chi-squared distribution with one degree of freedom. The probability of a deviance statistic greater than a given constant does not depend on the fixed value of μ_0, because z and z^2 are pivotal quantities (Section 3.5.4). □

5.4.2 NORMAL APPROXIMATION

Let us now expand the log likelihood function as a Taylor series at $\psi = \hat{\psi}$:

$$\begin{aligned} l(\psi) = l(\hat{\psi}) + (\psi - \hat{\psi})^T \frac{\partial l(\hat{\psi})}{\partial \psi} \\ + \frac{1}{2}(\psi - \hat{\psi})^T \frac{\partial^2 l(\hat{\psi})}{\partial \psi \partial \psi^T}(\psi - \hat{\psi}) + \cdots \end{aligned} \tag{5.62}$$

Because $\partial l(\hat{\psi})/\partial \psi = \mathbf{U}(\hat{\psi}) = \mathbf{0}$, ignoring higher order terms, we obtain the approximation

$$\mathrm{D}_{N_2}(\psi) \stackrel{\text{def}}{=} (\psi - \hat{\psi})^{\mathrm{T}}\mathbf{J}(\hat{\psi})(\psi - \hat{\psi}) \tag{5.63}$$

the deviance for a multivariate normal distribution of $\hat{\boldsymbol{\Psi}}$: $\mathrm{N}[\psi, \mathbf{J}^{-1}(\hat{\psi})]$. From this result, we see why it is convenient to define the deviance as the log likelihood multiplied by minus two, as was done in Equation (3.25).

For n sufficiently large, $\hat{\psi}$ will be close to the 'true' value, ψ, and this should be a reasonable approximation. It is a second normal approximation to the deviance, differing from the normal approximation in Equation (5.32) in that here the observed information at the m.l.e. replaces the expected information, although both are asymptotically equivalent. However, we have seen above that, in small samples, they can give very different results.

Examples

1. For n i.i.d. observations from a normal distribution with σ^2 known, $\mathrm{J}(\hat{\mu}) = n/\sigma^2$, so that we have

$$\mathrm{D}_{N_2}(\mu) = \frac{n(\hat{\mu} - \mu)^2}{\sigma^2} \tag{5.64}$$

which is the exact deviance, as might be expected.

2. For n i.i.d. observations, the exponential distribution was given by Equation (5.34). This normal approximation gives

$$\begin{aligned} \mathrm{D}_{N_2}(\mu) &\doteq n\left[\frac{\mu}{\hat{\mu}} - 1\right]^2 \\ &= \frac{n(\hat{\mu} - \mu)^2}{\hat{\mu}^2} \end{aligned} \tag{5.65}$$

This is the same normal approximation to the deviance for the mean which we obtained indirectly above in Equation (5.40) by using the canonical parameter and, in Section 5.3.4, the standard error of the mean. We saw in Figure 5.2 that it did not perform very well. □

5.4.3 CHI-SQUARED DISTRIBUTION

From the normal approximation to the distribution of the m.l.e., we can show that

The deviance has an asymptotic Chi-squared distribution.

proof: From Equation (5.41),

$$(\psi - \hat{\boldsymbol{\Psi}})^{\mathrm{T}}\mathbf{J}(\hat{\boldsymbol{\Psi}})(\psi - \hat{\boldsymbol{\Psi}}) \sim \chi^2_p \tag{5.66}$$

Thus, asymptotically, from Equations (5.62) and (5.63),

$$\mathrm{D}(\boldsymbol{\psi}) \sim \chi^2_p \tag{5.67}$$

where $\boldsymbol{\psi}$ is the 'true' value, with dimension p fixed. (Wilks, 1938) □

Notice that this result will not necessarily hold for deviances based on a saturated model, as in Section 4.1.1, because p may not be fixed but may then depend on n, the sample size.

Now, suppose that we wish to compare our model with p parameters to some submodel, $\boldsymbol{\psi}_1$, of dimension $r < p$, *nested* in Ψ, i.e. where $\Psi_1 \subset \Psi$. We have

$$\begin{aligned} -2[\mathrm{l}(\hat{\boldsymbol{\psi}}_1) - \mathrm{l}(\hat{\boldsymbol{\psi}})] &= -2\{[\mathrm{l}(\boldsymbol{\psi}) - \mathrm{l}(\hat{\boldsymbol{\psi}})] - [\mathrm{l}(\boldsymbol{\psi}_1) - \mathrm{l}(\hat{\boldsymbol{\psi}}_1)] - [\mathrm{l}(\boldsymbol{\psi}) - \mathrm{l}(\boldsymbol{\psi}_1)]\} \\ &= \mathrm{D}(\boldsymbol{\psi}) - \mathrm{D}(\boldsymbol{\psi}_1) + 2[\mathrm{l}(\boldsymbol{\psi}) - \mathrm{l}(\boldsymbol{\psi}_1)] \end{aligned} \tag{5.68}$$

The first term has an asymptotic χ^2_p distribution, the second, χ^2_r, and the third is a positive constant, near zero if the 'true' model is indexed by $\boldsymbol{\psi}_1$. Then, asymptotically,

$$\mathrm{D}(\boldsymbol{\psi}) - \mathrm{D}(\boldsymbol{\psi}_1) \sim \chi^2_{p-r} \tag{5.69}$$

because sums of independent Chi-squared variables are Chi-squared (Section C.1).

Examples

1. Consider a normal linear regression model, with $\xi_i = \mu_i = \beta_0 + \beta_1 x_i$. Let $\boldsymbol{\psi} = (\beta_0, \beta_1, \sigma^2)^{\mathrm{T}}$ and $\boldsymbol{\psi}_1 = (\beta_0, \sigma^2)^{\mathrm{T}}$. The normed profile likelihood, maximized over σ^2, similar to Equation (3.75), is

$$\mathrm{R}_P(\beta_0, \beta_1) = \left[\frac{\sum(y_i - \hat{\beta}_0 - \hat{\beta}_1 x_i)^2}{\sum(y_i - \beta_0 - \beta_1 x_i)^2}\right]^{\frac{n}{2}} \tag{5.70}$$

so that

$$\begin{aligned} \mathrm{D}(\beta_0 = \bar{y}_\bullet, \beta_1 = 0) &= -2 \log \mathrm{R}_P(\beta_0 = \bar{y}_\bullet, \beta_1 = 0) \\ &= n \log \left[\frac{\sum(y_i - \bar{y}_\bullet)^2}{\sum(y_i - \hat{\beta}_0 - \hat{\beta}_1 x_i)^2}\right] \\ &\sim \chi^2_1 \end{aligned} \tag{5.71}$$

asymptotically.

2. Consider a logistic linear regression model for a set of binomial observations, y_i successes out of n_i trials, with

$$\xi_i = \log\left(\frac{\pi_i}{1 - \pi_i}\right)$$

$$= \beta_0 + \beta_1 x_i \tag{5.72}$$

Let $\psi = (\beta_0, \beta_1)^{\mathrm{T}}$ and $\psi_1 = \beta_0$. The likelihood function is

$$\mathrm{L}(\beta_0, \beta_1) = \frac{\mathrm{e}^{\beta_0 \Sigma y_i + \beta_1 \Sigma x_i y_i}}{\prod (1 + \mathrm{e}^{\beta_0 + \beta_1 x_i})^{n_i}} \tag{5.73}$$

so that

$$\begin{aligned} &\mathrm{D}(\hat{\beta}_0^*, \beta_1 = 0) \\ &= 2\left\{ (\hat{\beta}_0 - \hat{\beta}_0^*) \sum y_i + \hat{\beta}_1 \sum x_i y_i + \sum n_i \log \left[\frac{1 + \mathrm{e}^{\hat{\beta}_0^*}}{1 + \mathrm{e}^{\hat{\beta}_0 + \hat{\beta}_1 x_i}} \right] \right\} \\ &\sim \chi_1^2 \end{aligned} \tag{5.74}$$

asymptotically, where $\hat{\beta}_0^*$ is the m.l.e. with $\beta_1 = 0$. □

As mentioned above, the deviance statistic of a generalized linear model, with p location parameters, compared to a saturated model with n parameters, will not necessarily be close to having a Chi-squared distribution if $s = n - p$ and is not fixed but increases with n.

Examples

1. Consider n Bernoulli observations, y_i, with probabilities π_i. Our fixed model has $\pi_i = \pi$, a constant. The deviance was given in Equation (4.7),

$$\mathrm{D}(\pi) = -2n[\bar{y}_\bullet \log(\pi) + (1 - \bar{y}_\bullet) \log(1 - \pi)]$$

Here $s = n - 1$ and this deviance statistic depends too heavily on π, rather than on the individual observations, to be asymptotically Chi-squared.

2. Consider now two binomial observations, with y_i successes out of n_i, and probabilities π_i. Our fixed model is again $\pi_i = \pi$. The deviance is

$$\mathrm{D}(\pi) = 2\sum_{i=1}^{2} \left[y_i \log\left(\frac{y_i}{\pi n_i}\right) + (n_i - y_i) \log\left(\frac{n_i - y_i}{n_i(1 - \pi)}\right) \right] \tag{5.75}$$

which does not collapse, as in Equation (4.7). Here, $s = 1$ and the asymptotic approximation holds because we can allow the individual n_i to increase in size, while the number of binomial observations stays constant at two. This result is generally true for logistic and log linear models for contingency tables of fixed size. □

The Chi-squared distribution is commonly used for calibrating the deviance, i.e. for comparing deviances with different degrees of freedom, in model selection problems. However, in contrast to the approach through a smoothing factor, proposed in Section 3.6.1, the Chi-squared distribution does not provide even approximately compatible inferences for different degrees of freedom. For example, for the small degrees of freedom (< 7) most often used, it will provide narrower interval estimates (Section 7.5) than does the AIC; it does not allow for the uncertainty in model selection.

Directed deviance statistic In certain contexts, for a scalar parameter, it is more appropriate to use the directed deviance, $\mathrm{D}_d(\psi)$, of Equation (3.26), as a statistic. It will have an asymptotic standard normal distribution, $\mathrm{N}(0,1)$, showing why Equation (3.68) is an approximate pivotal quantity. The distributional properties of this approximation may be improved by taking

$$\mathrm{D}_d(\psi) - \frac{1}{\mathrm{D}_d(\psi)} \log\left[\frac{\mathrm{D}_d(\psi)}{(\hat{\psi}-\psi)\sqrt{\mathrm{J}(\psi)}}\right] + \frac{1}{\mathrm{D}_d(\psi)} \log\left[\frac{\mathrm{J}(\psi)}{\mathrm{J}(\hat{\psi})}\right] \tag{5.76}$$

where the denominator of the first logarithm is the Wald statistic (Pierce and Peters, 1992). To obtain parametrization invariance, the latter can be replaced by the sample space derivative of Equation (5.46). However, the cost of these corrections to improve asymptotic distributional properties is obviously to create a divergence from the form of the exact log likelihood function which will generally not be desirable.

Bartlett's adjustment Further useful results can be obtained by studying the first moment of the deviance statistic. By Taylor series expansion, it can be shown that

$$\mathrm{E}[\mathrm{D}(\psi)] = p\left[1 + \frac{b(\psi)}{n} + \mathrm{O}(n^{-2})\right] \tag{5.77}$$

where p is the number of degrees of freedom of $\mathrm{D}(\psi)$ and $b(\cdot)$ is some function to be determined. On the other hand, from Section C.1,

$$\mathrm{E}[\chi^2_p] = p \tag{5.78}$$

The asymptotic distribution of the deviance statistic, which is to order $\mathrm{O}(n^{-1})$, can, then, be improved to $\mathrm{O}(n^{-2})$ by using

$$D_B(\psi) \stackrel{\mathrm{def}}{=} \frac{D(\psi)}{1 + \frac{b(\psi)}{n}} \tag{5.79}$$

where $b(\psi)$ is a known constant or can be estimated.

This is known as *Bartlett's adjustment* (Bartlett, 1937). It is related to the modified profile likelihood of Section 5.3.5 above:

$$b(\psi) \doteq \frac{n}{2\pi}\left[\frac{c(\psi_0)}{c(\psi)}\right]^{\frac{2}{p}} - n \tag{5.80}$$

where ψ_0 is a fixed value. Because c is a normalizing constant, it will often be difficult to calculate.

Example

Suppose that we have n i.i.d. normal observations with mean μ and variance σ^2 and we are interested in μ. From Equation (5.56), we have

$$\frac{c_0}{c} = \frac{\sqrt{n\pi}\Gamma\left(\frac{n-1}{2}\right)}{\Gamma\left(\frac{n}{2}\right)} \tag{5.81}$$

with $p = 1$. From this, we obtain

$$b = \frac{3}{2} + \mathrm{O}(n^{-1}) \tag{5.82}$$

so that the adjustment to improve the distributional properties of the deviance statistic is

$$\mathrm{D}_B(\mu = 0) = \frac{\mathrm{D}(\mu = 0)}{1 + \frac{3}{2n}} \tag{5.83}$$

(Barndorff-Nielsen and Cox, 1984) □

Bartlett's adjustment generally only improves the Chi-squared approximation for continuous distributions. Our conclusion to this section is that corrections to the deviance statistic to improve the asymptotic distribution may only be useful for approximations to significance tests, at the expense of increased divergence from (the information in) the likelihood function.

5.4.4 LIKELIHOODS AS MARTINGALES

Most of the asymptotic results in this chapter require the assumption that the observations are independent. For stochastic processes, with dependence over time, this can be relaxed by the use of martingale theory (Section B.4).

If a series of random variables is observed (usually over time), then a sequence of likelihood ratios can be constructed for two models indexed by ψ_1 and ψ_2:

$$M_t = \frac{f(y_1|\mathcal{F}_0;\psi_1)f(y_2|\mathcal{F}_1;\psi_1)\cdots f(y_t|\mathcal{F}_{t-1};\psi_1)}{f(y_1|\mathcal{F}_0;\psi_2)f(y_2|\mathcal{F}_1;\psi_2)\cdots f(y_t|\mathcal{F}_{t-1};\psi_2)} \tag{5.84}$$

with $f(y_t|\mathcal{F}_{t-1};\psi_2) > 0\ \forall y$ and $\mathcal{F}_t$ the previous history or filtration. Because the observations are conditionally independent,

$$\begin{aligned}\mathrm{E}[M_{t+1}|\mathcal{F}_t] &= \mathrm{E}\left[M_t \left.\frac{f(y_{t+1}|\mathcal{F}_t;\psi_1)}{f(y_{t+1}|\mathcal{F}_t;\psi_2)}\right| \mathcal{F}_t\right] \\ &= M_t\mathrm{E}\left[\frac{f(y_{t+1}|\mathcal{F}_t;\psi_1)}{f(y_{t+1}|\mathcal{F}_t;\psi_2)}\right]\end{aligned}$$

but if the expectations are with respect to the distribution indexed by ψ_2, i.e. if we assume that the data were, in fact, generated by this model, then

$$\begin{aligned}\mathrm{E}\left[\frac{f(y_{t+1}|\mathcal{F}_t;\psi_1)}{f(y_{t+1}|\mathcal{F}_t;\psi_2)}\right] &= \int \frac{f(y_{t+1}|\mathcal{F}_t;\psi_1)}{f(y_{t+1}|\mathcal{F}_t;\psi_2)} f(y_{t+1}|\mathcal{F}_t;\psi_2)dy_{t+1} \\ &= 1\end{aligned} \tag{5.85}$$

so that $\{M_t\}$ is a martingale. Obviously, this result also holds for a series of completely independent observations.

Example
Consider a normal distribution where $\psi_1 = (0, \sigma^2)$ and $\psi_2 = (\mu, \sigma^2)$. Then,

$$\frac{f(y;\psi_1)}{f(y;\psi_2)} = \exp\left(-\frac{2\mu y - \mu^2}{2\sigma^2}\right) \tag{5.86}$$

and

$$M_n = \exp\left(-\frac{\mu\sum y_i}{\sigma^2} + \frac{n\mu^2}{2\sigma^2}\right) \tag{5.87}$$

forms a martingale if the data were generated by $f(y;\psi_2)$. □

Martingale limit By martingale convergence (Section B.6.3), the martingale, $\{M_n\}$, converges to a random variable, say M_∞, almost surely. Now,

$$\log(M_n) = \sum_{i=1}^{n} \log\left[\frac{f(y_i;\psi_1)}{f(y_i;\psi_2)}\right] \tag{5.88}$$

the sum of n i.i.d. random variables. The logarithm function is concave, so that, by Jensen's inequality,

$$\mathrm{E}\left[\log\left(\frac{f(y_i;\psi_1)}{f(y_i;\psi_2)}\right)\right] < \log\left\{\mathrm{E}\left[\frac{f(y_i;\psi_1)}{f(y_i;\psi_2)}\right]\right\} \tag{5.89}$$

$$= 0$$

Then, from the strong law of large numbers, $\log(M_n)/n$ converges, almost surely, to a point in $[-\infty, 0)$. Thus, $M_\infty = 0$.

From these results, we know that $M_n < c$ for $c > 0$ when n is sufficiently large. Indeed, from the maximal inequality,

$$P(M_n \geq c) \leq \frac{1}{c} \qquad \forall n \geq 1, c > 0 \tag{5.90}$$

This provides an asymptotic martingale justification for our preferring a model with a larger likelihood, as in Chapter 3.

5.5 Score statistic

The final statistic which we shall look at is based on the score. We already studied its mean and variance in Section 5.1.2.

5.5.1 EXACT DISTRIBUTION

For a given value of $\boldsymbol{\psi}$, in certain special simple cases, the exact distribution of the score statistic can be calculated, in the same way as for the deviance statistic.

5.5.2 ASYMPTOTIC DISTRIBUTION

Because we know the mean and variance of the score statistic, and because it is a sum, when the observations are independent, by the central limit theorem, asymptotically

$$\mathbf{U}(\boldsymbol{\psi}) \sim \mathrm{N}[\mathbf{0}, \mathbf{I}(\boldsymbol{\psi})] \tag{5.91}$$

under the usual regularity conditions. This justifies using Equation (3.67) as an approximate pivotal quantity. If $\boldsymbol{\psi}$ is far from the 'true' value, $\mathbf{U}(\boldsymbol{\psi})$ will differ from zero provided that the quadratic approximation is reasonable. It follows that, asymptotically,

$$\mathbf{U}^{\mathrm{T}}(\boldsymbol{\psi})\mathbf{I}^{-1}(\boldsymbol{\psi})\mathbf{U}(\boldsymbol{\psi}) \sim \chi^2_p \tag{5.92}$$

This is called the *score statistic*.

The same result can be obtained in another way. From the demonstration of the asymptotic normality of the m.l.e., in Equation (5.30), we know that

$$\hat{\boldsymbol{\Psi}} - \boldsymbol{\psi} \doteq \mathbf{I}^{-1}(\boldsymbol{\psi})\mathbf{U}(\boldsymbol{\psi})$$

Substituting this into Equation (5.41), we obtain an approximate deviance,

$$\mathrm{D}_U(\boldsymbol{\psi}) \stackrel{\text{def}}{=} \mathbf{U}^{\mathrm{T}}(\boldsymbol{\psi})\mathbf{I}^{-1}(\boldsymbol{\psi})\mathbf{U}(\boldsymbol{\psi}) \tag{5.93}$$

which will have an asymptotic Chi-squared distribution. Similar results can be obtained for dependent observations using martingale convergence.

Examples

1. For the binomial distribution, with

$$\mathrm{U}(\pi) = \frac{y - n\pi}{\pi(1-\pi)}$$
$$\mathrm{I}(\pi) = \frac{n}{\pi(1-\pi)}$$

we have the score statistic

$$\begin{aligned}\mathrm{D}_U(\pi) &= \frac{(y-n\pi)^2}{n\pi(1-\pi)} \\ &= \sum_{j=1}^{2} \frac{(y_j - n\pi_j)^2}{n\pi_j}\end{aligned} \tag{5.94}$$

where $\pi_1 = \pi$, $\pi_2 = 1 - \pi$, $y_1 = y$, and $y_2 = n - y$.

This is a simple case of the Pearson (1900) *Chi-squared goodness of fit statistic*, which we first encountered in Equation (4.4). It is the score statistic approximation to the exact deviance,

$$\mathrm{D}(\pi) = 2\sum_{j=1}^{2} y_j \log\left(\frac{y_j}{n\pi_j}\right) \tag{5.95}$$

Both have an asymptotic Chi-squared distribution. However, this approximate deviance, D_U, can differ substantially from the exact one.

Suppose that we observe $y = 8$ with $n = 10$. The 'exact' and approximate deviances are plotted in Figure 5.4, showing fairly close agreement.

2. For the exponential distribution, the score and expected information are

$$\begin{aligned}\mathrm{U}(\mu) &= \frac{\sum y_i - n\mu}{\mu^2} \\ &= \frac{n(\bar{y}_\bullet - \mu)}{\mu^2} \\ \mathrm{I}(\mu) &= \frac{n}{\mu^2}\end{aligned}$$

so that the score statistic,

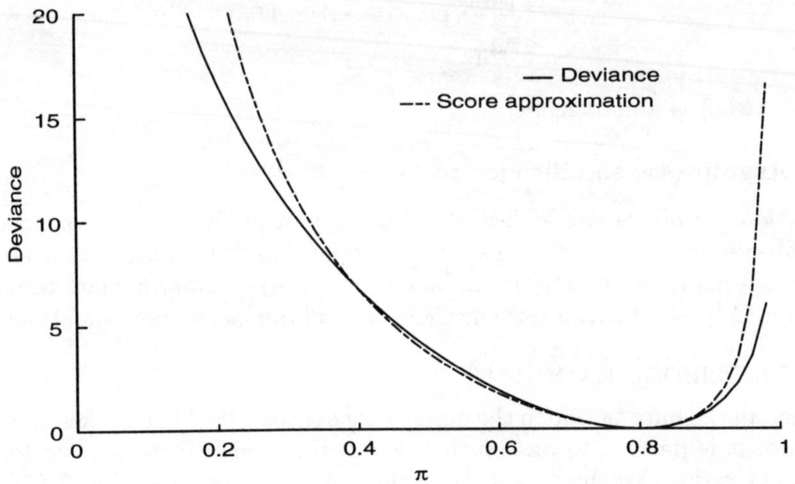

Fig. 5.4. 'Exact' deviance and the score approximation for the binomial probability.

$$D_U(\mu) = \frac{n(\bar{y}_\bullet - \mu)^2}{\mu^2} \tag{5.96}$$

is the same as our first normal approximation to the deviance, D_{N_1} in Equation (5.32). □

Although we have now seen that several of the approximations to the deviance provide the same function for the exponential distribution, this need not be the case in general for other distributions, especially those outside the exponential family.

The advantage of the score statistic, as compared to the deviance and its normal approximation, is that it does not require the calculation of $\hat{\psi}$; it depends only on the fixed value, ψ. However, it uses a quadratic approximation to the log likelihood function, like the Wald statistic. It looks at the squared slope of the likelihood function at ψ, multiplied by the inverse of the curvature, which will be close to zero if ψ is near $\hat{\psi}$.

5.5.3 SCORE STATISTICS AS MARTINGALES

If a series of random variables is observed over time, then a sequence of score statistics can be constructed using, successively, the different (cumulating) amounts of information available at each point in time. Let $\mathbf{M}_n = \mathbf{U}_n(\psi; y_n|\mathcal{F}_{n-1})$ be the score vector for the first n observations. Then, because $\mathbf{M}_n$ is known, given $\mathcal{F}_{n-1}$,

$$\mathrm{E}[\mathbf{M}_{n+1}|\mathcal{F}_n] = \mathrm{E}[\mathbf{M}_{n+1} - \mathbf{M}_n|\mathcal{F}_n] + \mathbf{M}_n$$

$$= \mathrm{E}[\mathbf{U}_{n+1}(\psi; y_{n+1}|\mathcal{F}_n)|\mathcal{F}_n] + \mathbf{M}_n$$
$$= \mathbf{M}_n$$

so that $\{\mathbf{M}_n\}$ is a martingale.

5.6 Asymptotic significance tests

In Section 4.2 above, we looked at some ways of performing significance tests for goodness of fit. Because exact tests of significance are only available in special cases, it is useful to be able to construct approximate tests. One common use of asymptotic methods is to obtain such approximations.

5.6.1 LIKELIHOOD RATIO TEST

Because all the information in the observations is contained in the likelihood function, it is natural to base such tests on the deviance statistic or log likelihood ratio. We have seen that this often has an asymptotic Chi-squared distribution, which we can now use to obtain a *likelihood ratio test.* We have also noticed that large values of the deviance statistic indicate the observation of rare events under a given fixed model. Thus, we look at the *one-tailed* Chi-squared probabilities of large deviances.

Example

Consider again the binomial distribution for the 2×2 contingency table, with frequencies n_{ij}. Because of the sum in the denominator of the conditional distribution, Equation (4.22), the deviance based on it is complex. Usually, the normed profile likelihood at the m.l.e.,

$$\mathrm{R}_P(\beta_0 = 0) = \frac{\left(\frac{n_{\bullet 1}}{n_{\bullet\bullet}}\right)^{n_{11}} \left(\frac{n_{\bullet 2}}{n_{\bullet\bullet}}\right)^{n_{12}} \left(\frac{n_{\bullet 1}}{n_{\bullet\bullet}}\right)^{n_{21}} \left(\frac{n_{\bullet 2}}{n_{\bullet\bullet}}\right)^{n_{22}}}{\left(\frac{n_{11}}{n_{1\bullet}}\right)^{n_{11}} \left(\frac{n_{12}}{n_{1\bullet}}\right)^{n_{12}} \left(\frac{n_{21}}{n_{2\bullet}}\right)^{n_{21}} \left(\frac{n_{22}}{n_{2\bullet}}\right)^{n_{22}}} \tag{5.97}$$

is used. Then, an asymptotic approximation to Fisher's exact test may be based on the corresponding deviance statistic

$$\begin{aligned} \mathrm{D}(\beta_1 = 0) = 2 \Bigg[& n_{\bullet\bullet} \log(n_{\bullet\bullet}) + \sum_i \sum_j n_{ij} \log(n_{ij}) \\ & - \sum_i n_{i\bullet} \log(n_{i\bullet}) - \sum_j n_{\bullet j} \log(n_{\bullet j}) \Bigg] \end{aligned} \tag{5.98}$$

For the first table in the example of Section 4.2.2, this gives a value of 4.73. The tail probability of a Chi-squared value as large as this is about 0.030, as compared to the exact value of 0.167. In this case, $n_{\bullet\bullet} = 9$ is too small for the approximation to be very good. For the second table, we obtain a deviance statistic of 4.76 with P-value of 0.029, as compared to the exact value of 0.039. With more observations this time, the approximation is better. □

Equation (5.98) is, in fact, applicable as an asymptotic test for independence in a two-way contingency table of any size.

5.6.2 WALD AND SCORE TESTS

Approximations to the log likelihood ratio test are sometimes used: the asymptotic normal distribution of the m.l.e. gives a *Wald test* (Wald, 1943) and the score statistic a *Rao score* or *Lagrange multiplier test* (Rao, 1948; Silvey, 1959).

Example
For the first table of Section 4.2.2, β_1 has an infinite estimate because of the zero frequency, so that the Wald statistic is not calculable. In the second table, $\hat{\beta}_1 = 1.404$ with s.e. 0.676, so that the Wald statistic is 2.076 with a two-tailed probability, or P-value, from the normal distribution of 0.038.

The Rao score approximation to the profile likelihood ratio test for a 2×2 contingency table is

$$\sum_i \sum_j \frac{\left(n_{ij} - \frac{n_{i\bullet} n_{\bullet j}}{n_{\bullet\bullet}}\right)^2}{\frac{n_{i\bullet} n_{\bullet j}}{n_{\bullet\bullet}}} \tag{5.99}$$

which is the Pearson Chi-squared statistic, similar to Equation (5.94). For the first table, this gives a value of 3.600 with P-value 0.058, and, for the second table, 4.591 with P-value 0.032. □

The Wald and Rao score tests may be thought to have certain (computational) advantages, especially when the model of interest sets certain parameters to zero. The Wald test is simple to apply, because, if the s.e. of an estimate is available, a rough rule of thumb is that a value of the m.l.e. at least twice as large as the s.e. will only occur with about probability 0.05 if the 'true' value is zero. However, as we have seen, this test is particularly sensitive to changes in parametrization and can give very bad results in many small sample situations. The log likelihood function needs to be close to quadratic for this normal approximation to hold. The other two tests are both parametrization invariant.

A likelihood ratio test requires estimation of both models to be compared. The Rao score test has the advantage, already mentioned, that it can be obtained from the simpler model, with the parameters set to zero, without calculating the m.l.e. of those parameters. The Wald test has a similar advantage, that only the more complex model need be estimated, although this generally requires more calculations than for the simpler model. These arguments have little weight when appropriate computer software is widely available.

In modern statistical practice, reliance should be placed, as far as possible, on the likelihood function, which contains all of the information about a model, and not on approximations to it. If the likelihood function is complex, say, with several local maxima of about the same height, with zero value for a certain range of the parameter, or with a supremum on the boundary of the parameter space, then only the likelihood ratio test, based on the deviance, may give sensible answers. Generally, when an exact test is unavailable, the deviance should be preferred to any of the other approximations, except as a rapid rule of thumb, because it is based directly on the likelihood function (Cox, 1988).

5.7 Appendix: large sample approximations

For more ample details on the material in this appendix, see the references at the end of the chapter.

5.7.1 LAPLACE APPROXIMATION

Because we shall often be working with complex integrals, we first provide a useful approximate formula for these. Other techniques are given in Appendix D. As we have seen, many density functions are such that their integral has a form somewhat similar to

$$g_n = \int_a^b e^{nt(u)} h(u) du \tag{5.100}$$

Suppose that we wish to evaluate such an integral, where the function $t(u)$ has a maximum at $u_m \in (a, b)$, so that

$$\left.\frac{\partial t(u)}{\partial u}\right|_{u=u_m} = 0$$

and

$$K(u_m) = -\left.\frac{\partial^2 t(u)}{\partial u^2}\right|_{u=u_m} > 0$$

These two quantities are obviously closely related to the score and observed information.

Expanding the log of the integrand as a quadratic about u_m and exponentiating yields an approximation with the form of a normal density having mean u_m and variance $K(u_m)$. Integrating this approximate integrand gives the *Laplace approximation* to the integral,

$$g_n = \mathrm{e}^{nt(u_m)} h(u_m) \sqrt{\frac{2\pi}{nK(u_m)}} [1 + \mathrm{O}(n^{-1})] \tag{5.101}$$

This approximation works especially well if the integrand is highly peaked around the maximum, which is often the case in statistical applications.

Example
The Laplace approximation to a gamma function is called Stirling's approximation:

$$\int_0^\infty \mathrm{e}^{-u} u^n du \doteq n^n \mathrm{e}^{-n} \sqrt{2\pi n} \tag{5.102}$$

where $t(u) = \log(u) - u/n$ and $h(u) = 1$, so that $u_m = n$. □

5.7.2 EDGEWORTH EXPANSIONS

Asymptotic expansions can involve writing the density function in a series in terms of its moments. Note that a normal distribution only requires the first two moments to be completely specified. We may ask how much error such series expansions will involve in other cases, and what improvement follows on adding further terms. Suppose, then, that we have n i.i.d. random variables, such that

$$\begin{aligned} \mathrm{E}[Y_i] &= 0 \\ \mathrm{E}[Y_i^2] &= \mu_2 \\ &= \sigma^2 > 0 \\ \mathrm{E}[|Y_i|^3] &= \mu_3 < \infty \end{aligned} \tag{5.103}$$

Let the cumulative distribution of the standardized mean statistic, $T_n = \sum_{i=1}^n Y_i/\sigma\sqrt{n}$, be given by $F_n(t) = \Pr[T_n < t]$, with corresponding density, $f_n(t)$. Then, the Berry–Esseen bound on the possible difference between this arbitrary cumulative distribution and that for a normal distribution is given by

$$\sup_t |F_n(t) - \Phi(t)| \leq \frac{1.332\mu_3}{\sigma^3\sqrt{n}} \tag{5.104}$$

where $\Phi(\cdot)$ is the standard normal cumulative distribution function. Thus, we have a limit to the error involved in approximating the distribution of this mean statistic by a normal distribution, when, for example, the central limit theorem is applied. However, according to this criterion, n needs to be about 10 000 before we know that the distribution is normal to two decimals and yet, in many situations, $n = 50$ may be sufficient to make the judgement by other criteria (Hoffmann-Jørgensen, 1994, p. 399)!

This suggests that an improvement can be obtained by adding further terms in an asymptotic series expansion in $1/\sqrt{n}$. In this way, we obtain the *Edgeworth expansion* for the density, $f_n(t)$, of this statistic for n observations, given by

$$f_n(t) = \phi(t) \left\{ 1 + \frac{1}{\sqrt{n}} \frac{\kappa_3}{6\sigma^3} H_3(t) + \frac{1}{n} \left[\frac{\kappa_4}{24\sigma^4} H_4(t) + \frac{\kappa_3^2}{72\sigma^6} H_6(t) \right] + \mathrm{O}(n^{-\frac{3}{2}}) \right\} \tag{5.105}$$

where $\phi(\cdot)$ is the standard normal density, κ_r is the rth cumulant, and $H_r(t)$ is the Hermite polynomial,

$$\begin{aligned} H_3(t) &= t^3 - 3t \\ H_4(t) &= t^4 - 6t^2 + 3 \\ H_6(t) &= t^6 - 15t^4 + 45t^2 - 15 \end{aligned} \tag{5.106}$$

The Edgeworth expansion is generally in powers of $1/\sqrt{n}$, where the leading term is the normal density. However, when $t = 0$, all coefficients for odd powers disappear and this is an expansion in $1/n$. This expansion provides a good approximation in the centre of the density, but can even be negative in the tails. Thus, it is rarely used in statistical applications.

5.7.3 SADDLEPOINT APPROXIMATIONS

An alternative series expansion, more closely connected with the exponential family, provides much better approximations. By means of a Fourier inversion of the cumulant generating function, the density of the standardized mean, T_n, for n observations can be written as an integral, with integrand of the form $\exp[n\mathrm{K}(ut)]$, where $\mathrm{K}(u)$ is the cumulant generating function. Then, one can derive a saddlepoint expansion in powers of $1/n$ in a neighbourhood of the saddlepoint, u_0, which will be a zero of $\partial\mathrm{K}(u)/\partial u$, by the method of steepest descent through that point. The result is

$$f_n(t) = \left[\frac{n}{2\pi \frac{\partial^2 \mathrm{K}(u)}{\partial u^2}} \right]^{\frac{1}{2}} \mathrm{e}^{n[\mathrm{K}(ut) - ut]} \left. \left\{ 1 + \frac{1}{n} \left[\frac{\frac{\partial^4 \mathrm{K}(u)}{\partial u^4}}{8 \left[\frac{\partial^2 \mathrm{K}(u)}{\partial u^2} \right]^2} - \frac{5 \left[\frac{\partial^3 \mathrm{K}(u)}{\partial u^3} \right]^2}{24 \left[\frac{\partial^2 \mathrm{K}(u)}{\partial u^2} \right]^3} \right] + \cdots \right\} \right|_{u=u_0} \tag{5.107}$$

with the saddlepoint, u_0, given by

$$\left.\frac{\partial \mathrm{K}(u)}{\partial u}\right|_{u=u_0} = t \tag{5.108}$$

The leading term of this expansion is called the *saddlepoint approximation*, which is $\mathrm{O}(1/n)$ and is always positive (Daniels, 1954).

This approximation may also be obtained in another way, by exponential tilting. We can associate a formal exponential family,

$$f_n(t;\theta) = \mathrm{e}^{t\theta-\kappa(\theta)} f_n(t) \tag{5.109}$$

with the density. To recentre it, we take $\theta = \hat{\theta}$, which turns out to be the saddlepoint given by Equation (5.108). Then, an Edgeworth expansion is used locally at that point, and the result back transformed to $f_n(t)$, yielding Equation (5.107).

Example
For the (k, k) linear exponential family, we take

$$\begin{aligned} f_n(\mathbf{t};\boldsymbol{\theta}) &= \mathrm{R}(\boldsymbol{\theta};\mathbf{y}) f_n(\mathbf{t};\hat{\boldsymbol{\theta}}) \\ &\doteq \frac{\mathrm{R}(\boldsymbol{\theta};\mathbf{y})}{\sqrt{(2\pi)^k |\mathrm{J}(\hat{\boldsymbol{\theta}})|}} \end{aligned} \tag{5.110}$$

using two terms of the Edgeworth expansion, but where the second vanishes and the normal density is evaluated at its mean. Notice the close relationship to the modified profile likelihood of Equation (3.77). □

Because the saddlepoint approximation does not integrate to one, it can be renormalized, which improves the order of approximation to $\mathrm{O}(n^{-3/2})$ for much of the range of t. For the normal, gamma, and inverse Gaussian distributions, the saddlepoint approximation is exact, the latter two only after renormalization. In general, this approximation gives very accurate results, even for small n.

This approximation is also applicable to other statistics than the mean. For the m.l.e., obtained from the score equations, $\mathrm{U}(\hat{\psi}) = 0$, we have

$$f_n(\hat{\psi};\psi_0) = \sqrt{\frac{n}{2\pi}} \frac{\mathrm{E}_{\psi_0}[\mathrm{J}(\hat{\psi})]}{\sqrt{\mathrm{E}_{\psi_0}[\mathrm{U}^2(\hat{\psi})]}} c^{-n}(\psi_0)[1+\mathrm{O}(n^{-1})] \tag{5.111}$$

where

$$c^{-1}(\psi_0) = \int \mathrm{e}^{u(\psi_0)U(\psi_0)} f(y;\psi_0) dy$$

and the saddlepoint, ψ_0, is the solution of

$$\int \mathrm{U}(\hat{\psi})\mathrm{e}^{u(\psi_0)U(\psi_0)} f(y;\psi_0)dy = 0 \tag{5.112}$$

for some function, $u(\psi_0)$. This, in turn, yields a new centred density

$$f_{\psi_0}(y) = c(\psi_0)\mathrm{e}^{u(\psi_0)U(\psi_0)} f(y;\psi_0) \tag{5.113}$$

Examples

1. For a one-parameter linear exponential family

$$f(y;\theta) = \mathrm{e}^{\theta t(y)-\kappa(\theta)-t_0^*(y)}$$

from Equation (3.107), the score equation gives

$$\begin{aligned} \mathrm{E}_{\hat{\theta}}[T(Y)] &= \frac{\partial\kappa(\hat{\theta})}{\partial\theta} \\ &= \mu(\hat{\theta}) \\ &= t(y) \end{aligned} \tag{5.114}$$

The saddlepoint, θ_0, is given by

$$\int [t(y) - \mu(\hat{\theta})]\mathrm{e}^{(\hat{\theta}-\theta_0)[t(y)-\mu(\theta_0)]} f(y;\theta_0)dy = 0 \tag{5.115}$$

with $u(\theta_0) = \hat{\theta} - \theta_0$, which is equivalent to recentring at $\hat{\theta}$:

$$f_{\theta_0}(y;\hat{\theta}) = \mathrm{e}^{\hat{\theta}t(y)-\kappa(\hat{\theta})-t_0^*(y)} \tag{5.116}$$

Then,

$$\begin{aligned} c^{-1}(\theta_0) &= \frac{\mathrm{e}^{\theta_0\mu(\theta_0)-\kappa(\theta_0)}}{\mathrm{e}^{\hat{\theta}\mu(\hat{\theta})-\kappa(\hat{\theta})}} \\ &= \mathrm{R}(\theta_0) \end{aligned} \tag{5.117}$$

$$\begin{aligned} \mathrm{E}_{\theta_0}[\mathrm{J}(\hat{\theta})] &= \frac{\partial^2\kappa(\hat{\theta})}{\partial\theta^2} \\ &= \mathrm{J}(\hat{\theta}) \end{aligned} \tag{5.118}$$

$$\begin{aligned} \mathrm{E}_{\theta_0}[\mathrm{U}^2(\hat{\theta})] &= \frac{\partial^2\kappa(\hat{\theta})}{\partial\theta^2} \\ &= \mathrm{J}(\hat{\theta}) \end{aligned} \tag{5.119}$$

so that the saddlepoint approximation is

$$f_n(\hat{\theta};\theta_0) = \sqrt{\frac{n}{2\pi}}\mathrm{R}(\theta_0)\sqrt{\mathrm{J}(\hat{\theta})} \tag{5.120}$$

but, because this is the canonical parametrization, $\mathrm{J}(\hat{\theta}) = \mathrm{I}(\hat{\theta})$, by Equation (3.110). (Field and Ronchetti, 1990, pp. 84–87)

2. For the (k, k) linear exponential family, we have Equation (5.110),

$$f(\mathbf{t};\boldsymbol{\theta}) \doteq \frac{\mathrm{R}(\boldsymbol{\theta};y)}{\sqrt{(2\pi)^k|\mathrm{J}(\hat{\boldsymbol{\theta}})|}}$$

which can be transformed, with Jacobian $|\mathrm{J}(\hat{\boldsymbol{\theta}})|$, to an approximation for the distribution of $\hat{\boldsymbol{\theta}}$:

$$f(\hat{\boldsymbol{\theta}};\boldsymbol{\theta}) \doteq \sqrt{\frac{|\mathrm{J}(\hat{\boldsymbol{\theta}})|}{(2\pi)^k}}\mathrm{R}(\boldsymbol{\theta};y) \tag{5.121}$$

which is parametrization invariant. Notice that Equation (5.121) is the approximation for the estimate of the canonical parameter, θ, while Equation (5.110) is that for the mean value parameter, μ, as follows from Equation (3.115).

If this formula is applied as an approximation outside the linear exponential family, where $\hat{\boldsymbol{\theta}}$ is not the sufficient statistic, as it may be in Section 6.3.2 below, it can be interpreted as conditioning on ancillary statistics. The usefulness of the approximation will depend on how close the m.l.e. is to being a sufficient statistic. □

If the conditional distribution of one statistic, given another statistic, is required, it can be approximated by finding the saddlepoint approximations to their joint distribution and to that for the conditioning statistic, and dividing the one by the other. This is known as a *double saddlepoint approximation.* We shall use this in Section 6.3.2.

The saddlepoint approximation applied to the distribution function yields

$$F(y;\theta) = F(\hat{\theta};\theta) \tag{5.122}$$

$$\doteq \Phi(z) + \phi(z)\left[\frac{1}{z} + \frac{1}{\theta_0\sqrt{\frac{\partial^2\mathrm{K}(\theta_0)}{\partial\theta^2}}}\right] \tag{5.123}$$

where $\phi(\cdot)$ and $\Phi(\cdot)$ are again the standard normal density and distribution functions, $z = \mathrm{sign}(\theta_0)\sqrt{2[\theta_0 y - \mathrm{K}(\theta_0)]}$, and the saddlepoint, θ_0, is given by $\partial\mathrm{K}(\theta_0)/\partial\theta = y$.

Further reading

Fisher (1915, 1922a) first introduced the mathematical concept of a hypothetical infinite population.

The reader may consult with profit the books by Barndorff-Nielsen and Cox (1989) and Field and Ronchetti (1990) and the review articles by Reid (1988) and Kass (1989) for further details about the techniques of asymptotic approximations which were outlined in the appendix (Section 5.7). Jørgensen (1987b) originally presented the bases of small dispersion asymptotics, while high precision asymptotics date back to Fisher (1922a). For the latter, the reader may also like to look at Heitjan (1989). It should be clear that large sample asymptotic justifications for maximum likelihood estimation are only valid if measurement precision, through Equation (3.4), is taken into account.

Distributions of statistics have primarily been the concern of frequentist statisticians. Thus, further details on the classical results presented in this chapter can be found in standard mathematical statistics texts such as Cox and Hinkley (1974). On the Bayesian side, the paper of Tierney *et al.* (1989) provides fundamental results on approximating distributions.

The use of such distributions as approximate likelihood functions seems to have been little discussed in the literature; see, however, Sprott (1973, 1975a) and Pierce and Peters (1992). The former was the first to point out that the accuracy of inferential statements, based on assumptions of asymptotic normality as the sample size tends to infinity, is not determined by the sample size so much as by the degree to which the parametrization allows the log likelihood to become quadratic; see also Slate (1994).

Tsou and Royall (1995) provide the set of criteria for asymptotic equivalence of approximate likelihood functions which were listed in Section 5.2.4. Generally, those approximations given in this chapter are support interval equivalent.

Buse (1982) provides a clear geometrical exposition of the ways in which the maximum likelihood and score approximations depend on the log likelihood being close to quadratic; see also Meeker and Escobar (1995).

The standard distributions derived from the normal distribution and presented in Appendix C are also discussed in all standard statistical texts. Fuller details can be found in Johnson and Kotz (1970).

5.8 Exercises

(1) (a) Write down the exact likelihood function for the normal distribution taking into account measurement precision.
  (b) Can you calculate the m.l.e. of the parameters?
  (c) Obtain the first few terms of its series expansion in Δ.
  (d) What can be said about high precision asymptotics for this distribution?

(e) Find some way to study the error in the usual approximate normal likelihood, as compared to the exact one.

(2) One sometimes hears that, in the limit for large n, all statistics tend to have normal distributions. Explain why this is not true in general, and specifically for the deviance statistic.

(3) As an example of the change point problem, consider a series of independent random variables, $Y_1, \ldots, Y_\psi, Y_{\psi+1}, \ldots, Y_n$, such that the first ψ have density f_1 and the last $n-\psi$ have density f_2. Both densities are from a (1,1) linear exponential family with canonical statistics y and known canonical parameters θ_1 and θ_2, but ψ is unknown.
   (a) Write down the likelihood function, score, and observed information for ψ.
   (b) What are the expectation and variance of the score?
   (Barndorff-Nielsen and Cox, 1994, p. 108)

(4) For the binomial distribution with $n = 10$,
   (a) compare the exact and asymptotic probabilities that the deviance statistic is greater than 2.706 for the models $\pi = 0.5$ and $\pi = 0.7$;
   (b) repeat for deviances greater than 3.841 and 6.635.
   (Kalbfleisch, 1985b, pp. 108–109)

(5) (a) Consider n i.i.d. observations from a Poisson distribution. Suppose that the parameter of interest is the reciprocal of the mean, $\lambda = 1/\mu$.
      i. What is the m.l.e.?
      ii. What is its variance?
      iii. What is its asymptotic variance?
      (Cox and Hinkley, 1974, p. 295)
   (b) Recall that $\sum y_i$ has a distribution with mean $n\mu$. Suppose that this mean is fixed at 10.
      i. Calculate the probability that the deviance statistic is greater than 2.706.
      ii. Repeat for deviances greater than 3.841 and 6.635.
      iii. Repeat for $n\mu$ fixed at 5, 20, and 40.
      iv. Construct a table to summarize your results.
      v. Compare them to the asymptotic probabilities.
      (Kalbfleisch, 1985b, pp. 109–110)

(6) (a) For the inverse Gaussian distribution of Equation (2.30), derive the p^* formula for the joint distribution of the m.l.e.s.
      (Barndorff-Nielsen and Cox, 1994, pp. 177–178)
   (b) Demonstrate that the p^* formula is exact for the location–scale family.
      (Barndorff-Nielsen, 1983)

(7) Consider n independent observations having a normal distribution with means $\mu_i = \beta_0 + \beta_1 x_i$, where $\sum x_i = 0$, and common variance, σ^2. This is the simple linear regression model for which we studied the residuals in Section 4.3.2.

(a) Write down the expected information matrix.

(b) For the statistic consisting of the sum of residuals cubed,

$$T = \frac{\sum(Y_i - \beta_0 - \beta_1 x_i)^3}{\sqrt{n}\sigma^3} \tag{5.124}$$

derive the asymptotic distribution.

(c) What role could this statistic play in goodness of fit?

(Barndorff-Nielsen and Cox, 1994, pp. 105–106)

(8) For the log normal distribution of Equation (1.6), the p quantile, $\lambda_p = \exp(\mu + z_p\sigma)$, is of interest, where z_p is the standard normal quantile.

(a) Write down the normed likelihood function.

(b) What is the m.l.e. of σ^2 for fixed λ_p?

(c) Write down the normed profile likelihood function for λ_p.

(d) For $\hat{\mu} = 1$, $\hat{\sigma}^2 = 4$, $n = 10$, plot the normed likelihood of $\lambda_{0.1}$ and σ, i.e. with $p = 0.1$.

(e) Derive the Wald statistics for λ_p and $\log(\lambda_p)$.

(f) Plot the approximate normed likelihoods for these two parametrizations for $\lambda_{0.1}$ on the same graph as the exact normed profile likelihood.

(g) Discuss their relative merits.

(Meeker and Escobar, 1995)

(9) Under the appropriate conditions, the deviance statistic has an asymptotic Chi-squared distribution, from Equation (5.67).

(a) Find a generalized linear model for which the deviance statistic does not have this distribution because the degrees of freedom increase with n.

(b) Find another example, similar to Equation (4.7), where the deviance statistic is not very useful.

(c) Find still another example, this time similar to Equation (5.75), where the deviance compares the model of interest to a saturated model, but nevertheless has an asymptotic Chi-squared distribution.

(10) As we saw in Section 4.1.1, the deviance can provide a measure of goodness of fit in certain circumstances. A small deviance indicates that the model is close to the data. We have now seen that the deviance statistic can have an asymptotic Chi-squared distribution.

From Figure C.1, we see that, according to this, small deviances are more probable only for small degrees of freedom. In other words, as the degrees of freedom become larger, deviances for models close to the data become rare.

(a) Explain why.

(b) Is this a useful characteristic of a tool commonly used for calibrating the deviance?

(11) Let $Y_1, \ldots, Y_n$ be independent random variables following a Poisson distribution with means $\lambda\phi_i$, the ϕ_i being known numbers.

(a) Suppose that $\sum_{i=1}^{\infty} \phi_i < \infty$. Show that λ cannot be estimated consistently.

(b) Now, let $\sum_{i=1}^{\infty} \phi_i = \infty$. Show that the m.l.e. of λ is consistent.

(Lehmann, 1983, pp. 447–448)

(12) (a) Calculate the score (Pearson Chi-squared goodness of fit) statistic for n Bernoulli observations, y_i, with probabilities π_i, where the fixed model of interest is $\pi_i = \pi$, a constant.

(b) Can this statistic be used in an asymptotic test for goodness of fit?

(13) In normal simple linear regression, the skew statistic is defined by

$$T = \sum [(Y_i - \boldsymbol{\beta}^{\mathrm{T}} \mathbf{x}_i)/\sigma]^3/n \qquad (5.125)$$

(a) What is the asymptotic distribution of $\sqrt{n}T$?

(b) What is the asymptotic distribution of $\sqrt{n}\hat{T}$, when the m.l.e. of the parameters is substituted into Equation (5.125)?

(c) How can this statistic be used for goodness of fit? (See also Exercise (4.8).)

(Pierce, 1982)

(14) The following 10 i.i.d. values of a random variable, Y, were observed: -5.0, 1.1, 1.5, 2.0, 4.9, 5.0, 3.6, 3.9, 4.5, 3.1. We are interested in some measure of location for these data.

(a) Assuming normality, derive a test that the mean is zero and give the resulting P-value.

(b) Try some other models, perhaps based on other distributions, testing for the location to be zero.

(c) If the above values were, in fact, obtained by drawing numbers, x_i, randomly from the interval $(-5, 5)$ and adding ψ to all of them, what is the distribution of $y_i = x_i + \psi$?

(d) What do we now know about the location parameter, ψ?

(Sprott, 1982)

(15) Pairs of independent observations have Poisson distributions with means μ_1 and μ_2, where the parameter of interest is their ratio, $\lambda =$

μ_1/μ_2. Show that the profile likelihood for λ is globally pointwise and information equivalent to the likelihood for λ with fixed μ_2, but not maximum likelihood, support interval, or locally pointwise equivalent. (Tsou and Royall, 1995)

(16) (a) Show that the approximate normal likelihood, from Equation (3.50),

$$\mathrm{L}_A(\mu) = \exp\left[-\frac{n(\bar{y}_\bullet - \mu)^2}{2s^2}\right] \tag{5.126}$$

is asymptotically support interval equivalent to any model which has $\hat{\mu} = \bar{y}_\bullet$ and $s^2 \xrightarrow{p} \mathrm{var}[Y]$, and hence is robust. (These include Poisson, gamma, normal, and negative binomial.)

(b) In contrast, show that the Poisson likelihood, which is only reasonable when $\mathrm{var}[Y] = \mu$, is not support interval equivalent to these other models.

(c) However, show that the approximate normal likelihood is not globally pointwise equivalent to a Poisson model.

(Tsou and Royall, 1995)

(17) Use Stirling's approximation to the gamma function in Equation (5.102) to obtain an approximate asymptotic density for the Poisson distribution.

(18) Derive the parameter transformation of the curved exponential family in Equation (2.55) using a saddlepoint approximation.

(19) Consider a compound distribution, where some parameter in a distribution is, itself, given a distribution, as we did in Section 2.6. Suppose that the latter is a normal distribution. Then, the marginal distribution of the observations, upon which a likelihood function can be based, is a random effects model (Section 6.4.1),

$$f_Y(y;\psi,\mu,\sigma) = \int_{-\infty}^{\infty} f_{Y|\theta}(y|\phi;\psi)p(\theta;\mu,\sigma)d\theta \tag{5.127}$$

where $p(\cdot)$ is a normal density with mean μ and variance σ^2.

(a) If $f_{Y|\theta}(\cdot)$ is a member of the linear exponential family, use a Laplace approximation to the integral to obtain a tractable approximate likelihood function.

(b) Take a set of m 2×2 contingency tables, with entries $y_{ijk}, i = 1,\ldots,m; j,k = 1,2$, and probabilities π_{i1} and π_{i2}, so that the log odds ratio of table i is

$$\theta_i = \log\left[\frac{\pi_{i1}(1-\pi_{i2})}{(1-\pi_{i1})\pi_{i2}}\right] \tag{5.128}$$

The conditional distribution for each table, given $y_{ij\bullet}$ and $y_{i\bullet k}$, is that of Equation (4.22). Suppose that this odds ratio has a normal distribution over the set of tables. Derive a suitable approximation to the marginal distribution of Y.

(c) Suppose, instead, that we have m pairs of counts, Y_{i1}, Y_{i2}, following a Poisson distribution.
   i. Show that the conditional distribution for each pair, given $y_{i\bullet}$, is binomial.
   ii. Derive a suitable approximation to the marginal distribution of Y.

(Liu and Pierce, 1993)

6
Factoring the likelihood function

The likelihood function provides an overall assessment of the relative merits of different members of a given family of statistical models, although this must be balanced against their relative complexity. However, as we saw in Section 3.6.3, we often require measures of precision of the estimates of individual parameters in the model. This is most often the case in a decision context (Part IV), where, for example, one might wish to decide which of two or more treatments provides the most positive effect, while controlling for various other influential factors. As in Section 3.6.3, the techniques to be discussed in this chapter should not be used for model selection, such as determining if a parameter could be set to zero. They are appropriate for interval estimation of individual parameters once an appropriate model has been selected.

When the parameter vector is of low dimension, the precision of a parameter can be determined by inspecting a plot of the (profile) likelihood function. However, when many parameters are present in a statistical model, the profile likelihood will give a poor measure of the uncertainty in any one parameter component, unless it is orthogonal to the others, because it effectively assumes that the other parameter components are known to be equal to their maximum likelihood estimates (for fixed value of the parameter of interest). Thus, in Section 3.6.3, we introduced a modified profile likelihood to meet this problem, with a further application in Section 4.2.2. In this chapter, we shall explore in more detail the validity of such approximations.

In many statistical problems, the parameter space can be partitioned according to two types of parameters, $\Psi = (\Lambda^{\mathrm{T}}, \Phi^{\mathrm{T}})^{\mathrm{T}}$, where Λ is the parameter vector of special *interest* and Φ a *nuisance* parameter. If the dimension of the nuisance parameter increases as a function of the sample size, let us call the latter an *incidental parameter.* As we shall see in some of the examples of this chapter, such parameters will usually correspond to aspects of the problem, such as explanatory variables describing varying conditions, which are not of interest. In the terms of model building described in Section 1.3, a model with incidental parameters is ill-specified. Such conditions should rather be made stochastic in the model.

If the likelihood function factors into two orthogonal parts, one a function only of $\boldsymbol{\lambda}$ and the other only of $\boldsymbol{\phi}$, then the problem is solved and inferences can be made about the parameter of interest independently of the values of the nuisance parameter. The goal of this chapter is to investigate under what conditions this can be done both exactly and approximately.

In Section 3.6.3, we saw how uncertainty about the values of nuisance parameters can, at least approximately, be taken into account by allowing for the width of the likelihood function in the direction of those parameters, creating a modified profile likelihood, and that, in the linear exponential family, it can even be done exactly. In Section 3.4.2, we saw that an ancillary statistic can determine the precision of inference about a parameter. Thus, for fixed values of the nuisance parameters, as in the construction of a profile likelihood, conditioning on ancillaries should make allowance for this lack of precision. Exact results are usually only available for certain models in the linear exponential and transformation families. In the examples of this chapter, we shall concentrate on those families, because other cases are often rather complex. Many results, however, may apply more widely, at least as reasonable approximations.

In general, as we saw in Section 3.6, one must face up to the fact that individual parameters in a parametric statistical model for a probabilistic data generating mechanism have little or no meaning if the likelihood function does not factor. In such cases, the parameters are so interlinked that they must be studied simultaneously. This chapter explores how closely one can come to unentwining them.

6.1 Parametrization orthogonality

Let us now begin our study of various ways in which nuisance and interest parameters can be separated within a likelihood function. We already saw, in Section 3.6.1, some of the complications which can result when parameters are not orthogonal. There, we also saw that the off-diagonal elements of the information matrix provide one approximate indication of this. Here, we must study in more detail the possible degrees to which parameters are inter-related in a likelihood function.

In the context of this chapter, parametrization invariance must have an additional characteristic to that described in Section 3.6.2. The separation between interest and nuisance parameters must be respected:

$$(\boldsymbol{\lambda}, \boldsymbol{\phi}) \rightarrow (\boldsymbol{\lambda}^*, \boldsymbol{\phi}^*)$$

such that

$$\begin{aligned}\boldsymbol{\lambda}^* &= \boldsymbol{\lambda}^*(\boldsymbol{\lambda})\\ \boldsymbol{\phi}^* &= \boldsymbol{\phi}^*(\boldsymbol{\lambda}, \boldsymbol{\phi})\end{aligned}$$

so that the new interest parameter depends only on the old one. Of course, it is preferable, and simpler, if $\boldsymbol{\phi}^* = \boldsymbol{\phi}^*(\boldsymbol{\phi})$ as well.

6.1.1 ESTIMATION ORTHOGONALITY

The simplest case of parameter orthogonality occurs if the m.l.e. of $\boldsymbol{\lambda}$ for $\boldsymbol{\phi}$ fixed does not depend on $\boldsymbol{\phi}$, so that $\hat{\boldsymbol{\lambda}}(\boldsymbol{\phi}) = \hat{\boldsymbol{\lambda}}_{\phi} = \hat{\boldsymbol{\lambda}}$ for all values of $\boldsymbol{\phi}$ and whatever observations are made. Then, those parameter components are *estimation orthogonal*. Obviously, this orthogonality is not symmetric: $\boldsymbol{\lambda}$ may be estimation orthogonal to $\boldsymbol{\phi}$, while $\boldsymbol{\phi}$ is not orthogonal to $\boldsymbol{\lambda}$. Nor, more importantly, does it ensure orthogonality of precision; interval estimates, of the interest parameter, will still generally depend on the nuisance parameters.

Examples

1. In the exponential family, the mean parameter of the mixed parameter representation is estimation orthogonal to the canonical parameter. In the most important case, a point estimate of a location parameter, which is of interest, can be obtained which does not depend on an incidental canonical scale or shape parameter.

2. For the normal distribution, $\hat{\mu} = \sum y_i/n$, which does not depend on σ^2. On the other hand, the m.l.e. of σ^2 does depend on μ. However, in Section 5.3.1, we saw that the likelihood function for the normal distribution can be factored into a part for $\hat{\mu}$, but containing both μ and σ^2, and a part for $\hat{\sigma}^2$ containing only σ^2, so that the precision of $\hat{\mu}$ will depend on σ^2. □

This definition is of limited use because it only refers to the point estimate. Measures of precision of individual parameters will involve more than just such orthogonality.

6.1.2 DESIGN ORTHOGONALITY

In an experimental situation, the values of the controlled explanatory variables are assigned, by the research worker, to the units under study. As we have seen, these define the design space. In survey sampling, some control can be had by stratifying on relevant explanatory variables. Thus, judicious choice of these values can provide improved separation among the parameters involved in the likelihood function, especially for linear regression models based on the normal distribution. In this way, factoring a likelihood function can, to some extent, be under the control of the research worker, before making the observations. In non-linear models, the effective design matrix of Equation (4.34) can provide similar information, but after the fact because the parameters are involved.

Consider a set of n independent random variables, each generated by a probability model from the same family, but with a different location parameter, η_i. Suppose that these location parameters are defined as a *linear* function of $p \leq n$ other regression parameters, β_j, such that $\eta_i =$

$\sum_j \beta_j x_{ij}$, where the x_{ij} are known constants, the explanatory variables fixed by the experiment. If the columns of the design matrix, $\mathbf{X} = (x_{ij})$, are orthogonal, the regression parameters, β_j, will be said to be *design orthogonal.*

Example
Take the univariate exponential dispersion family of Equation (2.58) and let $\xi_i = \beta_0 + \beta_1 x_i$, a simple canonical generalized linear model. Then, if the $n \times 2$ design matrix, $\mathbf{X}$, has a first column of ones, the second column, containing x_i, must sum to zero for β_0 and β_1 to be design orthogonal. If the second column contains equal numbers of -1 and 1 (n even), we have a balanced generalized one-way 'analysis of variance' model. □

More generally, some suitable arbitrary function of the mean, the link function, $g(\cdot)$, may be related to such a linear structure:

$$g(\mu_i) = \sum_j \beta_j x_{ij} \tag{6.1}$$

When $g(\mu) = \xi(\mu)$ in the exponential dispersion family of Equation (2.58), we have canonical linear regression with a canonical link function.

Example
For observations following a binomial distribution, if the link function is a logit, $\log[\pi/(1-\pi)]$, we have a canonical logistic regression model; if it is a standard normal probability integral, we have a *probit* model. In either case, the same design matrix as in the previous example will still be design orthogonal. □

Design orthogonality applies to all regression models, such as these, with a linear structure relating the known explanatory variables to the mean through a link function. Its existence obviously does not depend on what link function is used. However, its usefulness does, as we shall see in the next subsection. Nevertheless, in most cases, design orthogonality should reduce the strength of the relationships among parameters. Experimental designs yielding such orthogonality are especially important for linear normal models with an identity link, where they provide the greatest separation among parameters.

6.1.3 INFORMATION ORTHOGONALITY

In Section 3.5.2, we have seen how the observed information matrix provides an approximate measure of precision of estimates, i.e. of the width of the likelihood function. In addition, the off-diagonal elements show, again approximately, how far a multidimensional likelihood surface is from being oriented parallel to the axes. When it is not, in which case these elements will be non-zero, inference about one parameter will depend on the values of the other parameters. Because this is only a second-order approximation, the reverse will not generally be true: zero elements do not necessarily mean that parameters are unrelated (except perhaps asymptotically).

Two parameter components, $\boldsymbol{\lambda}$ and $\boldsymbol{\phi}$, are said to be *expected information orthogonal* if the corresponding off-diagonal elements of the expected information matrix are zero:

$$\begin{aligned}\mathrm{I}_{\lambda_i\phi_j}(\boldsymbol{\psi};\mathbf{y}) &= \mathrm{E}\left[-\frac{\partial^2 \log[\mathrm{L}(\boldsymbol{\psi};\mathbf{y})]}{\partial\lambda_i\partial\phi_j}\right] \\ &= \mathrm{E}\left[\frac{\partial \log[\mathrm{L}(\boldsymbol{\psi};\mathbf{y})]}{\partial\lambda_i}\frac{\partial \log[\mathrm{L}(\boldsymbol{\psi};\mathbf{y})]}{\partial\phi_j}\right] \\ &= 0\end{aligned} \tag{6.2}$$

for all i and j. This is called *global* (expected) *information orthogonality* if it holds for all $\boldsymbol{\psi} \in \Psi$. On the other hand, if it only holds at some specific fixed value, $\boldsymbol{\psi}_0$, the parameter components, $\boldsymbol{\lambda}$ and $\boldsymbol{\phi}$, are said to be *locally* information orthogonal at $\boldsymbol{\psi}_0$. If the information matrix is diagonal, the parameters are *totally* information orthogonal.

Information orthogonality, defined in this way, is of limited usefulness, except asymptotically, because it does not necessarily mean that we have orthogonality for given observed data (except for canonical parameters in the exponential family). Thus, such orthogonality should rather be defined with respect to the observed information matrix. Nevertheless, the term, information orthogonal, generally refers to the expected information, and we shall use it that way.

Because the expected information matrix is the variance of the score statistic, if parameter components are information orthogonal, the corresponding components of the score are uncorrelated. If $\boldsymbol{\lambda}$ and $\boldsymbol{\phi}$ are information orthogonal, the m.l.e. of $\boldsymbol{\lambda}$ when $\boldsymbol{\phi}$ is given, $\hat{\boldsymbol{\lambda}}_{\phi}$, will usually vary only slowly with $\boldsymbol{\phi}$ and vice versa (Cox and Reid, 1987). As we saw in Section 5.3.3, the m.l.e.s will be asymptotically independent. If parameters are estimation orthogonal, they are information orthogonal. The reverse is not necessarily true.

Examples

1. For n i.i.d. observations from a normal distribution,

$$\begin{aligned}\mathrm{J}_{\mu,\sigma^2} &= \frac{\sum_i(y_i-\mu)}{\sigma^4} \\ \mathrm{I}_{\mu,\sigma^2} &= \mathrm{E}[\mathrm{J}_{\mu,\sigma^2}] \\ &= 0\end{aligned}$$

so that μ and σ^2 are information orthogonal. Because this is a mixed representation, μ is also estimation orthogonal, but σ^2 is not.
2. Consider n independent observations from a normal distribution with canonical regression

$$\mu_i = \beta_0 + \beta_1 x_i \tag{6.3}$$

for known x_i. Then

$$\begin{aligned} \mathrm{J}_{\beta_0,\beta_1} &= \frac{\sum_i x_i}{\sigma^2} \\ &= \mathrm{I}_{\beta_0,\beta_1} \end{aligned} \tag{6.4}$$

If β_0 and β_1 are design orthogonal, they are information orthogonal.
3. Now take n independent observations from a Bernoulli distribution with the canonical logistic regression of Equation (4.16),

$$\begin{aligned} \theta_i &= \log\left(\frac{\pi_i}{1-\pi_i}\right) \\ &= \beta_0 + \beta_1 x_i \end{aligned} \tag{6.5}$$

for known x_i, the explanatory variables. We have

$$\begin{aligned} \mathrm{J}_{\beta_0,\beta_1} &= \sum_i \frac{x_i \mathrm{e}^{\beta_0+\beta_1 x_i}}{(1+\mathrm{e}^{\beta_0+\beta_1 x_i})^2} \\ &= \mathrm{I}_{\beta_0,\beta_1} \end{aligned} \tag{6.6}$$

Here, design orthogonality does not imply information orthogonality.
4. More generally, for any dispersion model, from Equation (2.56), the score is

$$\mathrm{U}_\mu = \frac{1}{\sigma^2}\frac{\partial t(y,\mu)}{\partial \mu} \tag{6.7}$$

$$\mathrm{U}_{\sigma^2} = -\frac{\partial \kappa(\sigma^2,y)}{\partial \sigma^2} - \frac{t(y,\mu)}{\sigma^4} \tag{6.8}$$

so that the relevant observed information is

$$\mathrm{J}_{\mu,\sigma^2} = \frac{1}{\sigma^4}\frac{\partial t(y,\mu)}{\partial \mu} \tag{6.9}$$

Because $\mathrm{E}(\mathrm{U}_\mu) = 0$, we also have $\mathrm{I}_{\mu,\sigma^2} = 0$, so that the location and dispersion parameters are information orthogonal. Because Equation (6.7) factors, we can also see that the location parameter is estimation orthogonal.□

The contrast between Examples 2 and 3 demonstrates why design orthogonality is so important for models based on the normal distribution, but of more limited usefulness in other contexts.

Transformation to information orthogonality When the parameter of interest is one dimensional, it is always possible to find transformations of the nuisance parameters such that it is information orthogonal to them. The corresponding elements of the information matrix are set to zero and the resulting differential equations solved:

$$\frac{\partial \boldsymbol{\phi}^{\mathrm{T}}}{\partial \lambda}\mathbf{I}_{\phi\phi} + \mathbf{I}_{\lambda\phi} = 0 \tag{6.10}$$

Example
Consider independent random variables, Y_1 and Y_2, from the exponential distribution of Equation (1.14) with means μ_1 and $\mu_2 = \lambda\mu_1$, where the ratio of means is the parameter of interest and $\phi = \mu_1$ is a nuisance parameter. We have

$$\mathrm{I}_{\lambda\phi} = \frac{1}{\lambda\phi} \tag{6.11}$$

so that the parameters are not information orthogonal. Suppose that the orthogonal nuisance parameter is ϕ^* so that $\phi = \phi(\lambda, \phi^*)$. Then, we require

$$\begin{aligned}
\mathrm{I}_{\lambda\phi^*} &= \left(\frac{\partial \phi}{\partial \lambda}\mathrm{I}_{\phi\phi} + \mathrm{I}_{\lambda\phi}\right)\frac{\partial \phi}{\partial \phi^*} \\
&= \left(\frac{2}{\phi^2}\frac{\partial \phi}{\partial \lambda} + \frac{1}{\lambda\phi}\right)\frac{\partial \phi}{\partial \phi^*} \\
&= 0
\end{aligned} \tag{6.12}$$

A possible solution is $\phi^* = \phi\sqrt{\lambda}$, which is information orthogonal to λ. □

6.1.4 LIKELIHOOD ORTHOGONALITY

The final, and strongest, condition is that the likelihood function factors into two distinct parts:

$$\mathrm{L}(\boldsymbol{\psi}; \mathbf{y}) = \mathrm{L}_1(\boldsymbol{\lambda}; \mathbf{t}_1)\mathrm{L}_2(\boldsymbol{\phi}; \mathbf{t}_2) \tag{6.13}$$

with one containing only $\boldsymbol{\lambda}$ and the other only $\boldsymbol{\phi}$. If, in addition, the two parameters are variation independent, then the two parameter vectors are said to be likelihood independent or *likelihood orthogonal*. This is the ideal, but rare, case where we can truly make inferences about the parameter of interest independently of what value the nuisance parameter might have.

In the linear exponential family, parameter components, $\boldsymbol{\theta}_1$ and $\boldsymbol{\theta}_2$, are likelihood orthogonal iff the cumulant function is a sum

$$\kappa(\boldsymbol{\theta}) = \kappa_1(\boldsymbol{\theta}_1) + \kappa_2(\boldsymbol{\theta}_2) \tag{6.14}$$

In the mixed representation, μ_1 and θ_2 are likelihood orthogonal if

$$\theta_1 = h_1(\mu_1) + h_2(\theta_2) \tag{6.15}$$

for some functions, $h_i(\cdot)$ (Barndorff-Nielsen, 1978, p. 148).

Example
In the normal simple linear regression of Equation (6.3), with design orthogonality and known variance, the likelihood function factors to allow separate inferences for β_0 and β_1. □

Other examples will be given in Sections 6.2.3 and 6.2.4 below.

As already stated above, if we have likelihood orthogonality, the problem of inference separately about the parameters of interest is solved. We need now to investigate under what conditions this can occur (unfortunately, very rarely), and how closely we can come to it in other cases.

In studying a parameter of interest in the presence of nuisance parameters, we have two goals which may be mutually conflicting: use the maximum available information about the interest parameter while allowing for the uncertainty about the nuisance parameters. In Section 3.6.3, we emphasized the latter aspect; here, we shall be more concerned with the former.

6.2 Partial sufficiency

In Section 2.2.3, we saw how, in the linear exponential family, the conditional distribution of one sufficient statistic, given another, only contains the parameters corresponding to the former statistic. We also saw, in Section 3.4.2, how ancillary statistics can contain information about the precision of a parameter. We shall now pursue these ideas further.

6.2.1 SUFFICIENCY AND ANCILLARITY

In order to be able to develop some appropriate new concepts, let us here call the sufficiency and ancillarity of Section 3.4 *full*: a statistic, $\mathbf{T}$, is F-sufficient for $\boldsymbol{\psi}$ if, from Equation (3.31),

$$\mathrm{L}(\boldsymbol{\psi};\mathbf{y}) = \mathrm{L}_1(\boldsymbol{\psi};\mathbf{t})\mathrm{L}_2(\mathbf{y}|\mathbf{t})$$

and it is F-ancillary if, from Equation (3.34),

$$\mathrm{L}(\boldsymbol{\psi};\mathbf{y}) = \mathrm{L}_1(\boldsymbol{\psi};\mathbf{y}|\mathbf{t})\mathrm{L}_2(\mathbf{t})$$

Suppose, now, that the likelihood function for $\boldsymbol{\psi} = (\boldsymbol{\lambda}^{\mathrm{T}}, \boldsymbol{\phi}^{\mathrm{T}})^{\mathrm{T}}$ factors such that

$$\mathrm{L}(\boldsymbol{\psi}; \mathbf{y}) = \mathrm{L}_1(\boldsymbol{\lambda}; \mathbf{t}_1)\mathrm{L}_2(\boldsymbol{\psi}; \mathbf{t}_2) \tag{6.16}$$

where $\mathbf{t}_1$ and $\mathbf{t}_2$ are observed values of the statistics $\mathbf{T}_1$ and $\mathbf{T}_2$, but where the factor L_1 may depend on $\boldsymbol{\phi}$ through $\boldsymbol{\lambda}$, if they are not variation independent. We would like to be able to make inferences about the parameter of interest, $\boldsymbol{\lambda}$, based on L_1 alone, there being little or no information about it in L_2.

Example
Suppose that two groups are selected and Bernoulli responses are recorded in each, with probability π_i of success for group i. The results may be represented in a 2×2 contingency table:

Group A	n_{11}	n_{12}	$n_{1\bullet}$
Group B	n_{21}	n_{22}	$n_{2\bullet}$
Total	$n_{\bullet 1}$	$n_{\bullet 2}$	$n_{\bullet\bullet}$

where the $n_{i\bullet}$ are fixed by the sample design. Interest centres on the odds ratio, so let us set

$$\lambda = \frac{\pi_1(1-\pi_2)}{\pi_2(1-\pi_1)} \tag{6.17}$$

where $\log(\lambda) = \beta_1$ in Equation (4.16). One possibly appropriate nuisance parameter is defined by

$$\phi = \frac{\pi_2}{1-\pi_2} \tag{6.18}$$

where $\log(\phi) = \beta_0$ in that same equation. Then, the likelihood function is

$$\begin{aligned} \mathrm{L}(\lambda, \phi; \mathbf{n}) &= \frac{(\phi\lambda)^{n_{11}}\phi^{n_{21}}}{(1+\phi\lambda)^{n_{1\bullet}}(1+\phi)^{n_{2\bullet}}} \\ &= \frac{\lambda^{n_{11}}\phi^{n_{\bullet 1}}}{(1+\phi\lambda)^{n_{1\bullet}}(1+\phi)^{n_{2\bullet}}} \end{aligned} \tag{6.19}$$

with sufficient statistics n_{11} for λ and $n_{\bullet 1}$ for ϕ. In Equation (4.22), we factored this by means of a conditional distribution involving only λ.

We may ask what information about λ is contained in the second factor, that involving $n_{\bullet 1}$, and the two fixed marginal totals. In Section 4.2.2, we already noticed that the likelihood function based on this latter factor alone is not necessarily flat in λ, and hence must contain some information about that parameter. □

A statistic, $\mathbf{T}$, for which

$$L(\boldsymbol{\psi}; \mathbf{y}) = L_1(\boldsymbol{\lambda}; \mathbf{t}) L_2(\boldsymbol{\psi}; \mathbf{y}|\mathbf{t}) \tag{6.20}$$

is *sufficient* (in some way) with respect to $\boldsymbol{\lambda}$ if L_2 is non-informative with respect to $\boldsymbol{\lambda}$. In the same way, a statistic, $\mathbf{T}$, for which

$$L(\boldsymbol{\psi}; \mathbf{y}) = L_1(\boldsymbol{\psi}; \mathbf{t}) L_2(\boldsymbol{\lambda}; \mathbf{y}|\mathbf{t}) \tag{6.21}$$

is *ancillary* (in some way) with respect to $\boldsymbol{\lambda}$ if L_1 is non-informative with respect to $\boldsymbol{\lambda}$ (see Barndorff-Nielsen, 1978, pp. 46–49, and Zhu and Reid, 1994). We know (Section 3.2.3) that a flat likelihood function indicates non-information; thus, we would like these functions to be as flat as possible in $\boldsymbol{\lambda}$.

Example
Such conditions will generally hold in the exponential family if $\boldsymbol{\lambda}$ is a canonical parameter and $\boldsymbol{\phi}$ a mean value parameter, with the corresponding sufficient statistics, $\mathbf{T}_1$ and $\mathbf{T}_2$, being respectively sufficient and ancillary for $\boldsymbol{\lambda}$.
□

Thus, ideally, we seek to have $L_2(\boldsymbol{\psi}; \mathbf{y}|\mathbf{t})$ in Equation (6.20) or $L_1(\boldsymbol{\psi}; \mathbf{t})$ in Equation (6.21) containing as little information as possible on $\boldsymbol{\lambda}$. However, these definitions do not tell us how to derive such ancillary and sufficient statistics.

6.2.2 L-SUFFICIENCY

A statistic $\mathbf{T}_1$ is said to be *L-sufficient* (L for likelihood) for $\boldsymbol{\lambda}$ if

(1) the set (or, more rigorously, σ-algebra) of events generated by the normed profile likelihood for $\boldsymbol{\lambda}$,

$$R_P(\boldsymbol{\lambda}) = \frac{L(\boldsymbol{\lambda}, \hat{\boldsymbol{\phi}}_\lambda)}{L(\hat{\boldsymbol{\lambda}}, \hat{\boldsymbol{\phi}})} \tag{6.22}$$

is contained in that generated by $\mathbf{T}_1$, and

(2) the marginal distribution of $\mathbf{T}_1$ depends on $\boldsymbol{\phi}$ only through $\boldsymbol{\lambda}$.

If the two sets are equal, $\mathbf{T}_1$ is minimal L-sufficient. If $\mathbf{T}_1$ is L-sufficient, the set of events generated by the m.l.e., $\hat{\boldsymbol{\lambda}}$, is contained in that generated by $\mathbf{T}_1$ (Rémon, 1984).

Examples
1. The normed profile likelihood for the mean of a sample of n i.i.d. normally distributed random variables can be written

$$\mathrm{R}_P(\mu) = \left(\frac{nt_2^2}{nt_2^2 + (t_1 - \mu)^2} \right)^{\frac{n}{2}} \tag{6.23}$$

where $t_1 = \sum y_i/n$ and $t_2 = \sum(y_i - \bar{y}_\bullet)^2/n$, so that the minimal L-sufficient statistic for μ is (t_1, t_2).

2. The normed profile likelihood for the variance of a sample of n i.i.d. normally distributed random variables can be written

$$\mathrm{R}_P(\sigma^2) = \exp\left(\frac{n}{2} - \frac{t_2}{\sigma^2} \right) \tag{6.24}$$

so that the minimal L-sufficient statistic for σ^2 is t_2. □

The results in these examples correspond to our factoring of the normal likelihood function in Section 5.3.1.

6.2.3 CUTS

If the marginal distribution, f_T, of the statistic $\mathbf{T}$, combined with the conditional distribution, $f_{Y|t}$, yields a probability distribution in $\mathcal{P} = \mathcal{P}_T \times \mathcal{P}_{Y|t}$, then $\mathbf{T}$ is called a *cut*. If $\mathbf{T}$ is neither F-sufficient nor F-ancillary, it is called a *proper cut*. For a suitable parametrization of $\mathcal{P}$, with $\mathbf{T}$ a cut, we have $\boldsymbol{\psi} = (\boldsymbol{\lambda}^{\mathrm{T}}, \boldsymbol{\phi}^{\mathrm{T}})^{\mathrm{T}}$, where $\boldsymbol{\lambda}$ and $\boldsymbol{\phi}$ are likelihood orthogonal (and variation-independent) components of $\boldsymbol{\psi}$, so that Equation (6.13) holds, in the form

$$\mathrm{L}(\boldsymbol{\psi}; \mathbf{y}) = \mathrm{L}_1(\boldsymbol{\lambda}; \mathbf{t})\mathrm{L}_2(\boldsymbol{\phi}; \mathbf{y}|\mathbf{t}) \tag{6.25}$$

This is a stronger condition than likelihood orthogonality alone.

Example

Consider the $p+1$ category multinomial distribution, for n events,

$$f(y_1, \ldots, y_p; \pi_1, \ldots, \pi_p) = \frac{n!}{y_1! \ldots y_p!(n - y_\bullet)!} \pi_1^{y_1} \ldots \pi_p^{y_p} (1 - \pi_1 - \cdots - \pi_p)^{n - y_\bullet} \tag{6.26}$$

Let us reparametrize it as

$$\lambda_1 = \pi_1$$

$$\lambda_2 = \frac{\pi_2}{1 - \pi_1}$$

$$\vdots$$

$$\lambda_p = \frac{\pi_p}{1 - \pi_1 - \cdots - \pi_{p-1}}$$

The distribution may, then, be rewritten as a product of binomial distributions with parameters λ_i and binomial totals $(n, n-y_1, \ldots, n-y_1-\cdots-y_p)$. The sufficient statistic is a proper cut among all of the new parameters. (Barndorff-Nielsen, 1978, pp. 26–29) □

For the exponential family with the mixed parameter representation (μ_1, θ_2), and sufficient statistics (t_1, t_2), t_1 is a proper cut iff θ_1 has the form

$$\theta_1 = h_1(\mu_1) + h_2(\theta_2) \tag{6.27}$$

for some functions, $h_i(\cdot)$ (Barndorff-Nielsen, 1978, pp. 196–202).

Example
For the normal distribution, with mixed representation (μ, σ^2),

$$\begin{aligned}\theta_1 &= \frac{\mu}{\sigma^2} \\ &= -2\mu\theta_2\end{aligned}$$

which cannot be written in the required form, so that $t_1 = \sum y_i$ is not a proper cut. □

Note that, if T_1 is a cut, then the marginal distribution, given by Equation (2.35), is in the exponential family.

6.2.4 S-SUFFICIENCY AND S-ANCILLARITY

If the statistic $\mathbf{T}$ is a proper cut for the parameters $\boldsymbol{\psi} = (\boldsymbol{\lambda}^{\mathrm{T}}, \boldsymbol{\phi}^{\mathrm{T}})^{\mathrm{T}}$, as in Equation (6.25), then $\mathbf{T}$ is *S-sufficient* for $\boldsymbol{\lambda}$ and *S-ancillary* for $\boldsymbol{\phi}$ (S for Sandved, 1967). A statistic which is S-sufficient is L-sufficient.

Example
Take n independent observations from a Poisson distribution in a canonical linear regression model, where $\theta_i = \log(\mu_i) = \beta_0 + \beta_1 x_i$, often called a log linear model. The probability distribution is

$$f(\mathbf{y}; \beta_0, \beta_1) = \frac{\mathrm{e}^{-\Sigma \mathrm{e}^{\beta_0 + \beta_1 x_i}} \mathrm{e}^{\beta_0 y_\bullet + \beta_1 \Sigma x_i y_i}}{y_1! \ldots y_n!} \tag{6.28}$$

The sufficient statistic for β_0, $T = \sum Y_i$, follows a (marginal) Poisson distribution with mean parameter $\mu_0 = \mathrm{e}^{\beta_0} \sum \mathrm{e}^{\beta_1 x_i}$. Then, the conditional distribution, given this statistic, is

$$f(\mathbf{y}|t; \beta_1) = \frac{y_\bullet!}{y_1! \ldots y_n!} \prod_i \left(\frac{\mathrm{e}^{\beta_1 x_i}}{\sum \mathrm{e}^{\beta_1 x_i}} \right)^{y_i} \tag{6.29}$$

which is a multinomial distribution not depending on β_0 or μ_0. Thus, T is a cut for (μ_0, β_1), S-sufficient for μ_0, and S-ancillary for β_1. Note also that

$$\beta_0 = \log(\mu_0) - \log\left(\sum e^{\beta_1 x_i}\right) \tag{6.30}$$

where (β_0, β_1) are the canonical parameters, as required by Equation (6.27). (Barndorff-Nielsen, 1978, p. 51) □

6.2.5 MARGINAL AND CONDITIONAL LIKELIHOODS

Likelihood orthogonality and cuts are strong conditions, rarely fulfilled. When available, they guarantee that all information in the complete likelihood function is contained in the factor for the parameter of interest, and, thus, is all used. Because of cancellation, the profile likelihood provides suitable inferences for the parameter of interest. However, we shall often require weaker, but more widely applicable, conditions for approximately factoring the likelihood function in the desired way. Unfortunately, these can no longer guarantee that all available information about a parameter is used.

If $\mathbf{T}$ is (approximately) sufficient, in some sense, for the parameter of interest, $\boldsymbol{\lambda}$, the *marginal likelihood*, based on the marginal distribution of $\mathbf{t}$, might be used for inference. If $\mathbf{T}$ is (approximately) ancillary, in some sense, for the parameter of interest, $\boldsymbol{\lambda}$, the *conditional likelihood*, based on the conditional distribution of the observations, given $\mathbf{t}$, might be used.

In the linear exponential family, conditioning on the sufficient statistics for the nuisance parameters is identical to conditioning on the m.l.e. of the corresponding mean value parameters (Section 3.7.1), which are estimation orthogonal to the canonical parameters of interest. As we saw in Section 2.2.3, this conditional distribution is also in the exponential family. For those normal models in which the conditioning statistics are linear, conditional and marginal likelihoods are identical.

Notice that, for fixed values of the nuisance parameters, their sufficient statistics become ancillary for the parameter of interest. Conditioning on them takes into account the lack of precision, which, thus, varies as a function of the nuisance parameter. Hence, a conditional likelihood is more appealing than a marginal one because the latter may not allow as much for the imprecision in information about the nuisance parameters.

Examples

1. In the logistic regression of Equation (4.16) and in Section 6.2.1 above, we saw how the nuisance parameter, β_0, could be eliminated by conditioning to provide a conditional likelihood for β_1 alone in Equation (4.19). However, we also noticed that the resulting conditional likelihood does not contain all of the information about β_1. The two parameters are not likelihood orthogonal.

2. For n i.i.d. normal random variables, suppose that the parameter of interest is the variance, σ^2. The mean of the observed values, $t_1 = \sum y_i/n$, is ancillary for this parameter and $t_2 = \sum(y_i - t_1)^2/n$ is L-sufficient, so that inference can be based on the conditional or marginal likelihood, both of which are identical:

$$\begin{aligned} \mathrm{L}_C(\sigma^2; y_1, \ldots, y_n | t_1) &= \mathrm{L}(\sigma^2; t_2) \\ &= \frac{\mathrm{e}^{-\frac{nt_2}{2\sigma^2}}}{\sigma^{n-1}} \end{aligned} \tag{6.31}$$

This is just the conditional density of Equation (2.38) which was also obtained as a modified profile likelihood in Equation (3.80). It is the likelihood function for a Chi-squared distribution of nt_2/σ^2 with $n-1$ degrees of freedom. (The Jacobian has a factor of σ^{-2}.)

The score is

$$\mathrm{U}_{\sigma^2}(\sigma^2; y_1, \ldots, y_n | t_1) = \frac{\sum(y_i - \bar{y}_\bullet)^2}{2\sigma^4} - \frac{n-1}{2\sigma^2} \tag{6.32}$$

Then, the conditional or marginal m.l.e. is $\hat{\sigma}^2 = \sum(y_i - \bar{y}_\bullet)^2/(n-1)$. This is sometimes called the restricted maximum likelihood or REML estimate (Patterson and Thompson, 1971). An example deviance was plotted in Figure 3.7.

3. Consider now an example first mentioned in Section 5.3.2. To ascertain the precision of an instrument, pairs of measurements, y_{ij} $(j = 1, 2)$, are made on n objects, indexed by i. Suppose that the observations follow a normal distribution with means $\mu_1, \ldots, \mu_n$ and variance σ^2. The latter, measuring the precision, is the parameter of interest, while the former are incidental parameters, their number increasing with the number of observations.

The profile likelihood is

$$\mathrm{L}_P(\hat{\boldsymbol{\mu}}, \sigma^2) = \frac{\exp\left[-\frac{1}{2\sigma^2}\sum_i(y_{i1} - y_{i2})^2\right]}{\sigma^{2n}} \tag{6.33}$$

and the corresponding m.l.e. is $\hat{\sigma}^2 = \sum_i(y_{i1} - y_{i2})^2/(4n)$. The statistic, $\mathbf{t}_1 = (\bar{y}_{1\bullet}, \ldots, \bar{y}_{n\bullet})^{\mathrm{T}}$, is ancillary for this parameter and $t_2 = \sum_i(y_{i1} - y_{i2})^2/(2n)$ is L-sufficient. Then, the conditional and marginal likelihoods are both

$$\begin{aligned} \mathrm{L}_C(\sigma^2; y_1, \ldots, y_n | \mathbf{t}_1) &= \mathrm{L}(\sigma^2; t_2) \\ &= \frac{\mathrm{e}^{-\frac{nt_2}{4\sigma^2}}}{\sigma^n} \end{aligned} \tag{6.34}$$

Table 6.1. Joint and marginal distributions of two variables.

	Y_1				
Y_2	1	2	3	4	Total
1	$\frac{(1-\lambda)(1-\phi)}{12}$	$\frac{(1+\lambda)(1-\phi)}{12}$	0	0	$\frac{(1-\phi)}{6}$
2	$\frac{(1-\lambda)(1+\phi)}{12}$	$\frac{(1+\lambda)(1+\phi)}{12}$	0	0	$\frac{(1+\phi)}{6}$
3	0	0	$\frac{(2-\lambda)(2-\phi)}{24}$	$\frac{(2+\lambda)(2-\phi)}{24}$	$\frac{(2-\phi)}{6}$
4	0	0	$\frac{(2-\lambda)(2+\phi)}{24}$	$\frac{(2+\lambda)(2+\phi)}{24}$	$\frac{(2+\phi)}{6}$
Total	$\frac{(1-\lambda)}{6}$	$\frac{(1+\lambda)}{6}$	$\frac{(2-\lambda)}{6}$	$\frac{(2+\lambda)}{6}$	1

and the resulting conditional or marginal m.l.e. is $\hat{\sigma}_c^2 = t_2$, twice the ordinary m.l.e. (Neyman and Scott, 1948)

4. Consider the two random variables Y_1 and Y_2, with the joint distribution given in Table 6.1. For one observation pair, (y_1, y_2), y_1 is S-sufficient and y_2 S-ancillary for the parameter of interest, λ. For any given observed value of Y_2, the marginal distribution of Y_1 eliminates the nuisance parameter, ϕ, but is not sufficiently restrictive for inference. It gives positive probability to values of Y_1 which are impossible given the observed $Y_2 = y_2$. We must use the conditional distribution, $\{(1-\lambda)/2, (1+\lambda)/2, 0, 0\}$ when $Y_2 = 1, 2$ and $\{0, 0, (2-\lambda)/4, (2+\lambda)/4\}$ when $Y_2 = 3, 4$, which places a further restriction on the marginal distribution. (Basu, 1977) □

The third example has special importance for the theory of inference, because the dimension of the parameter vector, $\boldsymbol{\mu}$, increases with the sample size, n. However, according to our criteria of Chapter 1, it is an ill-specified model. The n objects measured are of no intrinsic interest when we wish to determine the precision of the measuring instrument for any typical object. And yet we include a parameter for each in the model; the pattern studied depends on sample size. The preferable approach would be to assume random variability in these objects (see Section 6.4.1 below).

In each case, inference from the profile likelihood can be misleading, because it assumes that the unknown nuisance parameters are fixed at their m.l.e. But, in addition, we have seen how the presence of incidental parameters in a model indicates that it is badly specified and how a conditional likelihood can be preferable to a marginal one.

As mentioned in Section 3.6.3, when conditioning is applied to allow for the uncertainty in nuisance parameters which, in fact, are not necessary in a model (because proper model selection had not been carried out), the conditional likelihood for the interest parameter may be narrower than the profile likelihood. (See Lindsey, 1975, for an example.) However, the likelihood function without the unneeded nuisance parameters will be narrower still for the parameter of interest.

When a sufficient statistic is not available for a nuisance parameter, one may obtain an approximate likelihood by conditioning on the m.l.e. If its distribution is not easily available, then a further approximation may be made by using asymptotic results from Section 5.3.

6.3 Correcting the profile likelihood

In many examples that we have considered, the dimensions of the parameters and the sufficient statistics are identical because we are working with a linear exponential family. However, because sufficient statistics, of a dimension suitable for conditioning out the nuisance parameters, are not more generally available, it is useful to have some more approximate methods. Hence, we shall now consider in more detail an approach we first encountered rather intuitively in Section 3.6.3, the modified profile likelihood.

6.3.1 p^* FORMULA

As we might suspect from the presentation in Section 3.6.3, an approximate formula for a conditional likelihood can be obtained by a modification of the profile likelihood. It is based on the modified profile likelihood approximation to the distribution of the m.l.e. in Equation (5.56), the p^* formula,

$$p^*(\hat{\boldsymbol{\psi}}|\mathbf{t}_2;\boldsymbol{\psi}) \doteq c(\mathbf{t}_2,\boldsymbol{\psi})\sqrt{|\mathbf{J}(\hat{\boldsymbol{\psi}})|}\mathrm{R}(\boldsymbol{\psi})$$

where $c(.)$ is a norming constant, $\mathbf{J}$ is the observed information matrix, R is the normed likelihood, and $\mathbf{t}_2$ is ancillary. This formula is parametrization invariant and also invariant under one-to-one transformation of the random variable. For certain families of models, it is exact.

Example
Let us look at a classical example, where the idea of conditioning was first explored, the location–scale family described in Section 2.7.1 and defined in Equation (2.90). For n i.i.d. observations, we saw, in Equation (3.35), that the configuration of the sample (Fisher, 1934),

$$\mathbf{t}_2 = \left(\frac{y_1-\hat{\mu}}{\hat{\sigma}},\ldots,\frac{y_n-\hat{\mu}}{\hat{\sigma}}\right)^{\mathrm{T}}$$

is distribution constant, with

$$f(\mathbf{t}_2) = \frac{1}{c_0(\mathbf{t}_2)\hat{\sigma}^{n-2}} \tag{6.35}$$

and, thus, is ancillary. The conditional distribution is

$$f(\hat{\mu}, \hat{\sigma}|\mathbf{t}_2; \mu, \sigma) = c_0(\mathbf{t}_2)\hat{\sigma}^{n-2} \prod f(y_i; \mu, \sigma) \tag{6.36}$$
$$= c_0(\mathbf{t}_2)\frac{\hat{\sigma}^{n-2}}{\sigma^n} \prod f\left(\frac{y_i - \mu}{\sigma}; 0, 1\right)$$

where $\mathbf{y} = (\hat{\mu} + \hat{\sigma}t_{21}, \dots, \hat{\mu} + \hat{\sigma}t_{2n})^{\mathrm{T}}$.

Now, the normed likelihood is

$$\mathrm{R}(\mu, \sigma) = \left(\frac{\hat{\sigma}}{\sigma}\right)^n \prod \frac{f\left(\frac{y_i - \mu}{\sigma}; 0, 1\right)}{f(t_{2i}; 0, 1)} \tag{6.37}$$

We can calculate that

$$|\mathbf{J}(\hat{\mu}, \hat{\sigma})| = \frac{1}{\hat{\sigma}^4}\left\{\left[\sum \mathrm{l}''(t_{2i})\right]\left[\sum t_{2i}^2 \mathrm{l}''(t_{2i}) - n\right] - \left[\sum t_{2i}\mathrm{l}''(t_{2i})\right]^2\right\} \tag{6.38}$$

where $\mathrm{l}(t) = \log[f(t)]$ and primes indicate derivatives, so that the term in brackets is $c_0^2(\mathbf{t}_2)$ and

$$p^*(\hat{\mu}, \hat{\sigma}|\mathbf{t}_2; \mu, \sigma) = c(\mathbf{t}_2)\sqrt{|\mathbf{J}(\hat{\mu}, \hat{\sigma})|}\mathrm{R}(\mu, \sigma) \tag{6.39}$$

exactly, with $c(\mathbf{t}_2) = \prod f(t_{2i}; 0, 1)$. □

6.3.2 MODIFIED PROFILE LIKELIHOOD

Suppose that both $(\hat{\boldsymbol{\lambda}}, \hat{\boldsymbol{\phi}})$ and $(\mathbf{t}_1, \hat{\boldsymbol{\phi}})$ are one-to-one transformations of the minimal sufficient statistic $\mathbf{T}$, such that the distribution of $\mathbf{T}_1$ depends only on $\boldsymbol{\lambda}$. (More generally, the minimal sufficient statistic will contain ancillary information, besides the m.l.e.) We have the marginal distribution for $\mathbf{T}_1$, given by

$$f_{T_1}(\mathbf{t}_1; \boldsymbol{\lambda}) = \frac{f_{T_1\hat{\Phi}}(\mathbf{t}_1, \hat{\boldsymbol{\phi}}; \boldsymbol{\lambda}, \boldsymbol{\phi})}{f_{\hat{\Phi}|t_1}(\hat{\boldsymbol{\phi}}|\mathbf{t}_1; \boldsymbol{\lambda}, \boldsymbol{\phi})} \tag{6.40}$$
$$= \frac{f_{T_1\hat{\Phi}}(\hat{\boldsymbol{\lambda}}, \hat{\boldsymbol{\phi}}; \boldsymbol{\lambda}, \boldsymbol{\phi})\left|\frac{\partial(\hat{\boldsymbol{\lambda}}, \hat{\boldsymbol{\phi}})}{\partial(\mathbf{t}_1, \hat{\boldsymbol{\phi}})}\right|}{f_{\hat{\Phi}|t_1}(\hat{\boldsymbol{\phi}}_\lambda|\mathbf{t}_1; \boldsymbol{\lambda}, \boldsymbol{\phi})\left|\frac{\partial\hat{\boldsymbol{\phi}}_\lambda}{\partial\hat{\boldsymbol{\phi}}}\right|}$$

where $\hat{\boldsymbol{\phi}}_\lambda$ is the m.l.e. of $\boldsymbol{\phi}$ for fixed $\boldsymbol{\lambda}$ and partial derivatives are taken with respect to it, as a function of $\hat{\boldsymbol{\phi}}$ and $\mathbf{t}_1$.

If we use the p^* formula of Equation (5.56) for the density of the maximum likelihood estimates, conditional on ancillary information, in both the numerator and denominator, we obtain

$$\mathrm{L}_{MP}(\boldsymbol{\lambda};\mathbf{t}_1) = \left|\frac{\partial\hat{\boldsymbol{\phi}}}{\partial\hat{\boldsymbol{\phi}}_\lambda}\right| \frac{\mathrm{R}_P(\boldsymbol{\lambda};\mathbf{t})}{\sqrt{|\mathbf{J}_{\phi\phi}(\boldsymbol{\lambda},\hat{\boldsymbol{\phi}}_\lambda)|}} \tag{6.41}$$

$$= \frac{\sqrt{|\mathbf{J}_{\phi\phi}(\boldsymbol{\lambda},\hat{\boldsymbol{\phi}}_\lambda)|}\mathrm{R}_P(\boldsymbol{\lambda};\mathbf{t})}{\left|\frac{\partial^2 \mathrm{l}(\boldsymbol{\lambda},\hat{\boldsymbol{\phi}}_\lambda;\hat{\boldsymbol{\phi}},\mathbf{t}_1)}{\partial\hat{\boldsymbol{\phi}}_\lambda \partial\boldsymbol{\phi}^{\mathrm{T}}}\right|_{\phi=\hat{\phi}}}$$

using the relationship obtained by differentiating the score equation for $\hat{\boldsymbol{\phi}}_\lambda$ with respect to $\hat{\boldsymbol{\phi}}_\lambda$ and assuming that the norming constants are independent of $(\boldsymbol{\lambda},\boldsymbol{\phi})$ (Barndorff-Nielsen, 1983).

We have modified the (normed) profile likelihood to obtain an approximation to the marginal likelihood. However, because of its approximate nature, it can also be shown to be close to a conditional likelihood. Thus, it should take into account at least some of the uncertainty in estimating the nuisance parameters, $\boldsymbol{\phi}$. In the linear exponential family, this result is the double saddlepoint approximation to the conditional likelihood (see Section 5.7.3 and Barndorff-Nielsen, 1983). This modified likelihood is just Equation (3.77) which we developed intuitively in Chapter 3.

Examples

1. For n i.i.d. observations from a normal distribution, $\hat{\mu} = \hat{\mu}_{\sigma^2} = \sum y_i/n$, the partial derivative does not depend on σ^2, and $|\mathbf{J}_{\mu\mu}(\hat{\mu},\sigma^2)| = n\sigma^{-2}$. Then, the modified profile likelihood for σ^2 is just Equation (3.80):

$$\mathrm{L}_{MP}(\sigma^2;t_2) = \frac{\exp\left[-\frac{1}{2\sigma^2}\sum(y_i-\bar{y}_\bullet)^2\right]}{\sigma^{n-1}}$$

which also corresponds to the exact conditional or marginal Chi-squared likelihood of Equation (6.31).

2. For n independent observations from a normal distribution with fixed covariates, $\mathbf{X}$, consider a linear multiple regression model, $\mu_i = \mu + \boldsymbol{\beta}^{\mathrm{T}}\mathbf{x}_i$, where $\sum \mathbf{x}_i = 0$ and $\mathrm{rank}[\mathbf{x}_1,\ldots,\mathbf{x}_n] = r$. The modified profile likelihood for the intercept, $\lambda = \mu$, with $\boldsymbol{\phi} = (\boldsymbol{\beta},\sigma^2)$ and $t_1 = \hat{\mu}$, is

$$\mathrm{L}_{MP}(\mu;t_1) = \left[1 + \frac{n(\bar{y}_\bullet - \mu)^2}{\sum(y_i - \hat{\mu} - \hat{\boldsymbol{\beta}}^{\mathrm{T}}\mathbf{x}_i)^2}\right]^{-\frac{n-r-2}{2}} \tag{6.42}$$

This contrasts with the ordinary profile likelihood which has exponent $-n/2$ (Barndorff-Nielsen, 1983).

For the 10 observations, with $\sum y_1 = 50, \sum y_i^2 = 300$, and $r = 0$, the plots of the deviances are given in Figure 6.1. The modified profile deviance intervals are larger, reflecting the lack of information about the 'true' value of the variance. Because the m.l.e. of the mean does not depend on the

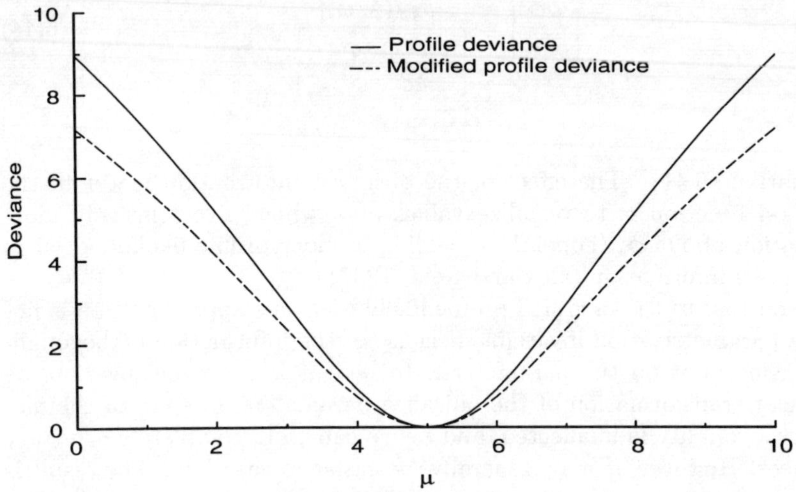

Fig. 6.1. Profile and modified profile deviances for the mean of a normal distribution.

variance (is estimation orthogonal), the point estimates are the same in the two cases. □

6.3.3 ADJUSTED PROFILE LIKELIHOOD

Because of the difficulty in obtaining the partial derivative in Equation (6.41), a number of further approximations have been proposed, called adjusted profile likelihoods. We shall only look at one example. Consider the special case where λ is a scalar. This is the most common situation of inference for a parameter of interest in the presence of nuisance parameters.

As we saw in Section 6.1.3, the parameters can be transformed so that λ and $\boldsymbol{\phi}$ are information orthogonal. We shall condition on the observed value of $\hat{\boldsymbol{\phi}}_\lambda$, the m.l.e. of $\boldsymbol{\phi}$, for a fixed value of λ. If we use the approximation to the marginal distribution of $\hat{\boldsymbol{\phi}}$ in Equation (6.40), we obtain

$$\begin{aligned} \mathrm{L}_{AP}(\lambda; \mathbf{t}|\hat{\boldsymbol{\phi}}_\lambda) &= \sqrt{\frac{|\mathrm{J}_{\psi\psi}(\hat{\lambda}, \hat{\boldsymbol{\phi}}_\lambda)|}{|\mathrm{J}_{\phi\phi}(\lambda, \hat{\boldsymbol{\phi}}_\lambda)|}}\mathrm{R}_p(\lambda) \\ &\propto \frac{\mathrm{L}(\lambda, \hat{\boldsymbol{\phi}}_\lambda; \mathbf{y})}{\sqrt{|\mathrm{J}_{\phi\phi}(\lambda, \hat{\boldsymbol{\phi}}_\lambda)|}} \end{aligned} \tag{6.43}$$

recalling that $\boldsymbol{\psi} = (\lambda, \boldsymbol{\phi}^{\mathrm{T}})^{\mathrm{T}}$ and again assuming the norming constant to be free of the parameters. We see that information orthogonality has permitted us to assume, at least asymptotically, that

$$\left|\frac{\partial\hat{\boldsymbol{\phi}}}{\partial\hat{\boldsymbol{\phi}}_\lambda}\right| = \frac{|\mathbf{J}_{\phi\phi}(\lambda,\hat{\boldsymbol{\phi}})|}{\left|\frac{\partial^2 \mathrm{l}(\lambda,\hat{\boldsymbol{\phi}}_\lambda;\hat{\boldsymbol{\phi}},\mathbf{t}_1)}{\partial\hat{\boldsymbol{\phi}}_\lambda \partial\hat{\boldsymbol{\phi}}^{\mathrm{T}}}\right|} = 1 \tag{6.44}$$

in Equation (6.41). The effect of the observed information factor in the likelihood function is to penalize values of λ which give relatively more information about $\boldsymbol{\phi}$, (hopefully) yielding a wider profile likelihood when there is less information (Cox and Reid, 1987).

In contrast to the modified profile likelihood, this approximation is not usually parametrization invariant, because of the information orthogonalization constraint on the parameters. Indeed, it is not even invariant to non-linear transformation of the nuisance parameters, in spite of the fact that orthogonality is unaffected, and hence can yield completely arbitrary inferences. However, it may, generally, be easier to calculate. The result is identical for the two examples of the previous section.

Example

In Section 6.2.5, we looked at an example of an experiment to measure the precision of one instrument. Suppose now that we want to obtain an accurate measure for one object. We shall use m instruments with differing unknown precisions. Each will be used to measure this same object n_j times, j being the instrument number. Then, we shall have m sets of n_j i.i.d. normal observations, so that $\bar{y}_{\bullet j} = \sum_i y_{ij}/n_j$ are independently normally distributed with mean μ, the parameter of interest, and variances σ_j^2/n_j, the incidental parameters.

Let $s_j = \sum_i (y_{ij} - \bar{y}_{\bullet j})^2$. The log likelihood function is

$$\mathrm{l}(\mu,\boldsymbol{\sigma}^2) = -\frac{1}{2}\sum_j n_j \log(\sigma_j^2) - \frac{1}{2}\sum_i\sum_j (y_{ij}-\mu)^2/\sigma_j^2 \tag{6.45}$$

and the m.l.e. is

$$\hat{\mu} = \frac{\sum(\bar{y}_{\bullet j} n_j/\hat{\sigma}_j^2)}{\sum(n_j/\hat{\sigma}_j^2)} \tag{6.46}$$

where $\hat{\sigma}_j^2 = \sum(y_{ij}-\hat{\mu})^2/n_j$. The conditional density of the means, $\bar{y}_{\bullet j}$, given the m $\hat{\sigma}_j^2$, is proportional to $\prod_j \hat{\sigma}_{j\mu}^{-(n_j-2)/2}$, where $n_j\hat{\sigma}_{j\mu} = s_j + n_j(\bar{y}_{\bullet j}-\mu)^2$. This gives

$$\mathrm{D}_{AP}(\mu) = \sum_j (n_j - 2)\log\left[\frac{s_j + n_j(\bar{y}_{\bullet j}-\mu)^2}{s_j + n_j(\bar{y}_{\bullet j}-\hat{\mu}_c)^2}\right] \tag{6.47}$$

where $\hat{\mu}_c$ satisfies the estimating equation

$$\sum_j \frac{(n_j - 2)n_j(\bar{y}_{\bullet j} - \hat{\mu}_c)}{s_j + n_j(\bar{y}_{\bullet j} - \hat{\mu}_c)^2} = 0 \tag{6.48}$$

(Barndorff-Nielsen and Cox, 1994, pp. 126–127) □

This is another example of an ill-specified model, where the number of parameters increases with the sample size, here with m. This time, the m hopefully typical instruments are of no intrinsic interest when we wish to obtain an accurate measure for one object. Again, the preferable approach would be to assume random variability among instruments (Section 6.4.1).

In models from the (k, k) linear exponential family, conditioning on the canonical statistic, $\mathbf{t}_2$, for $\boldsymbol{\phi}$, instead of on the m.l.e., $\hat{\boldsymbol{\phi}}_\lambda$, yields

$$\mathrm{L}_{AP}(\lambda; \mathbf{t}|\mathbf{t}_2) = \mathrm{L}(\lambda, \hat{\boldsymbol{\phi}}_\lambda; \mathbf{y})\sqrt{|\mathrm{J}_{\boldsymbol{\phi}\boldsymbol{\phi}}(\lambda, \hat{\boldsymbol{\phi}}_\lambda)|} \tag{6.49}$$

which we can contrast with Equation (6.43). This follows from the difference between Equations (5.110) and (5.121), and contrasts with the first example which was a mixed representation with information orthogonality (Davison, 1988).

In certain circumstances, the adjusted profile likelihood approximation of Equation (6.43) holds even without information orthogonality. In such cases, the modified and adjusted profile likelihoods are identical.

Example

Consider the extreme value distribution with

$$f(u) = \mathrm{e}^{u - \mathrm{e}^u} \tag{6.50}$$

where $u = (y - \phi)/\lambda$, a member of the location–scale family. For n observations, $t_i = (y_i - \hat{\phi})/\hat{\lambda}$ is ancillary and $\hat{\phi}_\lambda = \hat{\phi}$. The likelihood function is

$$\begin{aligned} \mathrm{L}(\lambda, \phi) &= \frac{1}{\lambda^n}\prod_{i=1}^{n} f\left(\frac{y_i - \phi}{\lambda}\right) \\ &= \frac{1}{\lambda^n}\prod_{i=1}^{n} f\left(\frac{\hat{\phi} - \phi + t_i\hat{\lambda}}{\lambda}\right) \end{aligned} \tag{6.51}$$

From this, we have

$$\mathrm{J}_{\phi\phi}(\lambda, \phi) = -\frac{1}{\lambda^2}\sum \frac{\partial^2 \log(f)}{\partial \phi^2}$$

$$= \frac{\partial^2 \log(\mathrm{L})}{\partial\hat{\phi}_\lambda \partial\phi} \tag{6.52}$$

so that, from Equation (6.44), the modified and adjusted profile likelihoods are identical. However,

$$\begin{aligned} \mathrm{I}_{\lambda\phi}(\lambda, \phi) &\propto -\frac{1}{\lambda^2}\sum \frac{\partial^2 \log(f)}{\partial\phi\partial\lambda} \\ &\neq 0 \end{aligned} \tag{6.53}$$

so that the parameters are not information orthogonal. (Barndorff-Nielsen and McCullagh, 1993) □

6.3.4 DIRECTED DEVIANCE

If the scalar parameter, λ, has a sufficient statistic, t_1, further insight can be obtained by transforming from this statistic to the directed deviance statistic, D_d, derived from Equation (3.26),

$$\mathrm{D}_d(\lambda) = \mathrm{sign}(\hat{\lambda} - \lambda)\sqrt{\mathrm{D}_P(\lambda)}$$

For $\lambda = \theta$, a canonical parameter of the exponential family, this has Jacobian $\mathrm{D}_d(\lambda)(\hat{\lambda} - \lambda)$, so that Equation (6.43) becomes

$$\mathrm{L}_{AP}(\lambda; \mathbf{t}|\hat{\phi}_\lambda) \doteq \frac{\mathrm{D}_d(\lambda)}{(\hat{\lambda} - \lambda)\sqrt{\mathrm{J}_{\lambda\lambda|\phi}(\hat{\lambda}, \hat{\phi}_\lambda)}}\sqrt{\frac{|\mathrm{J}_{\phi\phi}(\lambda, \hat{\phi}_\lambda)|}{|\mathrm{J}_{\phi\phi}(\hat{\lambda}, \hat{\phi}_\lambda)|}}\mathrm{e}^{-\frac{1}{2}\mathrm{D}_d^2(\lambda)} \tag{6.54}$$

We have used Equation (5.76) and the relationship

$$|\mathrm{J}_{\psi\psi}(\hat{\lambda}, \hat{\phi}_\lambda)| = |\mathrm{J}_{\lambda\lambda|\phi}(\hat{\lambda}, \hat{\phi}_\lambda)||\mathrm{J}_{\phi\phi}(\hat{\lambda}, \hat{\phi}_\lambda)| \tag{6.55}$$

In Equation (6.54), $\mathrm{D}_d(\lambda)$ is the directed deviance statistic and

$$(\hat{\lambda} - \lambda)\sqrt{\mathrm{J}_{\lambda\lambda|\phi}(\hat{\lambda}, \hat{\phi}_\lambda)} \tag{6.56}$$

is the Wald statistic of Equation (5.45). The first factor, the ratio of directed deviance to Wald statistic, is a measure of how far the deviance is from being quadratic, i.e. from a normal distribution, while the second factor corrects for the nuisance parameter (Pierce and Peters, 1992).

The first factor is of wider use than its derivation for a canonical parameter suggests. Values differing from unity indicate that the normal approximation to the m.l.e. of Section 5.3.3, in the given parametrization, is poor for the observed sample. Because this index is positive, with one indicating normality, its logarithm is most usefully plotted.

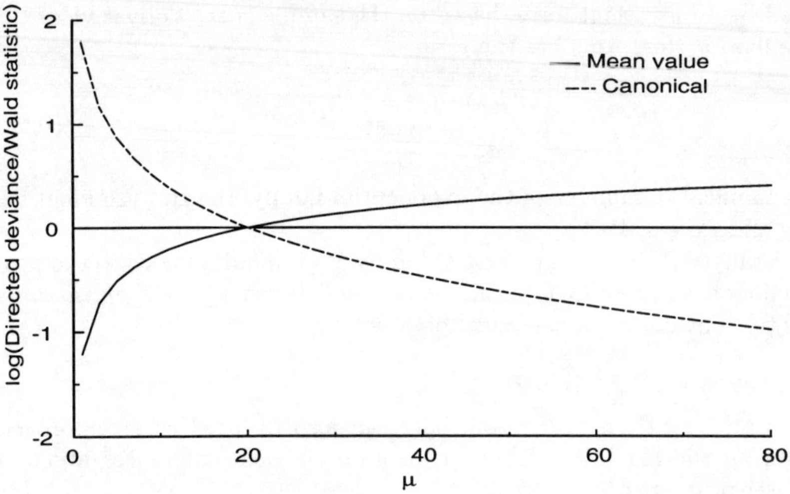

Fig. 6.2. Logarithm of the ratio of the directed deviance statistic to the Wald statistic for the normal approximations for the mean of an exponential distribution in mean value and canonical representations.

Example
Consider again the example in Section 5.3.3 of the exponential distribution. The logarithm of the ratio of the directed deviance statistic to the Wald statistic for the mean value representation approximation of Equation (5.36), and for the canonical representation approximation of Equation (5.40)

$$\log\left(\sqrt{\frac{\mathrm{D}(\mu)}{\mathrm{D}_{N_1}(\mu)}}\right) \tag{6.57}$$

are plotted in Figure 6.2. As might be expected, we see that such normal approximations are best when the 'true' value of the mean is near the m.l.e., here $\hat{\mu} = 20$. This also confirms what we found in Section 5.3.3: the canonical representation here gives an inferior approximation; it deviates further from zero as the mean increases. □

In certain situations, as for the example of the variance of a normal distribution above, the second factor of Equation (6.54) produces a translation which corrects the 'bias' (Section 7.4.1) of the unconditional m.l.e. It may also be more appropriate to inspect the logarithm of each of these quantities divided by D_p. A 'moderate'-sized correction factor, for each, can, then, be judged to be about 0.2.

Because the Wald statistic in the first factor of Equation (6.54) is not parametrization invariant, we have seen in Equation (5.46) that it can be replaced by an invariant form, based on the sample space derivative of the log likelihood. Here, this has the form

$$-\frac{1}{2}\frac{d\mathrm{D}(\lambda)}{d\hat{\lambda}}\sqrt{\mathrm{J}_{\lambda\lambda|\phi}(\hat{\lambda},\hat{\phi}_{\lambda})} \tag{6.58}$$

For a canonical parameter of the exponential family, the two are identical (Pierce and Peters, 1992).

Generally, within the exponential family, such modifications to the profile likelihood as presented in this section will be small for a mean value parameter, but not for a canonical parameter.

6.3.5 PIVOTAL LIKELIHOOD

The pivotal quantities of Section 3.5.4 can also be used as an approach to factoring the likelihood function in order to make inferences about a parameter of interest.

Example
Pivotal quantities for the location–scale family were given in Equations (3.63) to (3.65). Inferences about μ and σ can be based on the conditional distribution of Q_1 and Q_2, given the observed value, $\mathbf{Q}_3 = \mathbf{q}_3$, which has density function

$$f(q_1, q_2|\mathbf{q}_3; \mu, \sigma) \propto q_2^{n-1} f[(\mathbf{q}_3 + q_1)q_2] \tag{6.59}$$

Inference on μ alone might be obtained by integrating out q_2:

$$f(q_1|\mathbf{q}_3; \mu) = \int_0^{\infty} q_2^{n-1} f[(\mathbf{q}_3 + q_1)q_2]dq_2 \tag{6.60}$$

which does not depend on S.

For the normal distribution, this yields the Student t distribution. (Chamberlin and Sprott, 1989) □

Such inferences are closely related to the fiducial approach of Fisher (1930).

A pivotal quantity is *complete linear conditionally sufficient* for a parameter, $\boldsymbol{\lambda}$, if

(1) for every $\boldsymbol{\lambda}$, the pivotal quantity is a one-to-one differentiable transformation of a conditionally sufficient statistic for $\boldsymbol{\lambda}$, and
(2) every component of the pivotal quantity, which is a one-to-one differentiable transformation of some statistic, is also linear in $\boldsymbol{\lambda}$.

The parameter $\boldsymbol{\lambda}$, of $\boldsymbol{\psi} = (\boldsymbol{\lambda}^{\mathrm{T}}, \boldsymbol{\phi}^{\mathrm{T}})^{\mathrm{T}}$, is *separable* for a complete linear conditionally sufficient pivotal quantity, $(\mathbf{q}_1, \mathbf{q}_2)$, if

(1) $\mathbf{q}_1$ contains only $\boldsymbol{\lambda}$,
(2) $\mathbf{q}_2$ contains only ϕ,
(3) $\mathbf{q}_2$ is variation independent of $\mathbf{q}_1$ for fixed $\mathbf{y}$.

The *pivotal likelihood* of a separable parameter is proportional to the probability function for $\mathbf{q}_1$, as a function of $\boldsymbol{\lambda}$ (Sprott, 1990).

Example
For n i.i.d. normal random variables, the pivotal likelihood for μ can be based on the marginal distribution of $Q_1 = \frac{\sqrt{n}(\bar{Y}_\bullet - \mu)}{S}$, a Student t distribution with $n-1$ d.f.:

$$\mathrm{L}(\mu; \mathbf{q}) = \frac{\Gamma\left(\frac{n}{2}\right)}{\sqrt{(n-1)\pi}\,\Gamma\left(\frac{n-1}{2}\right)}\left(1 + \frac{q_1^2}{(n-1)}\right)^{-\frac{n}{2}} \tag{6.61}$$

The pivotal likelihood estimate is $\hat{\mu} = \bar{y}_\bullet$ and the normed pivotal likelihood is identical to the normed profile likelihood. For our 10 observations, this was plotted above in Figure 6.1. (Chamberlin and Sprott, 1989) □

6.4 Special models

As we have now seen several times, certain inference problems arise simply because a model has not been well-specified. This may remind us of certain paradoxes of likelihood inference, in Section 3.6.6, which arise in the same way. It is now time to look at some more appropriate models for such cases. All of the models which we shall study in this section involve isolating a parameter of interest from complex effects which do not concern us directly.

6.4.1 RANDOM EFFECTS

In certain models, nuisance parameters are incidental, so that their number increases with the size of the sample, n, as we have seen in some of our examples. We know that such models are ill-specified because conditions not of interest are being described in the systematic part of the model, whereas they should be stochastic; the patterns studied are changing with sample size. Let us now look more closely at this latter possibility for handling the study of such phenomena.

Instead of trying to factor the likelihood function to isolate the parameter of interest, we shall change our model to one which is more appropriate. Thus, suppose that these incidental parameters are themselves a random variable having a probability distribution, $p(\phi)$. Then, we may eliminate the nuisance parameter by obtaining the marginal distribution of the observations

$$f(\mathbf{y}; \boldsymbol{\lambda}) = \int_\Phi f(\mathbf{y}|\phi; \boldsymbol{\lambda})p(\phi)d\phi \tag{6.62}$$

a compound distribution. This is a procedure which we already studied in Section 2.6 for conjugate distributions, and which is also a direct extension of the finite mixture distributions, used in Section 4.1.3 for contaminated observations. It is often called the empirical Bayesian approach (Section 8.2.2), as well as multilevel or hierarchical modelling. (It is also, confusingly, sometimes called a 'marginal' likelihood, although it is the model, not the likelihood, which is marginal.)

Example
As we saw in Section 2.6, the exponential family, as well as the exponential dispersion family, have conjugate distributions which allow the necessary marginal distribution to be obtained in closed form. □

If the likelihood function does, in fact, factor, then we shall have

$$\begin{aligned} f(\mathbf{y};\boldsymbol{\lambda}) &= \int_{\Phi} f(\mathbf{t}_1;\boldsymbol{\lambda}) f(\mathbf{t}_2|\boldsymbol{\phi}) p(\boldsymbol{\phi}) d\boldsymbol{\phi} \\ &= f(\mathbf{t}_1;\boldsymbol{\lambda}) \int_{\Phi} f(\mathbf{t}_2|\boldsymbol{\phi}) p(\boldsymbol{\phi}) d\boldsymbol{\phi} \end{aligned} \tag{6.63}$$

Because the integral is a constant not containing the parameter of interest, $\boldsymbol{\lambda}$, the likelihood obtained from this marginal distribution of the random effects model is, in this case, proportional to the appropriate factor of the original likelihood function and is, thus, the same. In cases where approximate factorization of the likelihood function is possible, by conditional or marginal arguments, as previously in this chapter, the results should be similar to those obtained from the appropriate random effects model.

Example
Suppose that n subjects, indexed by i, each have a binary $(0, 1)$ response, y_{ij}, to m different questions, indexed by j. Let the binomial probability, π_{ij}, of a response be described by the logistic model,

$$\pi_{ij} = \frac{\mathrm{e}^{(\phi_i - \lambda_j) y_{ij}}}{1 + \mathrm{e}^{(\phi_i - \lambda_j)}} \tag{6.64}$$

with $\lambda_m = 0$. This is called the Rasch (1960) model, with ϕ_i the ability of the ith subject and λ_j the difficulty of the jth question. Suppose for the moment that the latter are of interest and the former are incidental parameters.

The likelihood function is

$$\mathrm{L}(\boldsymbol{\phi}, \boldsymbol{\lambda}) = \prod_i \prod_j \pi_{ij}^{y_{ij}} (1 - \pi_{ij})^{1 - y_{ij}} \tag{6.65}$$

and the sufficient statistics are $t_{1i} = y_{i\bullet}$ for ϕ_i and $t_{2j} = y_{\bullet j}$ for λ_j. Notice that abilities can only be estimated to have $m+1$ distinct values, the number of possible correct answers to the questions. The conditional likelihood, $L(\boldsymbol{\lambda}; \mathbf{y}|\mathbf{t}_1)$, is usually used for inference.

For the special case, $m = 2$, the data can be rearranged in a 2×2 contingency table with entries n_{kl} being the frequencies of both questions wrong, one or the other correct, and both correct. The appropriate conditional distribution is binomial for the two entries in this table, n_{12} and n_{21}, with only one correct answer, with the total number of such subjects, $n_{12} + n_{21}$, fixed. The conditional maximum likelihood estimate is $\hat{\lambda}_1 = \log(n_{12}/n_{21})$.

Consider now a random effects model for the parameter $\boldsymbol{\phi}$, such that $p(\boldsymbol{\phi})$ has a finite number of point mass probabilities. It can be shown that $(m+1)/2$ points will maximize the likelihood obtained from the random effects model. When $m = 2$, if the 2×2 contingency table shows positive association between the two items, in that there is a positive log odds ratio, then it can also be shown that the random effects model will give an estimate of the ability parameter identical to the conditional estimate. (Lindsay *et al.*, 1991)

Of course, the model of Equation (6.64) is symmetric and, in certain contexts, one may be interested in the abilities of subjects rather than the difficulties of questions. □

One important application of random effects models occurs when prior empirical information is available (Section 3.6.4). A series of studies may be similar in pertinent aspects, so that one wishes to combine the information in them to improve estimation. However, the studies may differ in other aspects, not of direct interest. If circumstances are stable, and not evolving over time, a reasonable model may allow the latter to vary randomly among studies (Fisher, 1958, pp. 99–101; Cochran, 1937).

Example

The results of the clinical study of a new medicine are often summarized as a 2×2 table, classifying the participants according to the numbers of successful cures and failures with this new medicine and with a control, usually either placebo or the current standard treatment. The log odds ratio of Equation (5.128), comparing the two treatments, is of interest. Previous studies with the same two treatments are available, but the other clinics involved will have had varying types of patients, some specializing in acute cases and others in much milder ones. If the odds ratio can be assumed constant for differing severity, variability in the latter could be modelled as a random effect:

$$\log\left(\frac{\pi_{ij}}{1-\pi_{ij}}\right) = \mu + \lambda_i + \phi_j \tag{6.66}$$

where i indexes treatment and j the clinic, and hence the severity, and the parameter, ϕ_j, is assumed to vary randomly among clinics. In this way, prior information is included with the current results to strengthen the conclusions. When such a model is applied to a set of studies already published in the medical literature, this is known as meta-analysis. □

Again, it is important that the individual raw data be available from each study, and not just summaries such as the 2×2 tables in this example, for proper modelling, including checking, to be carried out.

6.4.2 MISSING VALUES

Missing values are a practical possibility in any actual collection of data. (Indeed, all statistics might be said to be a missing data problem, because samples only contain part of the population of interest!) They will be especially likely when responses are durations or when they are taken several times on the same individuals. Some individuals in the study may be unavailable at certain points in time for some reason and their responses will not be recorded. A simpler special case involves drop-outs, where missing data are due to individuals disappearing completely from the study after some point in time. As we have already seen, for durations, this is known as censoring. Models for missing data are closely related to the theory for stopping rules (Section 3.3.2).

Responses are said to be *missing at random* (MAR) if, for all values of the parameter, the conditional probability of the missing responses, given the observed data, is the same for all possible values of these missing responses. If, in addition, this probability is the same for all possible values of the observed data, the responses are said to be *missing completely at random* (MCAR). Only the first is necessary for likelihood inference to be valid, without an explicit model for the distribution of the missing values (Rubin, 1976). Notice the close analogy to the requirement that a sequential stopping rule be a martingale stopping time (Section 3.3.2): for MAR, missingness cannot depend on unavailable data.

Let Z be a random indicator variable, taking values 1 or 0, showing whether or not the random variable, Y, is observed, and having conditional density, $f(z|y;\phi)$. Thus, the pair, (Y, Z), will be observed, where $Y = y_{miss}$ if $Z = 0$ and $Y = y_{obs}$ otherwise. The resulting likelihood will be

$$\mathrm{L}(\boldsymbol{\lambda}, \boldsymbol{\phi}; \mathbf{y}_{obs}, \mathbf{z}) = \int_{u=y_{miss}} f(\mathbf{u}; \boldsymbol{\lambda}) f(\mathbf{z}|\mathbf{u}; \boldsymbol{\phi}) d\mathbf{u} \qquad (6.67)$$

If the data are MAR and $\boldsymbol{\lambda}$ is unrelated to $\boldsymbol{\phi}$, then the likelihood factors, the relevant part containing the parameters of interest becoming

$$\mathrm{L}_1(\boldsymbol{\lambda}; \mathbf{y}_{obs}, \mathbf{z}) = \int_{u=y_{miss}} f(\mathbf{u}, \mathbf{z}; \boldsymbol{\lambda}) d\mathbf{u} \qquad (6.68)$$

Table 6.2. Conditional probability distribution of missing values.

$\mathbf{z}$	$f(\mathbf{z}\|\mathbf{y};\phi)$
(1,0)	ϕ
(1,1)	$(1-\phi)I(y_1 \geq 0)$
(0,1)	$(1-\phi)[1-I(y_1<0)]$
(0,0)	0

$$= \frac{f(\mathbf{y}^*, \mathbf{z}; \boldsymbol{\lambda})}{f(\mathbf{y}^*, \mathbf{z}|\mathbf{y}_{obs}; \boldsymbol{\lambda})}$$

because the conditional distribution is a constant in the parameter of interest, $\boldsymbol{\lambda}$. Here $\mathbf{Y}^* = (Y_{obs}^{\mathrm{T}}, Y_{miss}^{\mathrm{T}})^{\mathrm{T}}$ represents the complete data that would have been observed if none were missing. The latter factorization of the likelihood function is used in the EM algorithm for maximizing the likelihood function (Section D.1.3). Notice that the length of $\mathbf{y}_{obs}$ depends on the values in this vector, so that derivation of the distribution of any statistic calculated from it will be difficult, depending on the missing data mechanism.

Examples

1. Consider a bivariate sample of size n for random variables Y_1 and Y_2, where Y_1 is always observed. If Y_2 is observed iff $Y_1 > c$, a constant, the observations are MAR, but if it is observed iff $Y_2 > c$, they are not.

2. For $n = 2$, the conditional probabilities of missing data are given in Table 6.2. The data are MAR unless $z = (0,1)$, and MCAR unless $z = (0,1)$ or $(1,1)$. (Rubin, 1976) □

If data are not MAR, then some model will have to be elaborated to describe the generation of the missingness. (If distributions of statistics are to be used, as in significance tests, the data must be MCAR.) Thus, the distribution of the complete data can be factored in two ways:

$$\begin{aligned} f(\mathbf{y}^*, \mathbf{z}|\mathbf{x}; \boldsymbol{\lambda}, \boldsymbol{\phi}) &= f(\mathbf{y}^*|\mathbf{x}; \boldsymbol{\lambda}, \boldsymbol{\phi}) f(\mathbf{z}|\mathbf{y}^*, \mathbf{x}; \boldsymbol{\lambda}, \boldsymbol{\phi}) \\ &= f(\mathbf{y}^*|\mathbf{z}, \mathbf{x}; \boldsymbol{\lambda}, \boldsymbol{\phi}) f(\mathbf{z}|\mathbf{x}; \boldsymbol{\lambda}, \boldsymbol{\phi}) \end{aligned} \tag{6.69}$$

where $\mathbf{x}$ is a vector of explanatory variables and the conditional distribution of $\mathbf{z}$ gives the model for missingness. Generally, no information is available about $f(\mathbf{y}^*|\mathbf{z} = 0, \mathbf{x}; \boldsymbol{\lambda}, \boldsymbol{\phi})$ because it only concerns the missing data.

Recall also that missing data mechanisms can pose special conceptual and practical problems for constructing large sample asymptotic approximations (Chapter 5).

6.4.3 COARSENED DATA

As we have seen, all continuous variables can only be measured to finite precision, the unit of measurement. Thus, the usual likelihoods for such data, as in Equation (3.5), are numerical approximations. A similar example of grouping occurs when duration data are censored, so that the duration is only known to be longer than some fixed observation time. This concept of grouping can be generalized to coarsened data, where the grouping is stochastic (Heitjan and Rubin, 1991; Heitjan, 1994).

Let Y^* be the variable of interest, with density $f(y^*; \boldsymbol{\lambda})$, but where values of this variable cannot be observed exactly. Suppose that Z is a random variable whose conditional distribution, dependent on the variable of interest, Y^*, is $f(z|y^*; \boldsymbol{\phi})$. This variable determines the precision of reporting Y^*, in the sense that the value of Z determines which collection of possible mappings, $Y^* \rightarrow Y$, has coarsened Y^*, where Y is the observable random variable. The conditional distribution of Y, given the actual value, $Y^* = y^*$, is degenerate,

$$\begin{aligned} f(y|y^*, z) &= 1 \qquad Y = y \\ &= 0 \qquad Y \neq y \end{aligned} \tag{6.70}$$

because z determines which mapping is observed. Then,

$$f(y|y^*; \boldsymbol{\phi}) = \int f(y|y^*, z) f(z|y^*; \boldsymbol{\phi}) dz \tag{6.71}$$

Equation (3.5) ignores the coarsening completely, while Equation (3.4) treats the groups as fixed, not stochastic. The full model for stochastic coarsening, for the observable values, is

$$f(y; \lambda, \boldsymbol{\phi}) = \int_{y^* \in y} f(u; \lambda) f(y|u; \boldsymbol{\phi}) du \tag{6.72}$$

where integration is over all possible values of Y^* in the coarsened interval, Y. The complete likelihood function can then be constructed from this model.

Example

Age heaping occurs, especially for young children, because ages may be reported to the nearest year, half year, or month. Thus, a reported age of 24 months may imply any of three possible intervals. □

Data are *coarsened at random* (CAR) if, for all values of the parameters, the conditional probability of the observed category, $f(y|y^*; \boldsymbol{\phi})$, is the same for all possible values, Y^*, in that category. Then, the coarsening is ignorable and Equation (6.72) reduces to the likelihood of Equation (3.4) for the parameters of interest. The MAR condition of the previous section is a special case of CAR.

Example
In Type I censoring, the time to stop observation of each individual is fixed in advance; if the event has not occurred before this time, the observation is censored.

In Type II censoring, all individuals are placed on trial at the same time and observation ceases after a fixed number of events.

Both of these types of censoring are CAR and hence non-informative. (Heitjan, 1993) □

6.5 Pseudo-likelihood

Pseudo-likelihoods are sometimes used when an exact likelihood is too complex to derive and use. A complete probability model may involve a complex parametrization, part of which is not of direct interest. However, this terminology is rather misleading. Any model is only an approximation to some complex data generating mechanism, ignoring many factors, so that any likelihood based on it might be called pseudo, partial, or quasi!

6.5.1 PARTIAL LIKELIHOOD

Suppose that $\{Y_t, Z_t\}$ is a sequence of pairs of random variables in some stochastic process observed over time and that, for each t, the conditional distribution of Y_t, given the previous history, $z_1, y_1, \ldots, z_{t-1}, y_{t-1}, z_t$, depends on $\boldsymbol{\lambda}$ but not on other unknown aspects of the model. If the conditional distribution of Z_t, given $z_1, y_1, \ldots, y_{t-1}$, contains little available information about $\boldsymbol{\lambda}$, then we obtain a *partial likelihood*,

$$\mathrm{L}_{CP}(\boldsymbol{\lambda}; \mathbf{y}, \mathbf{z}) = \prod f(y_i | z_1, y_1, \ldots, z_i; \boldsymbol{\lambda}) \tag{6.73}$$

by ignoring the distribution of Z_t (Cox, 1975).

Example
Suppose that the random variables Y_i are survival times of n different individuals, following a *proportional hazards* model with intensity or failure rate given by

$$h(t; \boldsymbol{\beta}) = h_0(t)\mathrm{e}^{\boldsymbol{\beta}^{\mathrm{T}}\mathbf{x}} \tag{6.74}$$

where $\mathbf{x}$ is a vector of known explanatory variables and $h_0(t)$ is an unknown baseline failure rate of unspecified form. Consider the *ordered* failure times, $0 < y_{(1)} < \cdots < y_{(n)}$, and take the conditional probability of failure of individual i among the set of $R_{y_{(j)}}$ individuals still at risk at time $y_{(j)}$, given by

$$\frac{e^{\boldsymbol{\beta}^{\mathrm{T}}\mathbf{x}_j}}{\sum_{j \in R_{y_{(j)}}} e^{\boldsymbol{\beta}^{\mathrm{T}}\mathbf{x}_j}} \tag{6.75}$$

The partial likelihood is a product of these terms over all observed failure times. Thus, the risk set, $R_{y_{(j)}}$, summarizes all of the necessary information in the previous sequence of observations. (Cox, 1972) □

However, this classical case of what was thought to be a partial likelihood turns out, in fact, to be simply a profile likelihood (Whitehead, 1980). It is an example where the m.l.e. of $\boldsymbol{\beta}$ is consistent in the presence of incidence parameters. This arises because there is a proper cut between the nuisance parameters and $\boldsymbol{\beta}$: the partial likelihood can essentially be derived as was Equation (6.29), a conditional binomial distribution arising from a Poisson model.

6.5.2 COMPOSITE LIKELIHOOD

The conditions for a partial likelihood can be further weakened if the pairs of random variables are not a sequence. Then, one might base a *composite likelihood* on

$$\mathrm{L}_C(\boldsymbol{\lambda}; \mathbf{y}, \mathbf{z}) = \prod f(y_i|z_i; \boldsymbol{\lambda}) \tag{6.76}$$

In contrast to a partial likelihood, here the terms of the score statistic will generally be correlated.

6.5.3 QUASI-LIKELIHOOD

We saw, in Section 2.5.1, that members of the univariate exponential dispersion family are characterized by the strict relationship between their mean and variance. A quasi-likelihood can be defined uniquely in terms of this relationship (Wedderburn, 1974). In other words, it is defined by the first two moments only. An important application is to overdispersion in count data.

Examples

1. In the 'least squares' interpretation of linear regression, the variance is assumed not to depend on the mean without claiming any other distributional assumptions. (See, however, the discussion of the Gauss–Markov theorem in Section 7.4.1.)

2. Suppose that events are counted on various individuals. If these counts, say Y_i, had a Poisson distribution, with mean μ, then $\mathrm{var}[Y_i] = \mu$. If this relationship does not hold, so that there is under- or overdispersion (Section

1.2), an additional parameter may be introduced, to give $\text{var}[Y_i] = \phi\mu$, without further specification of the distribution. This can be estimated from the mean deviance or from the empirical second moment. The standard errors of the parameters will be correspondingly modified, and might be used to make approximate inferences. □

6.5.4 GENERALIZED ESTIMATING EQUATIONS (GEE)

We have seen, in Section 3.5.1, how the score or estimating equations can be solved to obtain the m.l.e. Such a score can be integrated, as in Equation (3.37), to obtain the corresponding likelihood function and probability model (the latter, if the sample space is known). However, occasionally, one may wish to solve a system of estimating equations,

$$g(\mathbf{t}, \boldsymbol{\psi}) = \mathbf{0} \tag{6.77}$$

which cannot be integrated to obtain a proper probability distribution (Zeger and Liang, 1986). A minimal requirement would be that the equations are unbiased (Section 5.1.2), but the expectation involved in this concept is undefined if a probability distribution is not available. Thus, such equations have no clear interpretation as corresponding to a stochastic model for a data generating mechanism. Pseudo-'standard errors' can be obtained, for example from the derivative of $g(\cdot)$, a pseudo-information matrix, yielding precision intervals useful for a measurement error problem. However, in most other situations where one might want to use such an approach, they will be unreliable, as described in Section 5.3.4. Hence, such equations are essentially only useful for obtaining point estimates without a trustworthy measure of precision.

Summary

As usual, suppose that $\boldsymbol{\psi} = (\boldsymbol{\lambda}^{\mathrm{T}}, \boldsymbol{\phi}^{\mathrm{T}})^{\mathrm{T}}$, where $\boldsymbol{\lambda}$ is the parameter of interest and $\boldsymbol{\phi}$ the nuisance parameter. Let k_ψ be the dimension of the parameter vector and k_T that of the sufficient statistic. Then, the following possibilities are available:

(1) The linear exponential family has $k_\psi = k_T = k$. If
- there is no nuisance parameter,
- the parameter of interest is a canonical parameter, or
- the parameter of interest is a ratio of canonical parameters,

then conditioning is available to produce a unique distribution depending only on $\boldsymbol{\lambda}$.

(2) If the above conditions do not hold, even in the (k, k) linear exponential family, approximate methods must be used.

(3) If $k_\psi < k_T$, in the curved exponential family, a partition of the sufficient statistic may sometimes be found, such that $\mathbf{T} = (\mathbf{T}_1^{\mathrm{T}}, \mathbf{T}_2^{\mathrm{T}})^{\mathrm{T}}$,

where $k_\psi = k_{T_1}$ and $\mathbf{T}_2$ is ancillary. Conditioning on this may lead to one of the first two cases.

(4) Similar special properties arise from the transformation characteristics of certain models, such as the location–scale family. Again, conditioning on ancillary statistics resolves ambiguity.

(5) If the number of incidental parameters depends on the sample size, a random effects model may be more appropriate.

(Cox, 1988)

The approximations in this chapter, as well as the profile likelihood, are most often globally pointwise equivalent.

Further reading

The seminal articles on factoring the likelihood function to make inferences about parameters of interest are Andersen (1970) and Kalbfleisch and Sprott (1970). These were closely followed by the article on REML by Patterson and Thompson (1971). However, the first uses of such likelihoods date back to Student (1908) and Fisher (1915).

The conditions for factoring the likelihood function exactly, within the exponential family, were set out by Barndorff-Nielsen (1978). Christensen and Kiefer (1994) provide asymptotic procedures for obtaining local factoring. Cox and Reid (1987) introduce the idea of information orthogonality as giving a usable approximate factorization in other simple cases. Willmot (1990) provides general conditions for an arbitrary parameter to be information orthogonal to the mean.

The basis of many of the approximate procedures in this chapter can be found in the fundamental papers of Barndorff-Nielsen (1983) and Tierney *et al.* (1989). However, these are usually interpreted as aids in eliminating nuisance parameters for making frequentist or Bayesian probability statements about the remaining parameters of interest, and not as a means of obtaining an optimal decomposition of the likelihood function.

The problem of the number of incidental parameters becoming infinite was first discussed by Neyman and Scott (1948) and the random effects models, as a solution, by Kiefer and Wolfowitz (1956).

Random effects models have long been used in normal theory linear models and are a standard aspect of these; see any text on analysis of variance, such as Searle (1971). Their extension to other contexts is more difficult, primarily because numerical integration is often required. The first example appears to be Pierce and Sands (1975). A large literature now exists; extensive references are given in Lindsey (1993).

The fundamental ideas for modelling missing data have been developed by Rubin (1976; see also Little and Rubin, 1987, and Little, 1995) and extended to coarsened data by Heitjan and Rubin (1991). For a survey of the analysis of grouped data, see Heitjan (1989).

The early work on pseudo-likelihoods was done by Besag (1974), Cox (1975), and Wedderburn (1974).

6.6 Exercises

(1) Suppose that we have data following a series of multinomial distributions such that $\sum_j \sum_k \pi_{ijk} = 1$ for all i. We want to study the influence of the explanatory variable, x_i, on the two binary responses, $j = 1, 2$ and $k = 1, 2$. For example, the responses might be presence or absence of two characteristics during medical diagnosis and the explanatory variable, age.

(a) For the (conditional) log linear regression model,

$$\begin{aligned} \log\left(\frac{\pi_{i11}\pi_{i12}}{\pi_{i21}\pi_{i22}}\right) &= \beta_{10} + \beta_{11}x_i \\ \log\left(\frac{\pi_{i11}\pi_{i21}}{\pi_{i12}\pi_{i22}}\right) &= \beta_{20} + \beta_{21}x_i \\ \log\left(\frac{\pi_{i11}\pi_{i22}}{\pi_{i12}\pi_{i21}}\right) &= \beta_{30} + \beta_{31}x_i \end{aligned} \tag{6.78}$$

what elements of the parameter vector, $\boldsymbol{\beta}$, are information orthogonal?

(b) For the corresponding marginal regression model,

$$\begin{aligned} \log\left(\frac{\pi_{i11}+\pi_{i12}}{\pi_{i21}+\pi_{i22}}\right) &= \beta_{10} + \beta_{11}x_i \\ \log\left(\frac{\pi_{i11}+\pi_{i21}}{\pi_{i12}+\pi_{i22}}\right) &= \beta_{20} + \beta_{21}x_i \\ \log\left(\frac{\pi_{i11}\pi_{i22}}{\pi_{i12}\pi_{i21}}\right) &= \beta_{30} + \beta_{31}x_i \end{aligned} \tag{6.79}$$

what elements are information orthogonal?

(2) Suppose that Y_1 and Y_2 are independent random variables from an exponential distribution with means μ_1 and $\mu_2 = \lambda + \mu_1$. Find a transformation of μ_1 making it information orthogonal to the parameter of interest, λ.
(Cox and Reid, 1987)

(3) For the location family, Equation (2.90) with $\sigma = 1$, introduce an arbitrary additional parameter, ϕ.

(a) Show that the transformation,

$$\mu^* = \mu + g(\phi) \tag{6.80}$$

for some suitable function, $g(\cdot)$, can be made information orthogonal to ϕ.

(b) If we have n samples with different means μ_i, but common ϕ, what transformation of the means is orthogonal?

(c) Apply this result in a regression context.

(Barndorff-Nielsen and Cox, 1994, p. 51)

(4) Demonstrate that the score statistic is unaffected by conditioning on an F-ancillary statistic.

(5) (a) Find the marginal likelihood for μ in the normal distribution.

(b) To what distribution does this likelihood correspond?

(Cox and Hinkley, 1974, p. 109)

(6) The papers of 200 candidates for an examination were marked separately by the examiners of two examining boards, who classified them as pass or fail, with the following results:

	Board B		
Board A	Pass	Fail	Total
Pass	136	2	138
Fail	16	46	62
Total	152	48	200

These data differ from those in previous examples and exercises in that they are not independent observations: each candidate is observed twice. Thus, interest centres on the conditional probability, given that a candidate was treated differently by the two boards, that he or she was failed by board A and not by board B. Construct a suitable model and draw the appropriate inferences about the parameter of interest.

(Lindley, 1965, pp. 183–184)

(7) (a) Consider a 2×2 contingency table of frequencies, with entries n_{ij}, where $n_{\bullet\bullet}$, the total number of observations, is fixed.

i. What is a reasonable model?

ii. What are the minimal sufficient statistics?

(b) Suppose that we are particularly interested in the log odds ratio, or interaction, parameter

$$\lambda = \log\left(\frac{\pi_{11}\pi_{22}}{\pi_{12}\pi_{21}}\right)$$

i. How should inferences be made?

ii. Is there a proper cut?

(c) Suppose now that we have m such 2×2 contingency tables with binary variables y_i and z_i, each table of the form

y_i	z_i	n_i
$1-y_i$	$1-z_i$	$2-n_i$
1	1	2

and that we are interested in the log odds ratio, which is assumed to be the same for all tables. This is a special case of the Mantel–Haenszel model for matched binary pairs. Consider only discordant pairs, where $n_i = 1$, summarized by the number of pairs with $(y_i, z_i) = (1, 0)$ and the number with $(y_i, z_i) = (0, 1)$.

i. What are the profile likelihood and the modified profile likelihood?
ii. Compare the m.l.e. and the width of the likelihood curve in the two cases.
iii. How does this compare to the random effects model developed in Exercise (5.19)?

(Barndorff-Nielsen and Blaesild, 1975; Barndorff-Nielsen, 1978, p. 208, 1988, p. 246)

(8) Capture–recapture sampling is widely used in wildlife surveys. Consider a geographically well-defined animal population, such as a lake containing a fixed unknown number of fish, say λ, which we wish to estimate. In a first sample of the lake, a fixed number, n_1 fish, are caught, marked, and returned to the lake. After sufficient time for mixing with the remaining fish and recovery from the trapping, a second fixed sample size of n_2 is taken. We assume that marked and unmarked fish have equal probability of capture the second time. Let Y be the random variable denoting the number of marked fish in the second sample.

(a) Write down the probability distribution of Y.
(b) What is an estimate of λ?
(c) Suppose now that each sample, $t = 1, 2$, was taken with N_t, the total catch, a random variable, perhaps by continuing netting a certain fixed time. Let π_t be the corresponding catch probability at each time point. What is the probability distribution for the four categories of fish, not caught, caught the first, second, or both times?
(d) What is the probability distribution of the two catch sizes, N_t?
(e) Derive a conditional distribution for Y, given n_1 and n_2, and compare it with the distribution for fixed sample sizes.

(Whittle, 1992, pp. 70–71)

(9) In Section 3.6.5, we looked at the prediction of future values of a random variable.

(a) Derive a conditional likelihood function for the example of the normal distribution in that section and compare it with the modified profile likelihood.

(b) Do the same for the binomial distribution of Exercise (3.32) of that chapter.

(c) Derive profile, modified profile, and conditional predictive likelihoods for a future observation from an exponential distribution, given n independent observations.

(10) Consider the three-category multinomial distribution, with observations $y_1 + y_2 + y_3 = n$ and parameters $\pi_1 + \pi_2 + \pi_3 = 1$.

(a) The Hardy–Weinberg model sets $\pi_1(\lambda) = \lambda^2$, $\pi_2(\lambda) = 2\lambda(1-\lambda)$, and $\pi_3(\lambda) = (1-\lambda)^2$. Find a suitable nuisance parameter and show how to make inferences about λ.

(b) A symmetry model is defined by $\pi_1 = \pi_3$. The parameter of interest is the conditional probability that an observation falls in the first category, given that it is in either the first or third: $\lambda = \pi_1/(\pi_1 + \pi_2)$. Find a suitable nuisance parameter and show how to make inferences about λ.

(Kass, 1989)

(11) (a) Determine whether the score equations for σ^2 of a normal distribution, derived from Equations (6.33) and (6.34), are unbiased.

(b) Repeat for the score equations for μ derived from Equation (6.45), with σ_i^2 replaced by s_j/n_j, and for Equation (6.48).

(Barndorff-Nielsen and Cox, 1994, pp. 124–128)

(12) Consider n independent pairs of Poisson random variables with means μ_{1i} and $\mu_{2i} = \lambda\mu_{1i}$.

(a) Derive a suitable likelihood function to make inferences for the parameter of interest, λ.

(b) Compare the resulting m.l.e. with the usual m.l.e.

(c) Repeat when the means are $\phi\lambda$ and $\phi(1-\lambda)$.

(Andersen, 1970; Barndorff-Nielsen and Cox, 1994, p. 39)

(13) Consider independent random variables Y_1 and Y_2, with gamma distributions of Equation (1.3), where $\lambda = (\alpha_1\mu_2)/(\alpha_2\mu_1)$ is the parameter of interest. Derive a suitable likelihood function for making inferences.

(Barndorff-Nielsen, 1978, p. 53)

(14) Suppose that $Y_1, \ldots, Y_n$ are independent random variables with an exponential distribution having means $\log(\mu_i) = \beta_0 + \beta_1 x_i$ with $\sum x_i = 0$, an exponential regression model.

(a) Give the profile likelihood function for β_1.

(b) Find the modified and adjusted profile likelihood functions for β_1.

(c) Simplify these when x_i takes only two values, -1 and 1.

(Cox and Reid, 1987; Barndorff-Nielsen and Chamberlin, 1994)

(15) For the inverse Gaussian distribution of Equation (2.30), we are interested in $\lambda = \alpha$ with $\phi = \mu$ a nuisance parameter.

(a) What is the m.l.e. of ϕ for known λ?

(b) Give the profile likelihood function for λ.

(c) Write down the modified profile likelihood.

(d) Why is this identical to the adjusted profile likelihood?

(e) What are the m.l.e.s from these likelihoods?

(Barndorff-Nielsen and Chamberlin, 1994)

(16) Consider a non-homogeneous Poisson process, as in Exercise (3.14), but with the intensity, $\rho(t)$, depending on time. An event is observed at time zero, but then only every κth event is observed, with times $y_1, \ldots, y_n$.

(a) Show that their joint density is

$$f_n(y_1, \ldots, y_n) = \frac{e^{-r(y_n)} \prod_i \rho(y_i)[r(y_i) - r(y_{i-1})]^{\kappa-1}}{\Gamma(\kappa)^n} \tag{6.81}$$

with $r(y) = \int_0^y \rho(u)du$ and $y_0 = 0$.

(b) Derive the conditional density of $y_1, \ldots, y_{n-1}$ given that total observation time is y_n.

(c) By transformation to $z_i = r(y_i)/r(y_n)$, show that this conditional distribution is a product of beta densities, as in Equation (C.20).

(d) Now suppose that the intensity depends on some explanatory variables, as

$$\log[\rho(t)] = \beta_0 + \boldsymbol{\beta}^{\mathrm{T}}\mathbf{z}(t) \tag{6.82}$$

Write down the likelihood function.

(e) In this model, $\boldsymbol{\beta}$ is the parameter of interest, while β_0 and κ are nuisance parameters. Derive a conditional likelihood function whereby β_0 is eliminated.

(f) Can a similar procedure be used to eliminate κ? Why?

(Berman, 1981)

(17) (a) Generalize the example of the precision of a measuring instrument, in Section 6.2.5, to the location–scale family, with common scale and the mean parameters being incidental.
(Barndorff-Nielsen and Cox, 1994, pp. 283–284)

(b) Can the same results be obtained using pivotal quantities?

(18) (a) Derive the modified profile likelihood, in general, for the linear exponential family, where the parameter of interest is canonical and the nuisance parameter is mean value.
(Barndorff-Nielsen and Cox, 1994, pp. 265, 284)

(b) Now take the (k, k) linear exponential family, where the interest parameter, $\boldsymbol{\lambda}$, has dimension d. Derive a general formula for the modified profile likelihood in this situation.
(Barndorff-Nielsen, 1994)

(19) For the precision of a measuring instrument, in Section 6.2.5, demonstrate that the m.l.e. from the modified profile likelihood is consistent as $n \to \infty$, while that obtained in the usual way is not.
(Neyman and Scott, 1948)

(20) (a) Derive the modified profile likelihood in general for the location parameter of the location–scale family.
(Barndorff-Nielsen, 1994)

(b) Demonstrate that the adjusted profile likelihood approximation of Equation (6.43) holds in general for the location–scale family when the scale is the parameter of interest, although the location and scale parameters are not necessarily information orthogonal.
(Barndorff-Nielsen and McCullagh, 1993)

(21) At the end of Section 6.1.3, we derived the transformation to orthogonality when the interest parameter is the ratio of two exponential means, a special case of the model in Exercise (6.14).

(a) Write down the profile likelihood for this ratio, λ.
(b) Using the orthogonality result, derive the adjusted profile likelihood of Equation (6.43).
(c) Repeat for a series of power transformations of orthogonalized ϕ^*, the nuisance parameter. Consider, for example, the family of Equation (2.54), and especially those giving a quadratic log density and a constant variance.
(d) Why might the latter two be especially important?
(e) Derive the modified profile likelihood.
(f) Plot your set of normed likelihoods when $y_1 = 2$, $y_2 = 3$.
(g) What will happen as the number of pairs of observations increases?
(h) Derive appropriate conclusions.

(22) Suppose that the random variables $X_1, \ldots, X_n$ have an arbitrary joint distribution, $f(\mathbf{x}, \boldsymbol{\phi})$, while the random variables $Y_1, \ldots, Y_n$ have independent normal distributions with means $\mu_i = \beta_0 + \beta_1 x_i$, and variance σ^2, where the parameter spaces of $\boldsymbol{\phi}$ and $(\beta_0, \beta_1, \sigma^2)$ are unrelated.

(a) What will be a suitable likelihood function for inferences about $(\beta_0, \beta_1, \sigma^2)$?
(b) What types of sufficient and ancillary statistics are present?
(c) What kind of orthogonality exists between $\boldsymbol{\phi}$ and $(\beta_0, \beta_1, \sigma^2)$?

(d) How does this differ from the case where the observed values, $x_1, \ldots, x_n$, are assumed to be known constants?

(e) Assume now that X_i and Y_i have a bivariate normal distribution with correlation ρ. How will inferences about ρ differ from those about β_1?

(Barndorff-Nielsen and Cox, 1994, pp. 39–40)

(23) (a) Show that the conditional likelihood for the variance of a normal distribution in Equation (6.31) is asymptotically equivalent to the usual likelihood for the variance for fixed values of the mean in all five ways listed in Section 5.2.4.

(b) Show that the conditional likelihood for the mean of a normal distribution, based on a Student t distribution, is maximum likelihood, support interval, and locally pointwise equivalent, but not globally pointwise equivalent to the usual likelihood for the mean for fixed values of the variance.

(Tsou and Royall, 1995)

(24) Consider the gamma distribution of Equation (1.3), where the location parameter, μ, is of interest.

(a) Use the directed deviance to study the validity of a normal approximation to the deviance.

(b) How does this change when the canonical parameter is used?

(c) How important is the correction for the nuisance parameter?

(25) Show that Equation (6.60) is a Student t distribution for the normal distribution member of the location–scale family.

(26) (a) For the example of measuring the precision of an instrument, in Section 6.2.5, develop a random effects model by assuming that the means are independent normal with mean ϕ and variance δ.

(b) What is the resulting distribution for each pair of measurements?

(c) Compare the m.l.e. to those obtained from Equations (6.33) and (6.34).

(Barndorff-Nielsen and Cox, 1994, pp. 128–129)

(27) As in the previous exercise, let us assume that n independent random variables, Y_i, have means μ_i and common variance σ^2. Again, let these means have a normal distribution with mean ϕ and variance δ. Now, however, interest centres on the estimation of the individual means.

(a) Develop estimates for the means in this random effects model. (A special case of this, the James–Stein estimate, will be presented in Section 7.4.2 from a different point of view.)

(b) Discuss the advantages and disadvantages of using such estimates as opposed to the usual m.l.e. for the means.

(Efron and Morris, 1973)

(28) Suppose that n independent random variables, Y_i, are normally distributed with common mean and variances, σ^2/ϕ_i. Let $\phi_i = \xi_i/\alpha$, where ξ_i has a gamma distribution, from Equation (1.3) with $\mu/\alpha = 1$.

(a) Develop a random effects model for this situation.

(b) What is the marginal distribution of Y_i?

(c) Relate these results to the models for robustness and contaminated observations in Section 4.1.

(Little and Rubin, 1987, pp. 210–211)

(29) For data which are MAR, use Equation (6.68) to show that the expected information of the observed data is equal to the complete information if all data had been observed minus the information in the missing data.

(Orchard and Woodbury, 1972)

(30) Demonstrate that Type I and Type II censoring are both CAR.

(Heitjan, 1993)

(31) An object is weighed 100 times on a precision scale with a normal error distribution having unit variance. Let μ be the true weight of the object.

(a) Randomly, with probability 0.1, the scale does not give the weight. Only 91 values are recorded in our case. Write down the likelihood function.

(b) As for the previous point, the scale randomly fails, but now with unknown probability π, which does not depend on the weight of the object. Give the likelihood function.

(c) Now suppose that $\pi = \mu/(1 + \mu)$. Give the likelihood function.

(d) Instead of randomly not giving the weight, the scale reports 'too heavy' for all values over 200 kg. The observed data are censored. Write down the likelihood function.

(e) Now, the censoring point, say ϕ, is unknown. Give the likelihood function.

(f) As in the second last point, the scale only measures to 200 kg. However, you only have the 91 observed values and do not know how many times the 200 kg limit was exceeded. The data are truncated. Give the likelihood function.

(g) Finally, the truncation point, ϕ, is unknown. Write down the likelihood function.

(Gelman *et al.*, 1995, pp. 192–193)

(32) We plan to make n observations from a normal distribution with known unit variance and unknown mean, actually equal to zero.

(a) Suppose that the observations are sequential but independent. Observation stops (because the machine breaks down) after the first value greater than one is recorded, giving a total of $m < n$ observations. This is a drop-out mechanism.
   i. Will these observations be MCAR? MAR?
   ii. What is the distribution of the observed data?
   iii. What is the distribution of the sample mean?
   iv. What asymptotic conclusions can be drawn about it?

(b) Now suppose the same type of observations, but recording stops after the first time the sample mean is positive. (It may stop after the first observation.) Repeat the questions of the previous section.

(c) Finally, suppose that the observations follow an autoregression of order one, an AR(1), with mean and variance as above and unknown autocorrelation ρ, actually equal to 0.9. When an observed value is positive, the next value is missing with probability 0.8, but subsequent values continue to be recorded (the machine does not break down).
   i. Again, repeat the questions of the first section of this exercise.
   ii. What can be said about the sample autocorrelation?

Part IV

Decisions

7 Frequentist decision-making

Up until now, we have been studying how to extract the maximum information from data about models, and how to check their appropriateness, without being concerned about the uses to which the conclusions might be put. In this chapter and the next, we look at various approaches to handling the latter problem. These involve, in different ways, the peculiar notion (to a scientist) that the model for the phenomenon under study can be true or correct. However, in a context where one *must* make decisions involving irrevocable concrete actions, this is usually a necessary assumption. To incorporate the information from a study into this context requires an additional new type of model, one describing how human beings make such decisions.

Three essential features will be involved in the decision-making situations which concern us:

(1) a definitive choice must be made among a set of enumerable possible alternative actions, based on data containing random variability;

(2) a precise measure of gain or loss can be associated with each such action;

(3) the choice is one of a reasonably long series, such that the expected value of gain or loss has meaning.

All three components are rarely simultaneously present (Barnard, 1954).

In certain circumstances, the most important aspect will be the series of repetitions, whereas, in others, it will be the optimal rational choice using all available information. We shall look at the former, as the frequentist model, in this chapter and the latter, as the Bayesian model, in the next. These certainly are not the only models of human decision-making. Others include fuzzy logic and neural networks, which are less closely related to statistical inference and will not be considered here.

7.1 Some principles for decision-making

In most cases, decision-making involves

- some means of quantifying the consequences of errors, in a utility function;

- additional information, other than that obtained directly from the data under analysis, perhaps as a prior probability distribution.

Example
In monitoring an industrial process to check for defects, the monetary costs of rejecting undamaged items and of letting defective items reach the market will be available. Previous information on the probability of defects under different adjustments of the machines may also have been recorded. □

7.1.1 UTILITY AND RISK

Many of the factors involved in a decision process can be formalized (Wald, 1949), as a model for human decision-making, where the results of various actions must be compared. Suppose that, in addition to a parametric statistical model, we have a set of possible decisions, $\mathcal{D}$, and a *utility function*, $u(\psi, d)$, defined for all possible pairs, $(\psi, d) \in (\Psi, \mathcal{D})$. This function gives the costs and benefits for each possible decision in each context. Notice that only *pairwise* comparisons of relative utility are involved.

Example
When a clinical trial is used to study a new medication, the short term relative costs and benefits of trying something unknown on a group of volunteers must be weighed against the longer term ones to the community as a whole. Human suffering and death are difficult to quantify! □

A *decision function*,

$$\delta : \mathcal{Y} \rightarrow \mathcal{D} \tag{7.1}$$

specifies what action to take for any possible observation. It is unbiased if, on average, it comes closer to the correct than to any wrong decision. The *loss function* can be defined as

$$\text{loss}(\psi, d_0) \stackrel{\text{def}}{=} \sup_{d \in \mathcal{D}} u(\psi, d) - u(\psi, d_0) \tag{7.2}$$

where d_0 is the decision to be taken and the supremum is the utility if we knew the true value of ψ. The (frequentist) *risk function* can be defined as

$$\text{risk}_F(\psi, \delta) \stackrel{\text{def}}{=} \text{E}[\text{loss}\{\psi, \delta(Y)\}; \psi] \tag{7.3}$$

the expected loss for fixed ψ.

7.1.2 REPEATED SAMPLING

In certain circumstances, one desires to reduce the long term risk of making mistakes in repeated application of a decision procedure. In quality control or acceptance sampling, a clearly defined population is available, from which samples are repeatedly to be taken. In contrast, in scientific inference, such a population is generally a mathematical fiction which has no unique definition. Various criteria have been developed to accomplish this long run goal, in what is known as the *Neyman–Pearson* or *frequentist school* of statistical decision-making.

This approach takes the following principle as fundamental to statistical procedures:

Repeated sampling principle
Decision-making procedures which would give misleading conclusions most of the time, in hypothetical repeated sampling, should not be used. □

This principle usually implies that the pertinent measure of uncertainty is a hypothetical frequency in the long run and that criteria of optimal decisions are formulated in terms of this behaviour in repeated sampling from some well-defined sample space. Thus, it can provide important criteria for judging the operating characteristics of decision-making procedures. In contrast to the asymptotic theory of Chapter 5 where the sample size was conceptually allowed to become very large, here sample sizes are generally fixed and the decisions are assumed to be repeated a large number of times.

Notice that the repetitions need not be performed by the same individual. Thus, this model of human behaviour could correspond to a certain view of scientific research, whereby the community checks the individual results by replicating them. When costs and benefits of the actions resulting from a decision are also taken into account in the model, these may correspond to those of an organization, such as a firm, or community, as a whole. But then the model distances itself from scientific behaviour.

Example
Faced with a multiple regression model, a scientist makes decisions about which variables to include. However, such decisions are provisional and may be reversed by further information; they do not imply specific actions with their costs and benefits. In contrast, an engineer, on a given job, must weigh the results of a regression analysis in terms of the final product and the long run performance. There will not usually be a second chance (if the bridge falls down). □

More concretely, the basic idea is

- to choose a function of the data, a statistic (not necessarily sufficient), which is related to a parameter of interest in an appropriate way;
- to derive the sample distribution of that statistic under the model; and

- to measure the acceptability of each possible parameter value by calibrating the observed value of the statistic against its long run behaviour under that model.

Thus, if the observed value of the statistic is within the region of high probability, given a parameter value, the latter is compatible with the data.

In general, the repeated sampling and sufficiency principles are in conflict, because the former uses criteria of what *might* happen, the sample space, while the latter requires that inferences use only what *did* happen. Decision-making procedures that follow the repeated sampling principle may often not give optimal or even reasonable results (in the sense of using all the information well; see Section 1.5) for any single sample. They are, rather, based on long term averages.

The situation is further complicated if the minimal sufficient statistic is of larger dimension than the parameter. Then, it is widely agreed that probabilities should be conditional on an appropriate ancillary statistic. In this way, probability calculations will be made more relevant to the data at hand, but the general long run interpretation of the repeated sampling principle is lost because it will only apply to observing data with the same value of the ancillary statistic, i.e. repetitions exactly like the observed data in all of these respects. Such conditioning may bring frequentist decision-making somewhat closer to the needs of the scientist, whose population is generally only theoretical in any case, but destroys its decisional basis for such goals as acceptance sampling.

Example

Suppose that random pairs of variables, (X, Y), are observed, where Y is the response of interest and X is the pertinent explanatory variable. If a normal distribution with a constant variance for all values of X can be assumed, the classical model is the usual 'least squares' simple linear regression, with a corresponding Student t test. However, this approach is conditional on the fixed observed values of X; any long run interpretation of that test is only valid under that condition: repetition with those exact values of the explanatory variables. If the members of the pair are both random, decision-making based on such a test will have no usable frequentist meaning. (Fisher, 1955, 1959, pp. 83-86) □

7.2 Tests: simple null hypothesis

One of the two basic ingredients of the frequentist approach is the *hypothesis test*, the other being the point estimate. In Section 4.2, we briefly mentioned hypotheses in the context of the Fisherian or pure significance test and, in Section 5.6, their asymptotic approximation. Here, we must study their construction, in detail, from a frequentist point of view, which is quite different: hypothesis testing can be interpreted in terms of a 0–1 loss function, zero if the null hypothesis is true, one otherwise.

7.2.1 TYPES OF HYPOTHESES

Suppose that we have a set of parametric statistical models, $\mathcal{P}_\psi$, and that we have some *hypothesis*, H, about which is the appropriate model or set of models. Such a hypothesis must be specified before observing the data; as should be clear, hypotheses derived from the data can rarely be rejected. Hypothesis *testing* is not designed for generating hypotheses! Nor is it appropriate for model selection in spite of its wide use in that context; the sequence of 'hypotheses' which would be 'tested' can never be specified in advance.

Let us suppose that $\mathcal{P}_\psi$ is the set of all possible models under consideration. A *null hypothesis*, $\mathcal{P}_{\psi_0} \subset \mathcal{P}_\psi$, is that subset of models of special interest and which is under test, usually denoted by H_0. As its name suggests, it is usually specified by setting some parameters to zero. The *alternative hypothesis*, denoted by H_A, which is, in fact, $\mathcal{P}_{\psi_A} = \mathcal{P}_\psi \setminus \mathcal{P}_{\psi_0}$, is the complementary subset of models which are taken to be competing, in some way, with H_0.

A hypothesis, i, is said to be *simple* if it completely specifies the model, so that $\mathcal{P}_{\psi_i}$ contains only one member and the probability of any possible outcome can be calculated. Otherwise, it is said to be *composite*. Usually, composite hypotheses specify the model completely except for the values of some of the unknown parameters.

Example

For the set of normal distributions with known variance, a simple null hypothesis would be that the observations have a normal distribution, with $\mu = 0$ and $\sigma^2 = 1$. In contrast, a composite alternative could be that $\mu > 0$ and $\sigma^2 = 1$.

For the set of all normal distributions, a composite null hypothesis would be that the observations have a normal distribution, with $\mu = 0$ and unknown σ^2, while a composite alternative could be that $\mu \neq 0$ with unknown σ^2. □

In hypothesis testing, the null and alternative are not treated in the same way. H_0 must be clearly specified and is the (set of) model(s) of interest; conclusions will be drawn under the assumption that it is true. In contrast, the alternative only serves to indicate possible directions of departure from the null. As well, it is not sufficient that $\mathcal{P}_0$ and $\mathcal{P}_A$ simply be disjoint subsets. $\mathcal{P}_0$ must be a subset of the models under consideration, defined by specifying certain parameter values of $\mathcal{P}$; it must be nested in the set (Section 3.6.1). Thus, this approach provides no easy means of directly comparing non-nested models, as can be done through direct likelihood methods (Sections 3.2.1 and 3.6.1).

Example
Suppose that the composite null hypothesis, H_0, is the set of regression models, $\mathcal{P}_0$,

$$\mu_i = \beta_0 + \beta_1 x_{1i} \tag{7.4}$$

nested in $\mathcal{P}$,

$$\mu_i = \beta_0 + \beta_1 x_{1i} + \beta_2 x_{2i} \tag{7.5}$$

If the composite alternative, H_A, thought to be suitable, is the set $\mathcal{P}_A$,

$$\mu_i = \beta_0 + \beta_2 x_{2i} \tag{7.6}$$

also nested in $\mathcal{P}$, then $\beta_2 = 0$ will not provide an appropriate test because $\mathcal{P}_0 \cup \mathcal{P}_A \neq \mathcal{P}$. □

7.2.2 SIMPLE ALTERNATIVE HYPOTHESIS

The goal of a hypothesis test is to define a set of observations, which, if a member occurs, places the null hypothesis in question because of its small probability. Thus, a hypothesis test is defined by a set of *critical regions*, $\mathcal{Y}_\alpha$, in the sample space such that

- $P(\mathbf{Y} \in \mathcal{Y}_\alpha; H_0) \stackrel{\text{def}}{=} \alpha \quad \forall\alpha$
- $\mathcal{Y}_{\alpha_1} \subset \mathcal{Y}_{\alpha_2}$ if $\alpha_1 < \alpha_2$
- the significance level of observation, $\mathbf{y}$, is $p_y = \inf(\alpha; \mathbf{y} \in \mathcal{Y}_\alpha)$.

Thus, for some α, with $0 < \alpha < 1$, $\mathcal{Y}_\alpha$ is the set of points in the sample space which, if observed, would be taken to contradict H_0 at the *significance level* α. It is usually specified by constructing a suitable test statistic. $\mathcal{Y}_\alpha$ is called a critical region of *size* α; there will usually be many such regions possible, primarily because no criteria have been specified as to which or how much information must be used, i.e. what test statistic to construct.

Up to this point, we have just a reformulation of the pure significance test of Section 4.2. Thus, the P-value of that approach is now called the significance level. However, we now reinterpret α as the long run frequency, in repeated sampling, of making observations which contradict H_0 when it is, in fact, true, rather than as a measure of the rarity of what we have actually observed.

Because $\mathcal{Y}_\alpha$ will not usually be unique, we require criteria for choosing an appropriate critical region in the sample space. Let us first look at the

Table 7.1. Numbers of significant tests with respect to true and false null hypotheses.

	Non-significant	Significant	Total
True null	M_{11}	M_{12}	m_0
False null	M_{21}	M_{22}	$m - m_0$
Total	$m - M_s$	M_s	m

simplest case, where the null and alternative hypotheses are both simple. In other words, we are comparing two completely specified models (as in direct likelihood inference).

Power Whereas the significance level involves the null hypothesis, the *power* of the critical region involves the alternative hypothesis. It is defined by

$$\text{pow}(\mathcal{Y}_\alpha) \stackrel{\text{def}}{=} P(\mathbf{Y} \in \mathcal{Y}_\alpha; H_A) \tag{7.7}$$

Suppose that more than one possible critical region of size α exists. If

$$P(\mathbf{Y} \in \mathcal{Y}_\alpha; H_A) > P(\mathbf{Y} \in \mathcal{Y}^*_\alpha; H_A) \qquad \forall \mathcal{Y}^*_\alpha \neq \mathcal{Y}_\alpha \tag{7.8}$$

the critical region, $\mathcal{Y}_\alpha$, has the largest power. Then, it is called the *best critical region* of size α. It is the size α critical region from which an observation has the largest probability of being observed, in the long run, if the alternative hypothesis is true. It yields the *most powerful test.* Thus, we are looking for a small significance level and a large power.

Suppose that some long run set of m tests has been performed. Of these, an unknown fixed number, m_0, will have true null hypotheses, while, for a random number, M_s, the null hypothesis is rejected, because the observations lie in the critical region. We can represent this in a two-way table, as shown in Table 7.1. Then, $\text{E}[M_{12}/m] = \alpha$ is the significance level, while $\text{E}[M_{22}/m]$ is the power. However, only m and M_s are observable. Correct conclusions will be drawn in the unobservable number, $M_{11} + M_{22}$ times out of m.

Likelihood ratios Now let us consider, as a possible test statistic, the usual likelihood ratio comparing the two simple hypotheses:

$$\begin{aligned} \text{LR}(H_A, H_0; \mathbf{y}) &\stackrel{\text{def}}{=} \frac{\text{L}(H_A; \mathbf{y})}{\text{L}(H_0; \mathbf{y})} \\ &= \frac{f_A(y)}{f_0(y)} \end{aligned} \tag{7.9}$$

For the comparison of these two hypotheses, all of the information in the data is contained in this ratio. As we know, the larger is the ratio, the more evidence there is in the data against H_0. Because the two models are fixed, $\mathrm{LR}(Y)$ is indeed a statistic with a distribution, so that there exists a unique value, c_α, for all α, $0 < \alpha < 1$, such that

$$P[\mathrm{LR}(\mathbf{Y}) \geq c_\alpha; H_0] \stackrel{\text{def}}{=} \alpha \tag{7.10}$$

The region defined by values $\mathrm{LR}(\mathbf{Y}) \geq c_\alpha$ is the size α *likelihood ratio critical region.* This region has certain optimal properties.

Neyman–Pearson fundamental lemma

For any fixed size α, the likelihood ratio critical region is the best critical region, i.e. has the largest power for simple null and alternative hypotheses. (Neyman and Pearson, 1933)

proof: Let $\mathcal{Y}_\alpha$ be the likelihood ratio critical region and $\mathcal{Y}_\alpha^*$ any other region, both of size α. Then,

$$\begin{aligned}\int_{\mathcal{Y}_\alpha} f_0(\mathbf{y})d\mathbf{y} &= \int_{\mathcal{Y}_\alpha^*} f_0(\mathbf{y})d\mathbf{y} \\ &= \alpha\end{aligned}$$

by definition, so that, subtracting the integral of the intersection of the two regions from both sides, we have

$$\int_{\mathcal{Y}_\alpha \setminus \mathcal{Y}_\alpha^*} f_0(\mathbf{y})d\mathbf{y} = \int_{\mathcal{Y}_\alpha^* \setminus \mathcal{Y}_\alpha} f_0(\mathbf{y})d\mathbf{y}$$

But in $\mathcal{Y}_\alpha \setminus \mathcal{Y}_\alpha^*$ which is inside $\mathcal{Y}_\alpha$, we have $\mathrm{L}(H_A; \mathbf{y}) \geq c_\alpha \mathrm{L}(H_0; \mathbf{y})$ whereas in $\mathcal{Y}_\alpha^* \setminus \mathcal{Y}_\alpha$, which is outside $\mathcal{Y}_\alpha$, we have $\mathrm{L}(H_A; \mathbf{y}) < c_\alpha \mathrm{L}(H_0; \mathbf{y})$. This implies that

$$\int_{\mathcal{Y}_\alpha \setminus \mathcal{Y}_\alpha^*} f_A(\mathbf{y})d\mathbf{y} \geq \int_{\mathcal{Y}_\alpha^* \setminus \mathcal{Y}_\alpha} f_A(\mathbf{y})d\mathbf{y}$$

Adding the integral over the intersection of the two regions to both sides, we have

$$\int_{\mathcal{Y}_\alpha} f_A(\mathbf{y})d\mathbf{y} \geq \int_{\mathcal{Y}_\alpha^*} f_A(\mathbf{y})d\mathbf{y}$$

or, equivalently,

$$\mathrm{pow}(\mathcal{Y}_\alpha) \geq \mathrm{pow}(\mathcal{Y}_\alpha^*)$$

the most powerful test, and the required result. □

In the pure significance test, we sought to judge a model in isolation, by some absolute criterion. When we reintroduce a second, alternative, model, with which our model of interest is to be compared, we discover, by a completely different route than in Chapter 3, that the comparison of a pair of likelihoods is an optimal procedure.

Examples

1. Take n i.i.d. random variables from the one-parameter linear exponential family, with canonical parameter θ and sufficient statistic $t = \sum y_i$. Then, directly from the log likelihood ratio, the best critical region has the form

$$(\theta_A - \theta_0)t - \kappa(\theta_A) + \kappa(\theta_0) \geq c_\alpha \tag{7.11}$$

for a suitable constant, c_α. If $\theta_A - \theta_0 > 0$, this is equivalent to

$$t \geq c_\alpha^* \tag{7.12}$$

(and inversely if the two equalities are reversed) so that the region is the same for all such θ_A.

2. Consider a normal distribution with known variance, where the value of the mean, $\mu = \mu_0$, is to be tested and such that $\mu_A > \mu_0$. The log likelihood ratio is

$$\log(\text{LR}) = \frac{\bar{y}_\bullet(\mu_A - \mu_0) - \frac{1}{2}(\mu_A^2 - \mu_0^2)}{\sigma^2/n} \tag{7.13}$$

From Equation (7.12), a critical region of the form $\text{LR} \geq c_\alpha$ is equivalent to one of the form $\bar{y}_\bullet \geq c_\alpha^*$.

We must now choose c_α^* to satisfy the size condition that the probability of the likelihood ratio being larger than this constant is α. But, under H_0, we have $\bar{y}_\bullet \sim \text{N}(\mu_0, \sigma^2/n)$. Then,

$$\begin{aligned} P\left(\frac{\bar{y}_\bullet - \mu_0}{\sigma/\sqrt{n}} \geq k_\alpha\right) &= P\left(\bar{y}_\bullet \geq \mu_0 + \frac{k_\alpha \sigma}{\sqrt{n}}\right) \\ &= \int_{k_\alpha}^{\infty} \phi(u) du \\ &= \alpha \end{aligned} \tag{7.14}$$

with $\phi(u)$ the density of the standard normal distribution, so that

$$c_\alpha^* = \mu_0 + \frac{k_\alpha \sigma}{\sqrt{n}} \tag{7.15}$$

Notice again that c_α^* and k_α do not depend on the value of $\mu_A > \mu_0$ chosen as alternative. □

Table 7.2. Critical regions for a Poisson distribution with $\mu_0 = 1$.

c^*_α	$\alpha = P(Y \geq c^*_\alpha; H_0)$
0	1.0000
1	0.6321
2	0.2642
3	0.0803
4	0.0190
5	0.0037
6	0.0006
7	0.0001
⋮	⋮

Discrete data For discrete distributions, critical regions only change by discrete steps, so that it will usually be impossible to find a region with exactly a given size, α. However, this is only of importance if we insist on using some fixed size, α, which must be attained, with attendant loss of information, instead of reporting the observed α size (see the discussion in Section 4.2.1). This may be important in a decision-making context.

Example
Suppose that we have a Poisson distribution, with the hypotheses $\mu_0 = 1$ and $\mu_A = 10$. The log likelihood ratio for one observation is

$$\log(\text{LR}) = -(\mu_A - \mu_0) + y[\log(\mu_A) - \log(\mu_0)] \tag{7.16}$$

so that, from Equation (7.12), the possible critical regions are as given in Table 7.2. □

If it is difficult to obtain the exact distribution, one of the normal approximations of Chapter 5 might be used.

Asymmetry In constructing critical regions, we are not treating the two models symmetrically; we are only concerned with testing the null hypothesis and not with which model fits the observations better!

Examples
1. For our Poisson example, suppose that we observe $y = 4$, so that $\hat{\mu} = 4$. For $\mu_0 = 1, \mu_A = 12$, Equation (7.16) gives $\log(\text{LR}) = -1.06$, indicating preference for μ_0 over μ_A. However, from Table 7.2, we see that H_0 is rejected with $\alpha = 0.019$. Because the only alternative is $\mu_A = 12$, this would imply accepting the latter, which goes against the evidence. Of course, if the alternative had been $\mu_A = 1000$, the conclusion would still

have been rejection of $\mu_0 = 1$ in favour of μ_A.

Let us also consider an asymptotic approximation. Under the null hypothesis, the m.l.e., $\hat{\mu}$, has an asymptotic normal distribution with mean and variance both μ_0. An asymptotic best critical region will be defined by

$$\frac{\hat{\mu} - \mu_0}{\sqrt{\mu_0}} \geq k_\alpha \tag{7.17}$$

where k_α is obtained from the standard normal integral, as above. For our data, $(\hat{\mu} - \mu_0)/\sqrt{\mu_0} = 3$. The probability of observing a value as extreme as this is 0.0013 so that we again reject the null hypothesis, although the approximation cannot be expected to be very good with only one observation.

2. For the normal distribution example, with known variance, $\sigma^2 = 1$, suppose that $\mu_0 = 0$ and $\mu_A = 1$. If we observed $\sum y_i = 10$ for $n = 25$, so that $\bar{y}_\bullet = 0.4$, the most powerful test of size $\alpha = 0.05$ rejects H_0 in favour of H_A if

$$\bar{y}_\bullet \geq \mu_0 + \frac{1.64\sigma}{\sqrt{n}}$$

from Equation (7.15). In other words, here we reject if $\bar{y}_\bullet > 0.328$, which is the case. However, the log likelihood ratio is -2.50, indicating support for μ_0 over μ_A. □

These examples illustrate how hypothesis testing is asymmetric. H_A only serves to indicate the possible direction of departure from the hypothesis of interest, H_0. In both examples, if we interchange the null and alternative hypotheses, the new null is also rejected, even more strongly, now being in agreement with the result obtained from examining the likelihood ratio.

Of course, a testing situation, where both hypotheses are simple, is not realistic. Unfortunately, as we shall see, with only one real exception, results as strong as the Neyman–Pearson lemma are only available in this case.

7.2.3 POWER FUNCTION

As we have seen briefly above in Equation (7.7), the probability, $P(\mathbf{Y} \in \mathcal{Y}_\alpha; H_A)$, that the observations fall in the critical region if the (simple) alternative hypothesis is true, is called the power of the test against that alternative. Power is used to choose among possible tests, before the observations are made (and sometimes to show that a test used was no good, after making the observations, perhaps because there were not enough).

Table 7.3. Two probability distributions.

	y_1	y_2	y_3	y_4
H_0	0.05	0.00	0.05	0.90
H_A	0.00	0.05	0.90	0.05

Table 7.4. Possible $\alpha = 0.05$ critical regions for the distributions in Table 7.3, and their power.

$\mathcal{Y}_{0.01}$	$P(Y \in \mathcal{Y}_{0.01}; H_A)$
$\{y_1\}$	0.00
$\{y_1, y_2\}$	0.05
$\{y_3\}$	0.90
$\{y_2, y_3\}$	0.95

The size, α, has sometimes been called the *Type I error* because it can be interpreted as the long run frequency of rejecting the null hypothesis when it is true. One minus the power, $\beta = 1 - P(\mathbf{Y} \in \mathcal{Y}_\alpha; H_A)$, is sometimes known as the *Type II error*, when it is interpreted as the long run frequency of accepting the null hypothesis when it is false. In Table 7.1, this is $\mathrm{E}[M_{21}/m] = \beta$, while $\mathrm{E}[M_{12}/m] = \alpha$. Although such acceptance does not occur in a scientific context, it will when decisions must be made, as in control of industrial processes.

Example
Let us imagine a random variable with four possible outcomes and consider the two probability distributions in Table 7.3 as hypotheses. According to our likelihood principle, we would prefer H_0 if y_1 or y_4 is observed and H_A otherwise.

Possible $\alpha = 0.05$ critical regions, with their power, are given in Table 7.4. The critical region with maximum power is $\{y_2, y_3\}$; we reject H_0 if either of these occurs. Not surprisingly, in this case with two simple hypotheses, we have agreement with the results using our likelihood principle. □

The power, taken as a function of the alternative hypothesis, is called the *power function*,

$$\mathrm{pow}(\psi_A; \mathcal{Y}_\alpha) \stackrel{\text{def}}{=} P(\mathbf{Y} \in \mathcal{Y}_\alpha; \psi_A) \tag{7.18}$$

Note that $\mathrm{pow}(\psi_0; \mathcal{Y}_\alpha) = \alpha$.

Because the power function is a function of the different possible models, given (a function of) the observations, it provides a type of likelihood comparison. However, it does not generally contain all of the information in the data, like a proper likelihood function. Thus, in the same way as we saw for significance tests in Equation (4.13), the power function, for fixed α, is identical to a likelihood function given, not the complete data, $\mathbf{y}$, but only the information whether or not $\mathbf{y}$ is in the critical region. The parametric statistical model, $(\mathcal{Y}, \mathcal{P}, \Psi)$, is reduced to the much less informative model, $(\mathcal{Z}, \mathcal{P}, \Psi)$, where the new sample space, $\mathcal{Z}$, consists of only two points, $z = 0$ corresponding to $\mathbf{y} \notin \mathcal{Y}_\alpha$ and $z = 1$ to $\mathbf{y} \in \mathcal{Y}_\alpha$:

$$\begin{aligned} \mathrm{L}_{PF}(\psi) &= \textstyle\int_{\mathcal{Y}_\alpha} f(\mathbf{y};\psi)d\mathbf{y} \qquad z = 0 \\ &= \textstyle\int_{\mathcal{Y}\setminus\mathcal{Y}_\alpha} f(\mathbf{y};\psi)d\mathbf{y} \qquad z = 1 \end{aligned} \tag{7.19}$$

The lost information will depend on what critical region is chosen. It can be represented by the likelihood function based on Y, given z (Barnard *et al.*, 1962).

7.2.4 SAMPLE SIZE

Besides its use in the comparison of possible tests to be used, the power function may serve as a guide to choosing an appropriate fixed sample size, n. For fixed α, the power generally increases as n increases so that n may be chosen to obtain some given level of power.

Example

For a normal distribution, with known variance, the power function is

$$\mathrm{pow}(\mu;\mathcal{Y}_\alpha) = P\left(\bar{y}_\bullet \geq \mu_0 + \frac{k_\alpha\sigma}{\sqrt{n}};\mu\right) \tag{7.20}$$

Suppose that $\mu_A = \mu_0 + \delta$ is an alternative of sufficient practical importance that we wish to obtain enough observations to be fairly sure of not being mistaken if this hypothesis is true. We shall have

$$\begin{aligned} \mathrm{pow}(\mu_0 + \delta;\mathcal{Y}_\alpha) &= P\left(\bar{y}_\bullet \geq \mu_0 + \frac{k_\alpha\sigma}{\sqrt{n}};\mu_A\right) \\ &= P\left(\bar{y}_\bullet \geq \mu_A - \frac{k_\beta\sigma}{\sqrt{n}};\mu_A\right) \\ &= 1 - \beta \end{aligned} \tag{7.21}$$

the latter by the symmetry of the normal distribution. From this, we see that

$$n = \frac{(k_\alpha + k_\beta)^2\sigma^2}{\delta^2} \tag{7.22}$$

is the required sample size. When $k_\alpha = k_\beta$, this is just Equation (3.28), which we obtained by likelihood methods in Section 3.3.1. □

A fixed sample size can only be calculated if both hypotheses are simple. Otherwise, it will be a function of the unknown parameters, such as the variance in Equation (7.22).

Tests based on a fixed sample size design may not necessarily yield the most efficient powerful test.

Example
In Equation (7.12), we saw that we might require $t \leq c^*_\alpha$. If the statistic, T, is a sum of positive functions of the individual observations, this sum may already be attained for $m < n$ observations. Thus, we could have stopped then, already knowing that we must reject the null hypothesis. □

In a testing situation, it is often possible, and preferable, to take observations one by one, or in small sets, accepting H_0, rejecting it, or continuing sampling, depending on the information so far available. This is a sequential design, such as we have already discussed several times. A sequential probability ratio test (SPRT), in fact a likelihood ratio test, is defined by a stopping time at sample size $N = m$, such that $\mathrm{R}_i \leq c_1$ or $\mathrm{R}_i \geq c_2$, for appropriately chosen constants, $0 < c_1 < c_2$, where R_i is the likelihood ratio of H_1 to H_0 at observation i. Then, H_0 is accepted in the first case, and rejected in the second (Wald, 1945; Barnard, 1946).

7.2.5 COMPOSITE ALTERNATIVE HYPOTHESIS

We shall now look at the more realistic situation where we have a collection of alternative hypotheses, any one of which is $H_A \in \mathcal{H}_A$.

This is a new situation, not covered by any of our previous results in this text. Two possible solutions, which we considered in Section 3.1.3, were to take an average of the likelihoods of models in the set and to take the maximum likelihood to represent the set. As we know, a likelihood ratio only allows direct, and symmetric, comparison of two models, or hypotheses, at a time. Here, we wish to test one hypothesis, as compared, simultaneously, to a collection of others. Thus, we shall require evidence against all of that collection, something which may be difficult both to define and to attain.

Uniformly most powerful One simple case can be handled immediately. If we obtain the same size α best critical region for all $H_A \in \mathcal{H}_A$, the results of Section 7.2.2 still apply. We do not need to specify which particular alternative hypothesis is of concern. Then, the critical region is called a *uniformly most powerful* (UMP) size α region for $\mathcal{H}_A$, and if the set of critical regions has this property for all α, the test is UMP.

Example
As we have already seen above, for the one-parameter exponential family, tests based on a one-sided alternative hypothesis, such that

$$t \geq c^*_\alpha \qquad \forall \theta_A > \theta_0 \tag{7.23}$$

and inversely if the two inequalities are reversed, where t is the sufficient statistic, are UMP. □

However, if alternatives in more than one direction from the null hypothesis are of interest, no UMP test can exist.

Example
In the exponential family, if, instead of, say, $\theta_A > \theta_0$, we want $\theta_A \neq \theta_0$ as the alternatives, as will usually be the case, then no UMP test exists. □

This linear exponential family, with a one-sided alternative, is a special case where a UMP test exists and, hence, where the Neyman–Pearson lemma can be directly applied for a composite alternative hypothesis.

If no UMP test exists, some rather arbitrary further criterion is needed to make the choice among possible tests. For example, one can

- choose a 'typical' H_A and use the most powerful test for that alternative,
- maximize the power locally, by taking H_A close to H_0,
- maximize some weighted average of power for the different alternatives.

Unbiasedness One solution is mathematically attractive and is widely used: a critical region of size α is said to be *unbiased* if

$$P(\mathbf{Y} \in \mathcal{Y}_\alpha; H_A) \geq P(\mathbf{Y} \in \mathcal{Y}_\alpha; H_0) \qquad \forall H_A \in \mathcal{H}_A \tag{7.24}$$

Then, among the unbiased regions, we choose the one with the greatest power. If all of these regions are UMP, they are called uniformly most powerful unbiased (UMPU) tests. Otherwise, the test is said to be biased. This ensures that there exist no alternatives for which acceptance of the hypothesis is more probable than in cases when the null is true. Note that, as is usual with power, unbiasedness is a likelihood, and not a probability, criterion, because the models are different on the two sides of the inequality.

This approach has at least two drawbacks:

- There may exist a biased test, which is, nevertheless, preferable for most alternatives.
- $P(\mathbf{Y} \in \mathcal{Y}_\alpha)$ does not take into account the direction of departure, so that evidence that $\psi > \psi_0$ could be used to conclude that $\psi < \psi_0$.

Table 7.5. Distribution of a random variable and one special case.

y	$f(y;\boldsymbol{\psi})$	$f(y;\psi_1=\alpha,\psi_2=\frac{1}{2})$
-2	$\psi_1(1-\psi_2)$	$\frac{1}{2}\alpha$
-1	$(\frac{1}{2}-\alpha)\left(\frac{1-\psi_1}{1-\alpha}\right)$	$\frac{1}{2}-\alpha$
0	$\alpha\left(\frac{1-\psi_1}{1-\alpha}\right)$	α
1	$(\frac{1}{2}-\alpha)\left(\frac{1-\psi_1}{1-\alpha}\right)$	$\frac{1}{2}-\alpha$
2	$\psi_1\psi_2$	$\frac{1}{2}\alpha$

Example
Suppose that a random variable has the probability density given in the second column of Table 7.5, where α is a known constant and $0 \leq \psi_1 \leq \alpha \leq 1/2$, $0 \leq \psi_2 \leq 1$.

Let H_0 be $\boldsymbol{\psi} = (\alpha, 1/2)^{\mathrm{T}}$, with probabilities given in the third column of Table 7.5, and H_A be $\boldsymbol{\psi} \neq (\alpha, 1/2)^{\mathrm{T}}$. A size α critical region is $\mathcal{Y}_\alpha = \{-2, 2\}$, which has power

$$
\begin{aligned}
f(y = \pm 2|H_A) &= \psi_1\psi_2 + \psi_1(1-\psi_2) \\
&= \psi_1 \\
&< \alpha
\end{aligned}
$$

Therefore, this is a biased test.

Another size α critical region is $\mathcal{Y}_\alpha = \{0\}$, which has power

$$
\begin{aligned}
f(y = 0|H_A) &= \alpha\left(\frac{1-\psi_1}{1-\alpha}\right) \\
&> \alpha
\end{aligned}
$$

According to the criterion of biasedness, this is better than the previous test. However, if $1/4 < \alpha < 1/2$, H_0 will be rejected when the most probable event under that hypothesis is observed, and accepted otherwise. □

A more satisfactory solution is to admit that we really require two tests, one for each direction, each with its own power function. Then, each region is usually taken to have size $\alpha/2$.

More generally, as a direct extension of the Neyman–Pearson lemma, in the set of tests of H_0 against $H_A \in \mathcal{H}_A$, the power is maximized, for a given maximum α, by keeping constant the likelihood ratio.

7.3 Tests: composite null hypothesis

7.3.1 TYPES OF NULL HYPOTHESES

In all but very simple problems, the null hypothesis is composite. Usually, it will be given by $P_0(\mathbf{y};\boldsymbol{\psi}_0)$, where $\boldsymbol{\psi}_0 \in \Psi_0$, and the alternative by $P_A(\mathbf{y};\boldsymbol{\psi}_A)$, where $\boldsymbol{\psi}_A \in \Psi_A$, with $\Psi_0 \cap \Psi_A = \emptyset$, $\Psi_0 \cup \Psi_A = \Psi$. However, in certain situations, P_0 and P_A may be different families.

Three common cases are:

(1) a single parametric model, with a null hypothesis that $\boldsymbol{\psi} \in \Psi_0 \subset \Psi$ and alternative, $\boldsymbol{\psi} \in \Psi \backslash \Psi_0$;
(2) a single parametric model, but where $\boldsymbol{\psi} = (\boldsymbol{\lambda}^{\mathrm{T}}, \boldsymbol{\phi}^{\mathrm{T}})^{\mathrm{T}}$ with $\Psi = \Lambda \times \Phi$; often, the null hypothesis is $\boldsymbol{\lambda} = \boldsymbol{\lambda}_0$ and the alternative, $\boldsymbol{\lambda} \neq \boldsymbol{\lambda}_0$, with the nuisance parameters, $\boldsymbol{\phi}$, unspecified in both cases;
(3) two different parametric models which we wish to compare, one being the null hypothesis.

This third case cannot really be solved by hypothesis testing in the frequentist context, but only by embedding both models in some wider model and applying the interval estimation techniques of Section 7.5 below. In this section, we shall be primarily concerned with the second case, tests in the presence of nuisance parameters, as an application of our results in Chapter 6 on factoring the likelihood function to the decision context.

One approach would be to fix the nuisance parameter at some value, $\boldsymbol{\phi}^*$, and then test the null hypothesis as previously. The significance level so obtained will be a function of $\boldsymbol{\phi}^*$. If this level does not vary very much for $\boldsymbol{\phi}^* \in \Phi$, the appropriate conclusion can be drawn about the null hypothesis. The drawback of this method is that the nuisance parameter is allowed to take any value in its parameter space, when, in fact, we have unused information about its value.

A second approach is to make the test at the m.l.e. of the nuisance parameter, in the same way as in the construction of a profile likelihood. As we have seen in Chapter 6, this assumes that we know exactly the value of the nuisance parameter, which, here, will lead to the true significance level exceeding the size α when the null hypothesis is true.

7.3.2 SIMILAR REGIONS

When $\boldsymbol{\phi}$ is a nuisance parameter, what we would like is that

$$P(\mathbf{Y} \in \mathcal{Y}_\alpha; \boldsymbol{\lambda}_0, \boldsymbol{\phi}) \stackrel{\text{def}}{=} \alpha \qquad \forall \boldsymbol{\phi} \in \Phi \tag{7.25}$$

A region satisfying this criterion is called a *similar region of size* α, with the corresponding *similar test* (Bartlett, 1937).

As we have seen in Chapter 6, the general approach is to condition on the sufficient statistic, $\mathbf{T}_2$, for $\boldsymbol{\phi}$. Here, the situation is somewhat simpler, because we only need require that $\mathbf{T}_2$ be sufficient for $\boldsymbol{\phi}$ when $\boldsymbol{\lambda} = \boldsymbol{\lambda}_0$. Then, the distribution of $\mathbf{Y}$, given $\mathbf{t}_2$, will not depend on $\boldsymbol{\phi}$ when H_0 is true, and the similar test is based on it:

$$P(\mathbf{Y} \in \mathcal{Y}_\alpha | \mathbf{t}_2; \boldsymbol{\lambda}_0) \stackrel{\text{def}}{=} \alpha \tag{7.26}$$

For any particular alternative, $\boldsymbol{\lambda} = \boldsymbol{\lambda}_A$, we can apply the Neyman–Pearson lemma to the conditional distribution to obtain the best critical region of size α. If the result applies for all possible alternatives, we have a *uniformly most powerful similar* (UMPS) region. Evidently, the existence of such regions will, as above, be restricted to the linear exponential family (for $\boldsymbol{\phi}$ with fixed $\boldsymbol{\lambda} = \boldsymbol{\lambda}_o$). When they do not exist, the other approaches of Section 7.2.5 can be applied.

Examples

1. Consider n i.i.d. random variables from a normal distribution with unknown mean and variance, where μ is the parameter of interest. For fixed $\mu = \mu_0$, the sufficient statistic for σ^2 is $S_2 = \sum(Y_i - \mu_0)^2$. In Section 6.3.5, we used a pivotal argument to obtain the conditional distribution, given s_2, a Student t distribution. The *Student t statistic* is

$$T_n = \frac{(\bar{Y}_\bullet - \mu_0)\sqrt{n(n-1)}}{\sqrt{\sum(Y_i - \bar{Y}_\bullet)^2}} \tag{7.27}$$

If the alternative is $\mu > \mu_0$, the critical region will be given by sufficiently large values of t_n, which yields a UMPS test.

2. Consider n independent binary random variables, Y_i, having possible values zero and one, each with corresponding fixed and known explanatory value, x_i. We want to fit a logistic regression, of Equation (4.16),

$$\log\left(\frac{\pi_i}{1-\pi_i}\right) = \beta_0 + \beta_1 x_i$$

which we have already looked at in Section 4.2.2 on exact tests. The sufficient statistic is $\mathbf{T} = (\sum Y_i, \sum x_i Y_i)^{\mathrm{T}}$.

If β_1 is the parameter of interest, for a fixed value, $\beta_1 = \beta_{10}$, t_0, the total number of ones, is sufficient for β_0. In Fisher's exact test, from Equation (4.19), we have seen that the conditional probability is

$$f(t_1 | t_0; \beta_1) = \frac{c(t_0, t_1)\mathrm{e}^{\beta_1 t_1}}{\sum_u c(t_0, u)\mathrm{e}^{\beta_1 u}}$$

As in the previous example, if the alternative to β_{10} is $\beta_1 > \beta_{10}$, the UMPS test is obtained by summing the upper tail probabilities.

3. Consider two independent random variables having Poisson distributions with means μ_i, where $\mu_1 = \lambda\mu_2$ and where we are interested in the ratio of the two means. We shall use the parameters λ and $\phi = \mu_2$, so that the joint distribution is

$$f(y_1, y_2; \lambda, \phi) = \frac{e^{-\phi(1+\lambda)}\phi^{y_1+y_2}\lambda^{y_1}}{y_1!y_2!} \tag{7.28}$$

For fixed $\lambda = \lambda_0$, the sufficient statistic for ϕ is $T_2 = Y_1 + Y_2$, which is independent of the value of λ_0 chosen.

Because T_2 has a Poisson distribution with mean $\mu_1 + \mu_2 = \phi(1 + \lambda)$, the conditional distribution is

$$f(y_1, y_2|t_2; \lambda, \phi) = \binom{t_2}{y_1}\frac{\lambda^{y_1}}{(1+\lambda)^{t_2}} \tag{7.29}$$

a binomial distribution. If the alternative to λ_0 is $\lambda > \lambda_0$, the UMPS test rejects H_0 for large y_1, with a significance level obtained by summing this conditional distribution, at λ_0, over values of y_1 greater than or equal to that observed, for fixed observed total, t_2.

4. Suppose now that we modify the previous example only slightly, so that $\log(\mu_1) = \lambda\log(\mu_2)$ or $\mu_1 = \mu_2^{\lambda}$. Here, the distribution is

$$f(y_1, y_2; \lambda, \phi) = \frac{e^{-\phi^{\lambda}-\phi}\phi^{\lambda y_1+y_2}}{y_1!y_2!} \tag{7.30}$$

so that the sufficient statistic, when $\lambda = \lambda_0$, is $T_2 = \lambda_0 Y_1 + Y_2$, which now depends on λ_0. Similar regions are obtained from the conditional distribution of all non-negative integer pairs, (y_1^*, y_2^*), such that

$$\lambda_0 y_1^* + y_2^* = \lambda_0 y_1 + y_2 \tag{7.31}$$

Because the y_i are integers, no useful similar test exists if λ_0 is irrational. In addition, the conditional probability distribution for constructing similar regions changes drastically as λ_0 varies. □

This last example shows that exact similarity is a very strong requirement. There may exist more powerful tests which are nearly similar, but which have, thus, been excluded by exactness.

When conditioning is used, application of the repeated sampling principle becomes more difficult. The idea of repetitions no longer applies to all observations, but only to those with the same value of the sufficient statistic. This approach has been developed primarily in an attempt to bring frequentist decision-making closer to scientific needs, i.e. to provide useful information from the sample actually observed, by severely restricting the class of what might have been observed. Thus, care must be taken with any long run interpretations, whether, for example, in quality control or in simulation studies.

7.3.3 INVARIANT TESTS

In order to proceed, let us now consider a special case of the sufficiency principle.

Invariance principle
If, for two values of the random variable Y, say y_1 and y_2,

$$f(y_1; \psi) = f(y_2; \psi) \qquad \forall \psi \tag{7.32}$$

then the conclusions to be drawn from y_1 and y_2 are the same. □

We may generalize this by taking a transformation of the sample space, with the associated transformation of the parameter space. Consider a model, $f(\mathbf{y}; \psi)$, and a group of transformations, $\{g : g \in \mathcal{G}\}$: $\mathbf{Y} \rightarrow g\mathbf{Y}$, giving unique transformed parameter value, $g^*\psi$, such that the distribution of the transformed random variable is $f(g\mathbf{y}; g^*\psi)$. Invariance is particularly important when the units of measurement have a physical meaning. As we might expect, it finds application primarily in the location–scale family.

Example
Suppose that the statistic $\mathbf{t}(\mathbf{y})$ is a point estimate of ψ. According to the invariance principle, we should have

$$\mathbf{t}(g\mathbf{y}) = g^*\mathbf{t}(\mathbf{y}) \tag{7.33}$$

□

An *invariant test* is such that the probability distribution of $g\mathbf{Y}$ belongs to the same subset of distributions as that of $\mathbf{Y}$, in the sense that either both satisfy H_0 or both satisfy H_A. The probability distribution of $g\mathbf{Y}$ is obtained from that of $\mathbf{Y}$ by replacing ψ by $g^*\psi$, such that the set, $\mathcal{G}^*$, of these induced parameter transformations is a group preserving both Ψ_0 and Ψ_A:

$$P(\mathbf{Y} \in \mathcal{Y}_1; \psi) = P(g\mathbf{Y} \in \mathcal{Y}_1; g^*\psi) \qquad \forall \mathcal{Y}_1 \subseteq \mathcal{Y} \tag{7.34}$$

for any given $g \in \mathcal{G}$ and for some $g^* \in \mathcal{G}^*$ with

$$\begin{aligned} g^*\Psi_0 &= \Psi_0 \\ g^*\Psi_A &= \Psi_A \end{aligned} \tag{7.35}$$

$$g^*\Psi = \Psi$$

H_0 remains either correct or incorrect, whether we observe $\mathbf{Y}$ or $g\mathbf{Y}$.

Example
Consider a normal distribution with mean μ and variance 1. Let $\mathcal{G}$ be the set of linear transformations, $gy = ay+b$, so that $g^*\mu = a\mu+b$. We wish to be able to draw the same conclusions for $H_0 : \mu = \mu_0$ against $H_A : \mu \neq \mu_0$ as for $H_0^* : \mu^* = a\mu_0 + b$ against $H_A^* : \mu^* \neq a\mu_0 + b$. □

A test with critical region $\mathcal{Y}_\alpha$ is an invariant test if

$$\mathbf{Y} \in \mathcal{Y}_\alpha \Rightarrow g\mathbf{Y} \in \mathcal{Y}_\alpha \qquad \forall g \in \mathcal{G} \tag{7.36}$$

For any group of transformations, $\mathcal{G}$, there will usually not be one unique test available. Thus, we should take $\mathcal{G}$ to be the largest group for which the problem is invariant, such that

$$t(y_1) = t(y_2) \Rightarrow y_2 = gy_1 \tag{7.37}$$

for some $g \in \mathcal{G}$. Then, t is called *maximal invariant* with respect to $\mathcal{G}$.

When all the transformations in $\mathcal{G}$ are applied to a sample point $\mathbf{y}$, they define an *orbit* in the sample space, within which the maximal invariant statistic is constant and unique to that orbit. The maximal invariant statistic creates a partition of the sample space, analogous to that of a sufficient statistic. For any maximal invariant, there will exist a sufficient reduction. However, usually, it is simplest to apply the invariance argument to the sufficient statistic directly.

Example
Consider n i.i.d. observations from a normal distribution. Suppose that the standard deviation is the parameter of interest, with unknown mean. The appropriate transformations are location shifts:

$$gy = y + \delta \tag{7.38}$$

where δ is an arbitrary fixed constant.

The sufficient statistic for (μ, σ^2) is $(\sum Y_i, \sum(Y_i - \bar{Y}_\bullet)^2)$. The orbits are lines parallel to the $\sum y_i$ axis, with constant values of $\sum(Y_i - \bar{Y}_\bullet)^2$. The latter indexes the orbits and is maximal invariant.

This result can also be obtained from the ancillary statistic $\mathbf{T}_3(Y) = (Y_2 - Y_1, \ldots, Y_n - Y_1)^{\mathrm{T}}$, which is a possible maximal invariant, because it can be reduced to the sufficient statistic $T_2 = \sum(Y_i - \bar{Y}_\bullet)^2$. □

Because ψ and $g^*\psi$ are equivalent for all transformations in the group $\mathcal{G}^*$, a maximal invariant parameter, say $\boldsymbol{\lambda} = \boldsymbol{\lambda}(\psi)$, is also produced. The distribution of the maximal invariant statistic depends only on $\boldsymbol{\lambda}$. In this way, the nuisance parameter is removed.

The best invariant test will be determined by the likelihood ratio with the critical region at level α, using the maximal invariant sufficient statistic:

$$\mathcal{Y}_\alpha = \left\{ s : \frac{f_S(t; \boldsymbol{\lambda}_A)}{f_S(t; \boldsymbol{\lambda}_0)} \geq c_\alpha \right\} \tag{7.39}$$

for a simple alternative, $\boldsymbol{\lambda}_A$. If this does not depend on $\boldsymbol{\lambda}_A$, we have a *uniformly most powerful invariant* or UMPI test.

7.3.4 LIKELIHOOD RATIO TESTS

We have now looked at a variety of rather *ad hoc* criteria for optimal tests, including unbiasedness, similarity, and invariance, without any real theory for generating them or means of selecting among them. The problem arises from the absence of a requirement to use all information in the data (Section 1.5) and from the lack of a universal criterion for comparing *sets* of models. In Section 3.1.3, we considered representing the set of models by its most likely member, given the observed data. This criterion is also often used in a testing context.

Thus, the idea of the Neyman–Pearson lemma can be extended in cases where similar or invariant regions are not available. When no points in the parameter space specified by H_0 are preferred to others, the likelihood function can be maximized under the null and alternative hypotheses:

$$\frac{\sup_{\Psi_0} \mathrm{L}(\boldsymbol{\psi})}{\sup_{\Psi_A} \mathrm{L}(\boldsymbol{\psi})}$$

or, equivalently, if $\Psi_A = \Psi \setminus \Psi_0$,

$$\mathrm{R} = \frac{\sup_{\Psi_0} \mathrm{L}(\boldsymbol{\psi})}{\sup_{\Psi} \mathrm{L}(\boldsymbol{\psi})} \tag{7.40}$$

Small values are significant.

However, the optimum properties of likelihood ratios for simple hypotheses, in general, no longer apply, except asymptotically. In addition, as we saw in Chapter 5, the required probabilities are often difficult to obtain without using asymptotic approximations. Thus, the fact that the deviance statistic has an asymptotic distribution which is well known is generally used to obtain significance levels. Wald and Rao score test approximations might also be used, with the same precautions noted in Section 5.6.

Examples

1. For the Weibull distribution of Equation (1.4), the power parameter,

Table 7.6. Deviances for the distribution of Table 7.5.

y	$f(y;\psi)$	$\hat{\psi}_1$	$\hat{\psi}_2$	$-2\log[\mathrm{R}(\psi_1=\alpha,\psi_2=\frac{1}{2})]$
-2	$\psi_1(1-\psi_2)$	α	0	$2\log(2)$
-1	$(\frac{1}{2}-\alpha)\left(\frac{1-\psi_1}{1-\alpha}\right)$	0	-	$-2\log(1-\alpha)$
0	$\alpha\left(\frac{1-\psi_1}{1-\alpha}\right)$	0	-	$-2\log(1-\alpha)$
1	$(\frac{1}{2}-\alpha)\left(\frac{1-\psi_1}{1-\alpha}\right)$	0	-	$-2\log(1-\alpha)$
2	$\psi_1\psi_2$	α	1	$2\log(2)$

$\lambda=\lambda_0$, may be of interest. Then, an appropriate test can be constructed from

$$\mathrm{R}=\frac{\sup_\theta\left[\lambda_0\theta y^{\lambda_0-1}\mathrm{e}^{-\theta y^{\lambda_0}}\right]}{\sup_{\lambda,\theta}\left[\lambda\theta y^{\lambda-1}\mathrm{e}^{-\theta y^{\lambda}}\right]} \tag{7.41}$$

2. Consider again the distribution of Table 7.5 above, with likelihood ratios or deviance statistics as in Table 7.6. The likelihood ratio test rejects H_0 for large values of the test statistic. Because $-2\log(1-\alpha)\leq 2\log(2)$, the size α region will be $\{-2,2\}$. This is the same region which we previously found to be biased. □

7.3.5 SEQUENTIAL TESTS

In a decision-making context, data are usually collected with one specific hypothesis in mind. For example, in pharmaceutical drug testing, this is called the primary end point of a clinical trial and will generally involve the existence of a treatment effect for some new medication. Often, because of the great expense involved in the data collection, they are analyzed for other purposes as well, to extract the maximum information. Here no *prior* hypotheses are usually available. Thus, in such contexts, additional testing on the given set of data is not justified.

Two distinct cases of multiple testing may be distinguished:

(1) as data accumulate, a series of tests of the same null hypothesis may be made, usually to determine if it is necessary to continue observation (Sections 3.3.2 and 7.2.4);

(2) on a given data set, a series of tests is made on different null 'hypotheses' in an attempt to simplify the model, i.e. model selection.

A central question is whether the error rate per test or per study should be held constant. The answer will depend on the type of decisions and on the context in which they are to be made. The combined size of several tests, each of size α, will usually be difficult to calculate. Globally, the probability of making observations which contradict the null hypotheses of such a series of tests will be much higher than α.

One common remedy is to apply a Bonferroni correction (Feller, 1957, pp. 100–101, 131), using a size α/m for each of m individual tests. If m is the total number of tests in Table 7.1, this procedure will guarantee that $P(M_{12} \geq 1) \leq \alpha$, whereas testing individually at level α has an average error rate, $\mathrm{E}[M_{12}/m] \leq \alpha$. Evidently, the Bonferroni correction is most suitable for the first case mentioned above, when one null hypothesis is repeatedly tested and the total probability of rejecting it when true must be α. Other, more complex corrections (for example, O'Brien and Fleming, 1979) are also used.

Model selection Sequential testing is commonly used for model selection, the second case above, especially in a regression context to determine which explanatory variables are unnecessary. However, as pointed out at the beginning of this chapter, this is not a legitimate use of testing; it necessarily leads to incompatible decisions (analogous to the incompatible inferences of Section 3.6.1). The sequence of specific hypotheses cannot generally be planned in advance, so that this is usually a hypothesis generating situation for which testing is not designed.

Example

Consider a model for the mean of a normal distribution, having unit variance, defined by

$$\mu_{ij} = \alpha_i + \beta_j \tag{7.42}$$

with each of two explanatory variables taking values one and minus one in a balanced factorial design. The parameters are design orthogonal. With a Wald test, the square of the ratio of each parameter estimate to its standard error will have a Chi-squared distribution with one degree of freedom and, because of orthogonality, their sum will have a Chi-squared distribution with two degrees of freedom under the null hypotheses that the appropriate parameters are zero.

Suppose that we perform separate tests for the two parameters and that each of the individual test statistics has a value of 3. This indicates that neither is significantly different from zero at the 5% level. Because of orthogonality, one parameter can be eliminated, then the other, conditional on the first being zero, in either order. However, if, instead, we performed a simultaneous test for both being zero, the statistic would have a value of 6 with two degrees of freedom, indicating, again at the 5% level, that neither could be eliminated.

Here, a Bonferroni correction makes matters worse. The two tests for individual elimination will be made at the 2.5% level, providing even less indication that either parameter is zero, while the simultaneous test would remain unchanged. □

Thus, we see that attempts to maintain a fixed long run probability level in multidimensional models aggravate the problem of compatibility. If one parameter is tested for elimination at a time, the resulting model will generally be much simpler than if a large number of parameters is tested for removal simultaneously.

7.4 Point estimation

The second fundamental basis of the frequentist approach is *point estimation*. A point estimator, instead of an interval estimate, is usually only required when some decision must be made, usually in an industrial setting.

Here, it is traditional to distinguish between a point *estimator*, say $T(Y)$, a function of a random variable, and its observed value, the point *estimate*, $t(y)$. A minimum requirement for a point estimator is usually that it be consistent. However, as we have seen in Section 5.3.2, the usual estimators, including those which will be discussed below, are not generally consistent for continuous distributions if based on the density approximation to the likelihood.

7.4.1 BIAS AND VARIANCE

Bias Suppose that T is a statistic which may be used as a point estimator for some parameter, ψ. Then, the *bias* of T is defined as

$$\text{bias}(\psi) \stackrel{\text{def}}{=} \text{E}[T] - \psi \tag{7.43}$$

We would hope for zero bias, called *unbiasedness*. When used repeatedly, such an estimator, in the long run, will estimate the true value on average.

Examples

1. For the exponential family, we have, from Equation (3.107),

$$\text{E}[\mathbf{T}] = \boldsymbol{\mu}$$

so that the canonical statistic is an unbiased estimator of the mean value parameter.

2. For the normal distribution, an unbiased estimator of the variance, σ^2, with unknown mean, is

$$S^2 = \frac{1}{n-1}\sum_i (Y_i - \bar{Y}_\bullet)^2 \tag{7.44}$$

which has $\text{E}[S^2] = \sigma^2$.

Suppose that $\sum_i (y_i - \bar{y}_\bullet)^2 = 125$ with $n = 50$. Then, the m.l.e. is $\hat{\sigma}^2 = 2.5$ while the unbiased estimate is 2.551. If the unit of measurement is $\Delta = 1$, Sheppard's correction is $\Delta^2/12 = 0.083$, which must be subtracted from the m.l.e. Thus, the unbiased estimator is correcting in the wrong direction for it to be a consistent estimator when there is finite measurement precision. In addition, for larger samples, the difference between the m.l.e. and the unbiased estimate quickly disappears, but Sheppard's correction does not because it does not depend on n.

3. For the normal distribution, the standard deviation, σ, is usually required; most often it is calculated from the unbiased estimator of the variance, S^2, in Equation (7.44). We, then, have

$$\mathrm{E}[S^2] = (\mathrm{E}[S])^2 + \mathrm{var}[S] \tag{7.45}$$

so that $\mathrm{E}[S] < \sigma$, a biased estimator.

4. For the geometric distribution,

$$f(y; \pi) = \pi(1-\pi)^{y-1} \qquad y = 1, 2, \ldots \tag{7.46}$$

an unbiased estimator of π, say $T(Y)$, must satisfy

$$\sum_{Y=1}^{\infty} T(Y)\pi(1-\pi)^{Y-1} = \pi \tag{7.47}$$

for any π. By polynomial expansion, the only estimator satisfying this condition is $T(1) = 1$, $T(Y) = 0, Y > 1$, which is neither very plausible, nor very useful. □

Any Fisher consistent estimator, including the m.l.e., will be asymptotically unbiased. As well, the approach which we studied in Chapter 6, using a conditional likelihood, can often reduce or remove the bias.

Example

We saw, in Equation (6.31) for the marginal or conditional likelihood of the variance from a normal distribution, that the resulting m.l.e. is just that of Equation (7.44). □

Mean squared error The *mean squared error* of a statistic T, used as an estimator, is

$$\begin{aligned} \mathrm{mse}(T; \psi) &\stackrel{\text{def}}{=} \mathrm{E}[(T-\psi)^2] \\ &= \mathrm{var}[T] + [\mathrm{bias}(\psi)]^2 \end{aligned} \tag{7.48}$$

We generally look for a small mean square error. As a criterion for choosing a point estimator, this is usually more important than unbiasedness. Thus, for example, a slightly biased estimator will usually be preferable if it has smaller mean squared error. However, imposing the criterion of unbiasedness greatly simplifies the task of minimizing mean squared error because then, from Equation (7.48), we only have to minimize var$[T]$.

Both of these criteria have been derived with symmetric distributions for independent observations in mind. Otherwise, for example, a definition of bias in terms of deviation from the mode might be more reasonable. Mean squared error entails the supplementary assumption that the variance does not depend on the mean (Section B.2.1). Note also that these both involve averages over the sample space of possible observations and, thus, violate the sufficiency principle.

We must now investigate under what conditions we can obtain point estimators with these characteristics. Let us first look at a well-known result.

Cramér–Rao lower bound

An estimator, T, has minimum variance if it reaches the Cramér–Rao lower bound:

$$\begin{aligned}\mathrm{var}[T] &\geq \frac{\left[1+\frac{\partial \mathrm{bias}(\psi)}{\partial\psi}\right]^2}{\mathrm{I}(\psi)} \\ &= \frac{\left[\frac{\partial \mathrm{E}[T]}{\partial\psi}\right]^2}{\mathrm{I}(\psi)}\end{aligned} \tag{7.49}$$

Equality holds iff the estimator is a linear function of the score. (Rao, 1945; Neyman and Scott, 1948)

proof: We have

$$\begin{aligned}\mathrm{E}[T] &= \int t(y)f(y;\psi)dy \\ &= \psi + \mathrm{bias}(\psi)\end{aligned}$$

by definition. Differentiating with respect to ψ, assuming suitable regularity conditions, we obtain

$$\begin{aligned}1+\frac{\partial \mathrm{bias}(\psi)}{\partial\psi} &= \int t(y)\frac{1}{f(y;\psi)}\frac{\partial f(y;\psi)}{\partial\psi}f(y;\psi)dy \\ &= \int t(y)\frac{\partial \log[f(y;\psi)]}{\partial\psi}f(y;\psi)dy \\ &= \mathrm{E}[T\mathrm{U}(\psi)]\end{aligned}$$

$$= \mathrm{cov}[T, \mathrm{U}(\psi)]$$

because the expected value of the score statistic is zero. But, by the Cauchy–Schwarz inequality of Equation (B.99),

$$\mathrm{var}[T]\mathrm{var}[\mathrm{U}(\psi)] \geq (\mathrm{cov}[T, \mathrm{U}(\psi)])^2$$

with equality iff T and $\mathrm{U}(\psi)$ are linear functions of one another. Then, because $\mathrm{var}[\mathrm{U}(\psi)] = \mathrm{I}(\psi)$ from Section 5.1.2,

$$\mathrm{var}[T]\mathrm{I}[\psi] \geq \left[1 + \frac{\partial \mathrm{bias}(\psi)}{\partial \psi}\right]^2$$

as required. □

An estimator which is unbiased and attains this lower bound,

$$\mathrm{var}[\mathbf{T}] = \mathbf{I}^{-1}(\psi) \tag{7.50}$$

is called *minimum variance unbiased.* If the estimator fulfils this condition for all values of the parameter, it is said to be uniformly minimum variance unbiased (UMVU). Notice that, in contrast to UMP tests, a UMV estimator will not exist because a constant will have zero variance when it is the true value. Thus, some criterion such as unbiasedness must be applied.

Example
As we saw in Section 3.7.6, Equations (3.107), (3.111), and (3.115), for the linear exponential family, with log likelihood function

$$\mathrm{l}(\boldsymbol{\theta}) = \boldsymbol{\theta}^{\mathrm{T}}\mathbf{t}(\mathbf{y}) - \kappa(\boldsymbol{\theta}) - t_0^*(\mathbf{y})$$

we have

$$\mathrm{E}[\mathbf{T}] = \boldsymbol{\mu} \tag{7.51}$$
$$\mathrm{var}[\mathbf{T}] = \mathbf{I}_{\mu}^{-1}(\boldsymbol{\theta}) \tag{7.52}$$

so that the sufficient statistics are the minimum variance unbiased estimators of the mean value parameters.

On the other hand, in most cases, $\boldsymbol{\theta}$ is a non-linear function of $\boldsymbol{\mu}$, so that no linear function of $\mathbf{T}$ can have expectation $\boldsymbol{\theta}$ and the lower bound cannot be attained: no minimum variance unbiased estimator exists for the canonical parameter. An exception is the normal distribution with known variance where $\theta_1 = \mu/\sigma^2$ is a linear function. □

Generally, the Cramér–Rao lower bound can only be achieved in exceptional circumstances. As well, decision-making with unbiased estimators is not parametrization invariant, because all one-to-one transformations of the parameter cannot have minimum variance unbiased estimators. Estimators with small, known, bias can have smaller variance than the unbiased estimator. Thus, such a criterion is essentially designed with the normal distribution and constant variance in mind and is of limited usefulness, except asymptotically, especially in asymmetric distributions where higher (odd) cumulants are non-zero.

These general results can be further clarified by the following two important theorems:

Rao–Blackwell theorem

Any function of a complete sufficient statistic is the unique minimum variance unbiased estimator of its expectation. (Rao, 1945)

proof: Suppose that T is the required sufficient statistic for ψ, T_u an unbiased estimator, and $S = \mathrm{E}[T_u|T]$, which will be a function only of T and not of ψ. Then,

$$\begin{aligned}\mathrm{E}[S] &= \mathrm{E}\{\mathrm{E}[T_u|T]\}\\ &= \mathrm{E}[T_u]\\ &= \psi\\ \mathrm{var}[T_u] &= \mathrm{var}\{\mathrm{E}[T_u|T]\} + \mathrm{E}\{\mathrm{var}[T_u|T]\}\\ &= \mathrm{var}[S] + \mathrm{E}\{\mathrm{var}[T_u|T]\}\end{aligned}$$

Thus, S is unbiased and

$$\mathrm{var}[S] \leq \mathrm{var}[T_u]$$

with equality iff $\mathrm{var}[T_u|T] = 0$, i.e. iff the unbiased estimator is a function of the sufficient statistic with probability one. Because this holds for any unbiased estimator of ψ, the conclusion follows. □

Example

We just saw that this is true for the mean value parameter in the exponential family. □

Gauss–Markov theorem

For the vector of random variables $\mathbf{Y}$, with means

$$\boldsymbol{\mu} = \mathbf{X}\boldsymbol{\beta} \tag{7.53}$$

known covariance matrix $\boldsymbol{\Sigma}$, and known values of $\mathbf{X}$, the weighted 'least squares' estimator of the p parameters

$$\hat{\boldsymbol{\beta}} = (\mathbf{X}^{\mathrm{T}}\boldsymbol{\Sigma}^{-1}\mathbf{X})^{-1}\mathbf{X}\boldsymbol{\Sigma}^{-1}\mathbf{y} \tag{7.54}$$

is the unbiased minimum variance linear estimator.

proof: Any linear estimator can be written in the form

$$\hat{\boldsymbol{\beta}} = [\mathbf{A} + (\mathbf{X}^{\mathrm{T}}\boldsymbol{\Sigma}^{-1}\mathbf{X})^{-1}\mathbf{X}\boldsymbol{\Sigma}^{-1}]\mathbf{y} \tag{7.55}$$

for some matrix $\mathbf{A}$. This has expectation

$$\mathrm{E}[\hat{\boldsymbol{\beta}}] = (\mathbf{A}\mathbf{X} + \mathbf{I}_p)\boldsymbol{\beta} \tag{7.56}$$

where $\mathbf{I}_p$ is the $p \times p$ identity matrix, so that $\mathbf{A}\mathbf{X} = \mathbf{0}$ if it is unbiased. It has variance

$$\begin{aligned}\mathrm{var}[\hat{\boldsymbol{\beta}}] &= [\mathbf{A} + (\mathbf{X}^{\mathrm{T}}\boldsymbol{\Sigma}^{-1}\mathbf{X})^{-1}\mathbf{X}\boldsymbol{\Sigma}^{-1}]\boldsymbol{\Sigma}[\mathbf{A} + (\mathbf{X}^{\mathrm{T}}\boldsymbol{\Sigma}^{-1}\mathbf{X})^{-1}\mathbf{X}\boldsymbol{\Sigma}^{-1}]^{\mathrm{T}} \\ &= \mathbf{A}\boldsymbol{\Sigma}\mathbf{A}^{\mathrm{T}} + (\mathbf{X}^{\mathrm{T}}\boldsymbol{\Sigma}^{-1}\mathbf{X})^{-1} \end{aligned} \tag{7.57}$$

because the cross-product is zero by the condition above. The first term is non-negative and the second positive definite, so that the minimum variance occurs when $\mathbf{A}$ is identically zero, the weighted 'least squares' estimator. □

Because no assumption of normality has been made, this theorem may appear to be quite general. However, it is difficult to find distributions such that variances and covariances are known in this way. Thus, it does not apply to the exponential family or generalized linear models (except the normal distribution), because the variance will depend on the mean, and hence on the unknown $\boldsymbol{\beta}$.

From this series of results, we see that unbiasedness is primarily useful for the exponential family, and often only for the normal distribution. For transformation families, such as the location–scale, invariance arguments can be applied (see Lehmann, 1983, pp. 154–235).

Asymptotic relative efficiency In evaluating point estimators, asymptotic criteria are often used. Then, they can be interpreted as the limit of a sequence of observable quantities. Suppose that the estimators, T_i, have asymptotic distributions,

$$\sqrt{n}(T_i - \psi) \sim \mathrm{N}(0, \sigma_i^2) \tag{7.58}$$

Then, the asymptotic relative efficiency of two such estimators is given by the ratio of variances, σ_i^2/σ_j^2. A value of 2, for example, would mean that the estimator, T_i, requires twice as many observations for the same asymptotic precision as does T_j.

7.4.2 DECISION THEORY

Mean squared error Specifying a loss function is a difficult practical problem. For a point estimator, $\mathbf{T}$, a common loss function is mean squared error,

$$\text{loss}(\boldsymbol{\psi}, d_t) \stackrel{\text{def}}{=} (\mathbf{t} - \boldsymbol{\psi})^{\mathrm{T}}(\mathbf{t} - \boldsymbol{\psi}) \tag{7.59}$$

not because it is generally thought to be a good approximation to actual losses, but because it simplifies the mathematical problem. Minimum variance estimation corresponds to a mean squared error loss function. More generally, convex loss functions lead to a number of simplifying properties.

Example
For a normal distribution, with known variance, we obtain n i.i.d. observations, and, in a particular context, wish to make a decision about whether the true mean is μ_0. Suppose that, for point estimator T, the utility function is

$$u(\mu, d_T) = a - (\mu - T)^2$$

so that the further the estimate is from the true value, the smaller the utility. If we know the value of μ, we take $T = \mu$ and have utility a. The loss function is

$$\text{loss}(\mu, d_T) = (\mu - T)^2$$

Two possible decision rules are to estimate the mean to be

$$\begin{aligned} \delta_1(Y) &= \bar{Y}_\bullet \\ \delta_2(Y) &= \mu_0 \qquad \forall Y \end{aligned}$$

Then, for a mean squared error loss function, the risk functions are

$$\begin{aligned} \text{risk}_F(\mu, \delta_1) &= \frac{\sigma^2}{n} \\ \text{risk}_F(\mu, \delta_2) &= (\mu - \mu_0)^2 \end{aligned}$$

The first is constant for all values of the mean, while the second is a parabola with zero risk when μ_0 is at the true mean, but very large risks as it deviates from that value. □

The faster a loss function increases, the more sensitive it will be to extreme values. Then, the performance of the resulting point estimators will be more strongly influenced by the tail behaviour of the distribution.

Example
Loss functions

$$\text{loss}(\mu, d_T) = |\mu - T|^p \qquad 1 < p < 2 \tag{7.60}$$

with a power between 1 and 2, will be more robust than a quadratic function.

□

Akaike information criterion A second important loss function is based on the Kullback–Leibler divergence of Equation (3.60):

$$\text{loss}(\psi, d_t) = 2 \int \log\left[\frac{f(y;\psi)}{f(y;\hat{\psi})}\right] f(y;\psi)dy \tag{7.61}$$

This can be estimated by the average deviance, D/n, so that the m.l.e. (approximately) minimizes this loss function. As we saw in Section 3.7.6, this will be exact for the exponential family. Then, using complex asymptotic approximations, one can show that the risk can be estimated by $\mathrm{D} + 2p$, where p is the dimension of the m.l.e. This is the Akaike information criterion (AIC), which can be used to compare nested models of different dimensions (Akaike, 1973). It can be obtained from the calibrated deviance of Equation (3.69) by setting $a = 1/e$.

Example
In quality control, a product, k, is weighed n times on the same precision balance. Different products are so weighed independently on different balances, all with known unit variance. Let us assume normal distributions with theoretical means, μ_k. For one product, a standard significance test compares $t_k = n(\bar{y}_k - \mu_k)^2$ to a Chi-squared distribution on one degree of freedom. Suppose that, for each of p such products, $t_k = 3.5$. Individually, we have no evidence, at the 5% level, that $\mathrm{E}[Y_k] \neq \mu_k$, so that we can accept them all.

However, now let us perform a combined test for the p products bought by a client. The test statistic will be $3.5p$, which, judged by the 5% level of the Chi-squared distribution with p degrees of freedom, provides strong evidence that the observed means are not equal to their hypothetical values. The whole batch of products must be rejected!

As the number of independent products grows, each with $t_k = 3.5$, so does the contradiction between the two methods of testing: 'evidence' accumulates that all observed means are different from their respective hypothetical values, although individually they are not. Compatible decisions will only be possible if the deviance is calibrated by the addition of some constant times the number of parameters involved. This is exactly what the AIC does. □

Thus, the AIC is one frequentist solution to the model selection problem, for which, as we saw in Section 7.3.5, testing does not provide suitable answers. However, this method will not be consistent in selecting the true model between two rivals with a difference of p parameters (when one exists), because the probability of rejection will be $P(\chi_p^2 > 2p)$, not zero, asymptotically.

James–Stein estimator The introduction of the principles of risk and loss is not neutral to the conclusions drawn.

Example
Consider an n-dimensional vector of independent random variables $\mathbf{Y}$, having a multivariate normal distribution with mean $\boldsymbol{\mu}$ and identity covariance matrix. The maximum likelihood (and 'least squares') estimate is $\hat{\boldsymbol{\mu}} = \mathbf{y}$. However, the estimator,

$$\tilde{\boldsymbol{\mu}} = \left[1 - \frac{n-2}{\sum Y_i^2}\right] \mathbf{Y} \tag{7.62}$$

known as the James–Stein estimator, has smaller mean squared error for $n \geq 3$ (James and Stein, 1961). Notice that the estimate for each individual μ_i uses all observed y_i, although they arise from *independent* (even unrelated) random variables. □

This paradoxical result arises because the same loss function is assumed for all n random variables, inducing a dependence among them. It increases the error in estimating certain individual values in order to reduce the overall average error, not a procedure one would want to use in testing people for AIDS! (Such an estimator is sometimes said to gain strength from the other observations, but might as easily be said to add weakness due to them.)

This estimator is closely related to those obtained from the random effects models of Section 6.4.1 (see Exercise 6.27). However, there the statistical model, not a model for human behaviour, i.e. the decision-making procedure, induced a dependence relationship among the observations.

The James–Stein estimator is known as a shrinkage estimator because it pulls all values towards the overall mean.

- If all observed values are close to the global mean, they will be shrunk considerably, typically improving accuracy.
- If there are some large deviations from the overall mean or a lot of moderate deviations, there will be little shrinkage.

- However, if most observations are close to the global mean, with a few moderately different, the latter will be greatly shrunk, although they represent the interesting cases.

(Lehmann, 1983, pp. 308–309).

Admissibility If there exists a decision function, δ, such that

$$\text{risk}_F(\psi, \delta_1) \geq \text{risk}_F(\psi, \delta) \qquad \forall \psi \tag{7.63}$$

then δ_1 is said to be *inadmissible* (Wald, 1939). If the set of possible risk functions is not closed, it may happen that no decision function is admissible.

Example
Neither of our decision functions based on mean squared error above is inadmissible with respect to the other, because δ_1 has smaller risk far from the true mean and δ_2 smaller near the true mean. □

Decision-making using a loss function is not generally parametrization invariant. In certain decision contexts, it might be argued that estimation of ψ and of $g(\psi)$ are different decision problems even when $g^{-1}(\cdot)$ exists. Having an estimate, t, close to ψ does not necessarily mean that $g(t)$ is considered to be close to $g(\psi)$. Different losses would then be involved in the two cases.

In applying the frequentist model for utility theory to statistical decision problems, the reader may like to reflect on the paradoxes of marginal probabilities in Section B.3.7.

7.5 Interval estimation

In the frequentist approach, interval estimation is a direct extension of hypothesis testing. Such intervals simply consist of the set of all null models not rejected at some significance level, i.e. an infinite number of hypothesis tests is performed. Thus, the results for hypothesis testing carry over directly to the construction of intervals for parameters. However, this is a somewhat surprising approach, given the problems which we have already encountered with applying multiple tests to the same data (Section 7.3.5).

7.5.1 CONFIDENCE LIMITS

For a scalar parameter, ψ, the $1-\alpha$ *upper* and *lower confidence limits* for the true value of that parameter are given by the statistics T^u and T_l such that

$$\left.\begin{aligned} P(T^u \geq \psi; \psi) = 1-\alpha \\ P(T_l \leq \psi; \psi) = 1-\alpha \end{aligned}\right\} \qquad \forall \psi \in \Psi \tag{7.64}$$

with the requirement that

$$T^{u_1} \leq T^{u_2} \text{ if } \alpha_1 > \alpha_2$$
$$T_{l_1} \geq T_{l_2} \text{ if } \alpha_1 > \alpha_2$$

Notice that these are probability statements about the statistic, as it will be observed in the long run, not about the parameter, which is taken to be fixed at its true value.

Example
For n i.i.d. observations from a normal distribution with known variance, the confidence limits for μ are

$$\bar{Y}_{\bullet} \pm \frac{k_\alpha \sigma}{\sqrt{n}} \tag{7.65}$$

where k_α is obtained from a standard normal probability integral, as defined in Equation (7.14) above. □

Often, an easy way to derive confidence limits is from the distribution of a pivotal quantity $Q(Y, \psi)$ (Section 3.5.4). Because this distribution does not depend on the parameter, a value, $p_{1-\alpha}$, such that

$$P[Q(y, \psi) \geq p_{1-\alpha}; \psi] = 1 - \alpha \tag{7.66}$$

can easily be found. If, for given y, Q is a monotonic decreasing function of ψ,

$$q(y, \psi) \geq p_{1-\alpha} \leftrightarrow \psi \leq T^u \tag{7.67}$$

where T^u is the upper limit. For an increasing function, we have lower limits.

Example
For the normal distribution example, $\bar{Y}_{\bullet} - \mu$ is an appropriate pivotal quantity, giving the above result, with $p_{1-\alpha}$ proportional to $k_{1-\alpha}$. □

When the pivotal quantity is based on a sufficient statistic, this is essentially the derivation of a fiducial probability. Because it must use all of the information in the data, Fisher (1930) claimed it to be a probability statement about the parameter, not the statistic. In contrast, confidence limits, which do not involve this requirement (although they will sometimes fulfil it), are probability statements about the statistic.

The same optimality criteria as for hypothesis testing, such as uniformly most powerful, may be applied. Confidence limits are formed from those values not rejected in a test at the required level. However, we also come up against the same problems and contradictions.

Example
In the exponential family, the directed deviance is monotonic with the sufficient statistic, so that

$$P(T(\mathbf{Y}) \leq t; \theta) = P(\mathrm{D}_d(\theta; T) \leq \mathrm{D}_d(\theta; t); \theta) \tag{7.68}$$

A first-order normal approximation is

$$P(T(\mathbf{Y}) \leq t; \theta) \doteq \Phi[\mathrm{D}_d(\theta; t)] \tag{7.69}$$

with $\Phi(\cdot)$ from Equation (7.14). □

Confidence limits obtained as in Equation (7.69) are parameter values where the log likelihood takes a value $[\Phi^{-1}(\alpha)]^2/2$ less than its maximum. This involves the assumption that the variance is constant for all values of the parameter of interest (Section 5.3.4). Thus, if the graph of the likelihood function is asymmetric, approximate normality of the m.l.e. is questionable and asymptotic standard errors will be misleading. Generally, limits based on the directed deviance are to be preferred. This was illustrated by our examples in Sections 5.3.3 and 6.3.3.

7.5.2 CONFIDENCE REGIONS

If upper and lower limits are taken in combination, this is called a *confidence interval.* Thus, a $1 - \alpha$ confidence interval, $[T_l, T^u]$, is defined by

$$P(T_l \leq \psi \leq T^u; \psi) = 1 - \alpha \qquad \forall \psi \in \Psi \tag{7.70}$$

Because any upper and lower limits at levels α_1 and α_2, such that $\alpha_1 + \alpha_2 = \alpha$, satisfy this requirement, there is no unique way of obtaining such an interval, without imposing further conditions. Usually, one selects the narrowest possible interval, which will generally be based on the likelihood function.

Example
For a scalar parameter, the confidence interval that is arbitrarily empty for 5% of data sets and the whole real line for the other 95% will always contain the true value with the required frequency. However, it will have infinite expected length. □

If the likelihood is multimodal, a confidence interval may include parameter values with lower likelihood than some values outside the interval.

Sometimes the data may be consistent with parameter values in two or more disjoint intervals, but not agree with values in between. In such cases, a disjoint *confidence region* will be required. The preferable criterion is to construct a region such that all parameter values within it have higher likelihood than those outside. This should yield the smallest region. (The proof is similar to that given for shortest credibility intervals in Section 8.5.1 below.)

Thus, if the asymptotic methods of Chapter 5 are to be used, those based on the deviance statistic are preferable to ones using the Wald or Rao score tests. Then, if the likelihood function has several maxima of roughly equal magnitude, a confidence region will consist of the appropriate disjoint parts. If the likelihood approaches a maximum only at a bound, for example when $\psi \rightarrow \infty$, as in Figure 3.1, the region will consist of all sufficiently large values. If the likelihood is zero for some values of the parameter, these will never be included in the region.

When nuisance parameters are present, similar or invariant arguments can be used to obtain confidence regions. If we require a confidence region for a vector of parameters, perhaps in the presence of other nuisance parameters, no uniformly most powerful critical regions will exist. The usual solution is a likelihood-based region.

Confidence regions are probability statements about the statistics defining the region and not about the unknown parameters. They state that, if the study is repeated, in the long run the corresponding series of regions will contain the true parameter value with frequency $1 - \alpha$, but, of course, the region changes with each repetition. They do not state that a region constructed from a given data set will contain the unknown parameters with the specified probability. However, if confidence regions are constructed after a model selection process, they will be too narrow, not allowing for the uncertainty in the latter and hence will not have the stated probability level under repeated sampling.

If properly constructed, confidence regions can also be given an interpretation analogous to that of pure significance tests. For some given data set, if a model outside the region has generated these observed data, then an improbable event has occurred under that model. Only models within the region make the data relatively probable. Nevertheless, being based on tests, regions for parameters of differing dimensions will generally be incompatible. They fix the probability of the multidimensional region excluding at least one parameter, whereas compatible regions fix the (average) probability per parameter.

Summary

Much of frequentist theory has been developed with only the symmetric normal distribution in mind. This is reinforced by the need for a heavy reliance on asymptotic procedures which use this distribution.

In a decision situation, parameters are often taken to have meaning out of the context of their probability model. Thus, pseudo-likelihood procedures (Section 6.5), such as estimating equations, have been developed by frequentists. Much of frequentist theory can appear *ad hoc* because it need not be model-based, not relying on the likelihood function nor any other single unifying principle, not even having the requirement that all of the information in the data must be used (see Section 1.5). This has led to the development of a set of criteria for judging the operating characteristics of decision-making procedures, which, however, do not take into account the uncertainty of any model selection performed and which may be mutually contradictory.

If long run properties are taken seriously, anything which might have happened in the study must be taken into account. For example, it is almost invariably the case that some data might have been missing in any study. Then the missing data mechanism (Sections 6.4.2 and 6.4.3) must be included in the model if it is not MCAR, modifying the definition of the sample space, even when no data are actually missing. Conditioning on a relevant subset alleviates certain of such problems, but destroys the long run properties of the decisions.

Frequentist testing is widely misused as an omnibus model selection procedure. In addition, few if any methods are available to allow for uncertainty in any selection process when confidence regions are constructed for the parameters of the final model (Chatfield, 1995).

If the chosen decision procedure uses the likelihood function, with appropriate conditioning, the conclusions can be similar to those obtained as inferences in previous chapters. (However, extreme care must be taken with long run interpretations when conditional methods are used.) In other cases, the conclusions from a given set of observations may be very wrong, although the long run frequency of error is being minimized. In certain situations, such as industrial quality control, this can be desirable. In a scientific context, it should be avoided.

Further reading

The frequentist decisional model originated with Neyman and Pearson, who denied that they were making inferences, in opposition to Fisher, whose scientific approach was developed in Part II above. All elements of this debate make fascinating reading; see, especially, Fisher (1958, originally published in 1925) and Neyman and Pearson (1933). Unfortunately, little is available in one central place, showing how the conflict evolved. Bartlett (1937) made important early contributions to the development of conditional testing.

Textbooks on frequentist decision-making abound. A standard reference in the decision-making context is Lehmann (1983, 1986). Cox and Hinkley (1974) provide the principal attempt to adapt frequentist procedures to scientific needs, through conditioning techniques.

7.6 Exercises

(1) Suppose that we obtain one observation from a normal distribution with mean μ. With probability 0.5, the variance is one and with probability 0.5, it is 10^6, depending on the machine used for measuring. It is known which machine was used.

  (a) Use the Neyman–Pearson lemma to construct a test for $\mu = 0$ against $\mu = \mu_A$.

  (b) Use an appropriate ancillary statistic to construct a better test.

(Cox and Hinkley, 1974, pp. 95–96)

(2) For one observation from a normal distribution with variance one, consider $H_0 : \mu = 0$ and $H_A : \mu = 2$. Two possible test statistics are Y and $|Y|$.

  (a) Construct size α critical regions for the two test statistics.

  (b) Compare the power of the two tests.

  (c) Is one of them UMP?

(Kalbfleisch, 1985b, p. 191)

(3) Consider a uniform distribution on $(\psi - 0.5, \psi + 0.5)$, as in Exercise (3.8).

  (a) What are the possible values for the likelihood ratio comparing ψ_0 to ψ_A?

  (b) With two observations, what is the minimal sufficient statistic?

  (c) What distribution is suitable for making decisions?

  (d) What levels of significance are possible?

  (e) What are the advantages and disadvantages of basing decisions only on the mean?

(Cox and Hinkley, 1974, p. 97; Barndorff-Nielsen and Cox, 1994, pp. 34–35)

(4) In a factory, a sample of n items is randomly selected from a lot of N and each inspected for defects. The unknown total number of defects in the lot is D, while d are found in the sample.

  (a) Develop a suitable model.

  (b) Find a UMP test for the null hypothesis that $D \leq D_0$ against the alternative, $D > D_0$.

(Lehmann, 1986, p. 80)

(5) For n i.i.d. random variables with density

$$f(y; \psi) \propto \mathrm{e}^{-\frac{y^2}{2} - \psi y^4} \tag{7.71}$$

find the UMP test for $H_0 : \psi = 0$ against $H_A : \psi > 0$.

(Cox and Hinkley, 1974, p. 44)

(6) Consider two independent random variables, Y_1 and Y_2, with gamma distributions, as in Equation (1.3) with $\alpha = 2$ and different means.
   (a) Derive a UMPU test of size α for H_0: $\mu_1 = \mu_2$ against H_A: $\mu_1 < \mu_2$.
   (b) Suppose now that observations are censored at y_c so that only the minimum of y_i and y_c is actually observed. What is the UMPU test now?
   (c) It may occur that censored data will produce observations where the uncensored test can be applied. Give an example.
   (d) In this latter case, will the two tests always provide the same result?

   (Sackrowitz and Samuel-Cain, 1994)

(7) For two independent samples of n_1 and n_2 i.i.d. random variables from the geometric distribution of Equation (7.46) with parameters π_1 and π_2, find a UMPS test for $H_0 : \pi_1 = \pi_2$ against $H_A : \pi_1 > \pi_2$. (Cox and Hinkley, 1974, pp. 54–55)

(8) For n i.i.d. random variables from an exponential distribution with lower bound ψ_1,

$$f(y;\psi) = \frac{1}{\psi_2} e^{-\frac{y-\psi_1}{\psi_2}} \qquad y \geq \psi_1 \tag{7.72}$$

   (a) Construct a similar or invariant test for $H_0 : \psi_1 = 0$ against $H_A : \psi_1 > 0$.
   (b) Construct a similar or invariant test for $H_0 : \psi_2 = 1$ against $H_A : \psi_2 \neq 1$.

   (Cox and Hinkley, 1974, pp. 58–59)

(9) Consider the exponential distribution of Equation (1.14).
   (a) Write down the deviance, score, and Wald statistics.
   (b) What would be appropriate tests for $\mu = \mu_0$?
   (c) What changes occur if the observed information is replaced by the expected information at μ_0?
   (d) Discuss the relative advantages of the different tests.

   (Barndorff-Nielsen and Cox, 1994, pp. 83–84)

(10) For y successes out of n i.i.d. Bernoulli trials, the m.l.e. of the binomial probability is $\hat{\pi} = y/n$. Suppose that n was fixed.
   (a) Is this an unbiased estimator?
   (b) Suppose that we are interested in π^2. What is the m.l.e.?
   (c) Is this an unbiased estimator?
   (d) If not, what is the bias?
   (e) Give an unbiased estimator.

(f) Find an unbiased estimator for $\pi(1-\pi)$ and compare it to the m.l.e.

(g) If we are interested in $1/\pi$, does an unbiased estimator exist?

(h) Show that a function of π only has an unbiased estimator if it is a polynomial of degree $\leq n$.

(Kalbfleisch, 1985b, p. 130; Lehmann, 1983, pp. 75, 81, 91–93)

(11) The random variable Y takes values $-1, 0, 1, \ldots$ with probability distribution

$$f(-1) = \pi$$
$$f(i) = (1-\pi)^2\pi^i \qquad i \geq 0$$

(a) What is an unbiased estimator of π?

(b) Of $(1-\pi)^2$?

(Lehmann, 1983, pp. 76–77)

(12) Suppose that Y has a Poisson distribution with mean μ. The parameter of interest is $\exp(-c\mu)$, with c a known constant.

(a) What is the UMVU estimator for one observation?

(b) Compare this to the m.l.e. for $c = 1, 2, 3$.

(c) Now suppose that $n > 1$ observations are made. Compare the behaviour of the two estimators.

(Lehmann, 1983, pp. 114–115; Bernardo and Smith, 1994, p. 463)

(13) For the normal distribution, in Equation (7.44) we gave an unbiased estimator of σ^2.

(a) Calculate the bias of the m.l.e. for σ^2.

(b) What is the variance of the unbiased estimator?

(c) What is the Cramér–Rao lower bound on the variance?

(Cox and Hinkley, 1974, p. 256)

(14) Suppose that we have 10 independent observations from the exponential distribution of Equation (1.14) with $\hat{\mu} = 28.8$.

(a) Find an approximate 95% confidence interval for μ.

(b) Try some other approximations and compare the results.

(c) Can an exact interval be derived?

(Kalbfleisch, 1985b, p. 117)

(15) Suppose that 17 successes are observed in 100 i.i.d. Bernoulli trials.

(a) Find an approximate 95% confidence interval.

(b) Try some other approximations and compare the results.

(Kalbfleisch, 1985b, p. 116)

(16) The model for a Brownian motion process over time t (Section 2.7.2) with drift is a normal distribution with $\mathrm{E}[Y_t] = \mu t$ and $\mathrm{var}[Y_t] = t$. Consider the null hypothesis, $\mu = \mu_0$.

(a) Write down the likelihood function and the m.l.e.
(b) Give the deviance statistic.
(c) Suppose that the stopping rule is that the process is observed until a fixed time, t_s. Then, the terminal position, Y_{t_s}, is a normal random variable with mean and variance as above. What is the distribution of the deviance statistic?
(d) Instead, suppose, for $\mu_0 > 0$, that observation continues until Y_t reaches $y_0 > 0$ for the first time. Now, T is a random variable with an inverse Gaussian distribution, Equation (2.30) with $y = t$, $\alpha = y_0^2$, and $\mu = y_0/\mu_0$, i.e.

$$f(t; \mu_0) = \frac{y_0}{\sqrt{2\pi t^3}} \mathrm{e}^{-\frac{(y_0 - \mu_0 t)^2}{2t}} \tag{7.73}$$

What is the distribution of the deviance statistic?
(Barndorff-Nielsen and Cox, 1994, pp. 241–242)

(17) We make observations on the number of people cured from an illness under placebo and under some medication. Suppose that n_1 patients are randomized to placebo and n_2 to the medication and that binomial distributions with probabilities π_1 and π_2 provide an adequate model. The medical doctors concerned are interested in the difference in probability of cure under the two treatments.
(a) Use the delta method (Section 5.3.4) to calculate the standard error of $\hat{\pi}_1 - \hat{\pi}_2$.
(b) Derive the Wald test for the null hypothesis that $\pi_1 = \pi_2$.
(c) What is the standard error in this case?
(d) Derive an asymptotic confidence interval for $\hat{\pi}_1 - \hat{\pi}_2$.
(e) The confidence interval and the hypothesis test may be in contradiction. For example, the null hypothesis may be rejected while the confidence interval with the same level of probability may contain $\pi_1 - \pi_2 = 0$. Explain why this is possible.

8
Bayesian decision-making

Observations of phenomena are always made in conditions where previous theoretical knowledge, or perhaps more informal ideas, about the data generating mechanism are available. As far as possible, these are embodied in the model for the data generating mechanism. The likelihood function contains all of the information in the (new) data for estimation of parameters in this model being entertained and, for this reason, is central to inference procedures. As we have seen in Sections 3.6.4 and 6.4.1, prior empirical information can be incorporated by combining likelihoods. Thus, where possible, any such prior knowledge and information should be used.

In a scientific context, only models which are thought to be relevant will be considered, but always with an eye to detecting the unexpected. They are to be ranked by the empirical information from the data, previous and present. However, when decisions, implying actions, must be made, the situation changes and an additional new model, for human behaviour, is required. Prior knowledge relating the models for the data generating mechanism to the decisions, and subsequent actions, will generally be available. If this can be formalized in a precise mathematical way, then it can be combined with the information in the likelihood function to yield a description of the posterior knowledge about the data generating mechanism after obtaining the observations. This model for human behaviour, based on changes in human beliefs, is known as a Bayesian approach.

Decision-making (about parameter values) in such a situation is conceptually very simple, often only requiring integration of a (posterior) probability distribution. This may, however, be analytically rather complex, a technical rather than statistical problem. One major benefit of this approach is that, in a certain sense, we shall be able to make absolute, rather than comparative, statements about the models. However, if this model of human behaviour is formalized, it will be found to exclude exploratory data analysis, the process which often leads to model development, and restructuring knowledge when models are discredited by the data, the basis of scientific discovery (Bernardo and Smith, 1994, pp. 4, 92–94).

In contrast to advocates of other methods of statistical inference and decision-making, those relying on this approach are in considerable disagreement among themselves. The proper way of mathematically formulating prior knowledge is a difficult problem, often involving deep philosophical aspects. Thus, it is necessary first to present some of the conflicting arguments in favour of the Bayesian approach, which imply somewhat different models of human decision-making.

8.1 Interpretation

The Bayesian model for human behaviour involves probability statements about unknown parameters. To maintain a minimum of rigour, we must employ certain elements of measure theory. Thus, for this chapter, we shall use the notation of Appendices A and B, so that it is clear to what probability measure we are referring.

8.1.1 PRIOR PROBABILITIES

Suppose (as in Section 2.6) that the parameter ψ in a parametric statistical model, $(\mathcal{Y}, \mathcal{P}_Y, \Psi)$, is itself a random variable on a statistical field, $(\Omega_\Psi, \mathcal{F}_\Psi, \mathcal{P}_\Psi)$, where $\Psi = \mathcal{F}_\Psi$. In other words, all uncertainty about the possible model, arising from sources other than the data, will be contained in the known function, P_Ψ. This assumption implies that one is capable of

(1) choosing a unique parametrization of interest,
(2) defining a measure on the corresponding parameter space, and
(3) delimiting all possible members of $\mathcal{P}_Y$, the *model space*.

The first point is necessary because, as we shall see, Bayesian procedures are generally not parametrization invariant. The second point can be difficult, especially if ψ is of high dimension, possibly indexing distinct functions. In contrast, the measure for an observable random variable is always clearly defined, because, in any empirical case, it is uniquely the unit of measurement. The final point is foreign to the scientific spirit, which must admit the possibility that the data can best be accounted for by a completely unforeseen model. Where the frequentist requires a clearly defined sample space to make decisions in terms of probabilities, the Bayesian requires a clearly defined model space.

Example
In the first two Bernoulli experiments that were described in Section 3.3.2, Bayesians apply the sufficiency principle to claim that the binomial and negative binomial likelihoods are identical. In doing so, they exclude the possibility of dependence among trials. □

Thus, we might assume that we have a two-stage process whereby a parameter value, $\psi \in \Psi$, is generated at random according to the known probability model, $P_\Psi \in \mathcal{P}_\Psi$, and, then, for this fixed value of ψ, the observation $\mathbf{y} \in \mathcal{Y}$ is generated at random according to the model, $P_{Y|\psi}(y|\psi)$, the latter being the 'true' model for the data generating mechanism. The first stage may correspond to an actual physical process and hence be a statistical model (Sections 6.4.1 and 8.2.2), or it may just be imagined by the statistician or decision-maker. We are primarily interested in the second case here.

We now have two distinct interpretations of probability, one used as a model to describe natural phenomena, the data generating mechanism, and the other as a model to describe states of the human mind. The first gives the probabilities of observable events, whereas the second involves meta-probabilities of unobservable probability models. In contrast to prior theoretical knowledge, these prior probabilities cannot be overturned in favour of a new model by new empirical information; they can only be down-weighted as the posterior probabilities.

Uncertainty about the sample data, arising from random variability in the data generation process, is being mixed with uncertainty of the postulated prior knowledge: the likelihood with the prior probability to obtain the posterior probability. There has been little discussion in the literature of the meaning in combining the two models in one mathematical equation (see the paradox of the probability of probabilities in Section B.3.7 and Barnard, 1949). Of course, it can be, and has been, argued that the data generating mechanism is also a state of mind! Often, in scientific work, a more reasonable procedure, instead of combining the prior and the likelihood, is rather to *compare* them (Spiegelhalter *et al.*, 1994).

At first sight, certain additional criteria for the construction of prior distributions might be considered desirable:

- Non-informative priors should be available for cases of complete prior ignorance about the parameters.
- In most situations, the information obtained from the sample would not be expected fundamentally to change the form of the distribution of the parameters from the prior to the posterior. This is called *closure under sampling* (Barnard, 1954; Wetherill, 1961). If the form of the prior knowledge *must* change whatever data are observed, something is wrong in most circumstances (except prior ignorance).
- As usual, parametrization invariance should hold.

Unfortunately, these criteria cannot always all be fulfilled by the same prior distribution, and the last rarely at all. For example, a non-informative prior cannot be expected to be closed under sampling, except as a limiting case.

8.1.2 BAYES' FORMULA

According to the Bayesian decision-making model, a parameter has a *prior probability distribution*, $P_\Psi(\psi|\mathcal{P}_Y)$, of knowledge or belief, conditional on some member of the model space, $\mathcal{P}_Y$ being true. Hence, Bayes' formula (Bayes, 1763) of Equation (B.38),

$$dP_{\Psi|y}(\psi|y, \mathcal{P}_Y) = \frac{dP_{Y|\psi}(y|\psi)dP_\Psi(\psi|\mathcal{P}_Y)}{dP_Y(y)} \tag{8.1}$$

can be applied to the likelihood function, $dP_{Y|\psi}(y|\psi)$, containing the information from the data, to obtain a *posterior probability distribution*, $dP_{\Psi|y}(\psi|y, \mathcal{P}_Y)$, for the parameters. To simplify notation, we shall drop explicit dependence on the model space, $\mathcal{P}_Y$, although this is an important restriction which should not be forgotten.

Bayes' formula can be written in several equivalent forms:

$$\begin{aligned} dP_{\Psi|y}(\psi|y) &= \frac{dP_{Y|\psi}(y|\psi)dP_\Psi(\psi)}{dP_Y(y)} \\ &= \frac{dP_{Y|\psi}(y|\psi)dP_\Psi(\psi)}{\int_\Psi dP_{Y|\psi}(y|\psi)dP_\Psi(\psi)} \\ &= \frac{\mathrm{L}_p(\psi)d\psi}{\int_\Psi \mathrm{L}_p(\psi)d\psi} \\ &\propto dP_{Y|\psi}(y|\psi)dP_\Psi(\psi) \end{aligned} \tag{8.2}$$

where $\mathrm{L}_p(\psi) = P_{Y|\psi}P_\Psi$ is called the *posterior likelihood*. Note that $dP_\Psi(\psi)$ need not necessarily be a proper density in order for $dP_{\Psi|y}(\psi|y)$ to exist; the latter may still be integrable, although the former is not (see Section 8.3.1 below).

The compound distribution,

$$dP_Y(y) = \int_\Psi dP_{Y|\psi}(y|\psi)dP_\Psi(\psi) \tag{8.3}$$

in the denominator of Equation (8.2) acts as the normalizing constant. This integration must be performed over the whole model space which thus must be explicitly defined for the posterior to exist. This distribution is also called the prior predictive distribution, because it can be used to predict a future value of Y.

An analogous posterior predictive distribution can also be formed to predict future observations after observing a sample. This can be used for checking goodness of fit (Gelman *et al.*, 1995, pp. 167–175); it generalizes the procedures of Chapter 4 by taking an average, weighted by the prior distribution, over the possible models.

If $\mathbf{T}$ is a sufficient statistic for ψ, then $dP_{\Psi|y}(\psi|\mathbf{y}) \propto dP_{\Psi|t}(\psi|\mathbf{t})$, from the factorization theorem (Section 3.4.1). Often, the latter is much easier to handle, allowing substantial simplification of calculations. Another possible simplification is to consider the linear form

$$\frac{\partial}{\partial\psi}\log[p(\psi|y)] = \frac{\partial}{\partial\psi}\log[\mathrm{L}(y|\psi)] + \frac{\partial}{\partial\psi}\log[p(\psi)] \tag{8.4}$$

where the first term on the right is just the score function.

Bayes' formula gives a weighting of the likelihood according to prior information or belief, a form of penalized likelihood. It provides an updating of prior to posterior knowledge or belief. Usually, a study is performed only when it can provide substantially more information than what is known *a priori*. Thus, the likelihood function should dominate the prior. As its name suggests, the prior should be specified before observing the data, although, as we shall see, this is not always the case in practice. For a given set of observations of random variables, the parameters are a corresponding realization from the prior distribution. In most cases, this prior distribution is taken to be completely known, because it represents prior belief or knowledge. We know which member of $\mathcal{P}_\Psi$ is involved.

Notice, however, that any model function, parameter, or parameter value which is given a prior probability (density) of zero will never have a positive posterior probability, no matter what data are observed. Thus, *all* possible models which might ever be considered must have positive prior probability (density) in P_ψ. One can only make choices among already known possibilities (de Finetti, 1974a, II, pp. 208–211). Unexpected models can never be detected by a strictly Bayesian procedure. In other words, scientific discovery is excluded from the Bayesian formalism: the completely unexpected always has a zero prior probability.

Example
In equating binomial and negative binomial likelihoods through the sufficiency principle, a Bayesian is placing a zero prior probability on the possibility of the Bernoulli probability changing over the course of the experiment. □

A Bayesian procedure is said to be calibrated if posterior probability statements have their asserted coverage in repeated experience. This can only occur if both the model function and the prior are correctly chosen. It assumes a full selection of possible statistical models, complete knowledge of the substantive area to be studied, and essentially unlimited resources (Dawid, 1982; Rubin, 1984).

Because proportional likelihoods give the same posterior distributions, for the same dominating measure on the parameter space, Bayesian decision-making obeys the sufficiency principle. Recall from Section 3.4.1 that, when studying the same model, experiments with different sample spaces have the same likelihood function. Following this, many Bayesians have argued that the way the data are obtained has no importance once they are observed. In particular, some claim that randomized designs are unnecessary. However, we have already seen, in Sections 3.3.2 and 3.6.6, that the appropriate likelihood function for the model can be modified by certain designs. A stopping rule which is not a martingale stopping time can have such an effect. In general, an important advantage of properly randomized designs is that they guarantee that the likelihood function does not depend on the method of collecting the data in undesirable ways, even unconsciously.

Let us now consider various schools of Bayesian thought; these are not mutually exclusive, nor necessarily exhaustive.

8.1.3 RATIONAL BAYES

One interpretation of a prior probability is as a representation of degrees of rational belief in possible values of the parameter. This objectivist view holds that any person given the generally available prior knowledge or beliefs should be able to accept these 'reasonable degrees of belief'. However, such rationality is individualistic, not usually implying that collective interests need to be taken into account.

In such a model, it is important to be able to represent coherently a state of initial ignorance. This is the problem of choosing a suitable measure for the parameter, that for which all members of $\mathcal{F}_\Psi$ are equally probable. Obviously, such a probability distribution of the parameters for this state will no longer represent ignorance for any non-linear one-to-one transformations of the parameters. Ignorance about one parametrization of the model, thus, must imply knowledge about other parametrizations!

Example

A natural form to express initial ignorance about a parameter might be to take a uniform prior density. However, any non-linear function of the parameter will not have a uniform distribution because of the Jacobian. □

As we shall see, this is not a problem specific to ignorance priors, but is general with all prior distributions, because the resulting credibility regions for the parameter will not be parametrization invariant under change of dominating measure (Sections 8.3.3 and 8.5.1).

8.1.4 SUBJECTIVE BAYES

For a subjective Bayesian, the statistician is always in a decision-making situation. This model of human behaviour is normative or prescriptive and does not attempt to describe how human decisions are actually made. (This contrasts with certain other decision-making models, such as fuzzy logic, which may be closer to how human beings actually act.) In many ways, it is closely related to the liberal economic theory of the rational human being. Thus, as in Section 7.1.1, we shall have a utility function, $u(\psi, d)$, and a decision, $d \in \mathcal{D}$, to be taken, in addition to a personal prior distribution for the parameters. However, here the utility will be strictly personal as well.

For statisticians to make individual rational decisions, given their individual knowledge or beliefs, certain (implicit) rules must be followed. These can be formalized as the

Bayesian coherency principle
Suppose a model of human decision-making behaviour such that

- each person can conceive of uncertainty about all events as probabilities, these, perhaps, being different for different people;
- for a given person, all such subjective probabilities are comparable;
- the person can determine them by imagining a decision in a betting situation.

Then, these subjective probabilities will ensure self-consistent betting behaviour. Let us denote the consequence of a decision by $c = (\psi, d) \in \mathcal{C}$. Now consider a situation where consequence c_i results if $\psi \in \Psi_i$. The selection of decision d_i is equivalent to the choice of a bet with consequence $c_i = (\psi, d_i)$ if ψ is true. Given the above suppositions about subjective probability, one can deduce the existence of the utility or value function.

Suppose, now, that there is a set of events, $\{E_\pi\}$, with known probabilities of occurrence, π, and that the person is equally prepared to bet on E_π as on ψ. Then, $P_\Psi(\psi) = \pi$. Thus, under the assumptions of this model of human behaviour, degrees of belief will follow the usual laws of probability. □

In determining the probabilities of more complex events, either the ordinary rules of probability are followed or the strategy is inconsistent, implying less than optimal decisions. The final decision problem is, then, solved by maximizing the *posterior expected utility*:

$$\int_\Psi u(\psi, d) dP_{\Psi|y}(\psi|y)$$

Here, in contrast to a rational Bayesian, a state of initial ignorance is excluded, because a decision would never be taken in such a state.

Coherency is formally equivalent to admissibility (Section 7.4.2; see Savage, 1976). It has only been demonstrated for finitely additive sets of probabilities (de Finetti, 1974a, I, pp. 32–35, 128-133, 245–246; Hill, 1988); its extension even to countably additive probabilities is questionable and is simply assumed for mathematical convenience (Bernardo and Smith, 1994, pp. 44–45). Unfortunately, Bayesian prior probabilities generally are not finitely additive, but have Lebesgue measure, so that coherency of the decisions is no more than an assumption.

Coherency applies only as a model of individual decisions of an informed rational human being; attempts to extend it to collective behaviour have been unsuccessful. Subjective Bayesians do, however, state, usually in passing, that the Bayesian reporting process is distinct from the individual coherent decision process, implying a wide range of possible belief mappings induced by a data set (Dickey, 1973). Obviously, coherency cannot apply to such a reporting process.

For the subjective Bayesian, probabilities are personal degrees of belief, conditional on the information currently available; they can *only* be modified by application of Bayes' formula. We may ask how the initial prior in such a sequence is obtained. Concrete examples in the literature generally involve an initial prior based on observed frequencies in the relevant population, such as the prevalence of a disease (Bernardo and Smith, 1994, pp. 44–45; see Section 8.2.1 below). However, if such a 'prior' arises from previous similar studies, the empirical information should be combined through the likelihoods (Section 3.6.4). On the other hand, priors elicited from a team conducting a study are generally overenthusiastic about positive results being obtained, rather than being centred around the empirical information of current evidence (Spiegelhalter *et al.*, 1994, and discussion).

8.1.5 DECISION THEORETIC BAYES

Closely related to the subjective Bayesian is the decision theoretic Bayesian, for whom the prior distribution is just a weight function. High prior probability is allocated to parameter values for which small risk is desired in some decision context, and low probability to values where the risk is not as important. As usual, decision problems are primarily concerned with point estimation. Bayes risk will be defined in Section 8.4.2 below.

8.1.6 ROBUST BAYES

Most Bayesians agree that determination of the prior distribution from prior belief is never easy. Thus, if the arguments of the subjective Bayesian were followed, it could never be done with complete accuracy in finite time. On the other hand, a quick guess at the prior may be worse than nothing, because further reflection might indicate that certain elements of it were in error.

In a robust analysis, one is concerned about the sensitivity of conclusions to prior assumptions and, thus, considers a class of plausible prior distributions felt certain to contain the correct one. The extremes can be called sceptical and enthusiastic priors. Unfortunately, once one works with a set of priors, the hope for coherency must be abandoned; in addition, there is the risk of dilation of the posterior probabilities (Section B.3.5).

Depending on the parametrization of a model, one may be inclined to choose one prior distribution over another. These will usually not be simply related through a Jacobian.

Example

The parameter π of the binomial distribution can only take values in the range $[0, 1]$, so that the beta distribution of Equation (C.20) may seem natural. However, if we work with thc canonical parametrization, θ, the parameter can take any value on the real line, and one might be more inclined to choose a normal distribution as appropriate. □

The robust Bayesian argues that, in such a situation, one should verify that the choice makes little difference to the final conclusions. Many robust Bayesians deem a conclusion to be clear cut only when any reasonable prior, when applied to the observed likelihood, gives essentially the same answer.

The class of plausible priors might also be gathered from various people. From this, one obtains a class of posterior distributions. Again, if the conclusion is essentially the same for all of the class, the problem is solved. If the conclusions are quite different for different members of the class, which might arise, for example, if the data and some priors are in contradiction, various options are available:

- attempt further refinement of the class of priors (but this will be difficult without being influenced by the data, which are now available),
- obtain more data,
- conclude that no answer exists,
- choose among the possible conclusions according to some other criterion.

The first two solutions may be excluded by time or cost factors.

The form of the likelihood function, after the data are observed, may mean that the prior can have virtually no effect over much of the range of parameter values. Thus, some argue that choice of the class of 'priors' should be considered only after examining the likelihood function. However, choosing the prior with the aid of the data requires caution. It can be justified in several ways. If the likelihood is small in a region of the parameter space, it is counterproductive to waste time trying to specify a prior there. From a robust perspective, some set of priors is reasonable, and one chosen from the data is that which is most plausible. But, the danger is that the posterior will be overconcentrated about the likelihood function, yielding estimates which are too precise (Berger and Berliner, 1986). It is the Bayesian analogue of a frequentist testing an hypothesis derived from the data.

8.2 Applications to statistical model building

In this section, we shall look at some applications of Bayes' formula which need not involve a Bayesian decision-making model.

8.2.1 POPULATION FREQUENCIES

In certain situations, such as the prevalence of a disease mentioned above, the exact population frequencies are available for the response variable under study. These may be used to construct the prior probability for Bayes' formula, avoiding the philosophical problems of mixing two diffcrent types of probability. The posterior probabilities will then also have a frequency interpretation which will be acceptable to a statistician of any inclination.

Example
In the prediction of election results on the evening of voting, the results of previous elections are generally known in great detail. As soon as the first new results are available, application of Bayes' formula yields the posterior probabilities of various possible proportions of votes for the parties. With more elaborate calculations, these can be converted to probabilities of seats in the parliament. Given the emotions connected with voting, this is a clear case where subjective priors should be avoided. □

8.2.2 EMPIRICAL BAYES

In many concrete situations, useful statistical models can be built by taking a prior distribution to have unknown parameters to be estimated. In other words, some of the parameters in the usual model can be considered to be generated by a physical random generating mechanism, itself described by a prior distribution with additional parameters. In standard regression problems, these are known as random effects models, and have already been treated in Section 6.4.1.

Suppose that a set of independent random variables, Y_i, are distributed as $dP_{Y_i|\beta_i}(y_i|\beta_i;\boldsymbol{\lambda})$ and that the β_i are i.i.d. with prior density $dP_B(\boldsymbol{\beta};\boldsymbol{\phi})$. On the other hand, let us assume that $\boldsymbol{\lambda}$ and $\boldsymbol{\phi}$ are not generated randomly, but are fixed unknown parameters. As we have seen, such a situation can happen, for example, when there is more variability among the individuals being observed than can be explained by the available explanatory variables. Then, the extra variability is absorbed by the random distribution of the parameter $\boldsymbol{\beta}$.

If we apply Bayes' formula, we obtain

$$dP_{B|y}(\boldsymbol{\beta}|y;\boldsymbol{\lambda},\boldsymbol{\phi}) \propto dP_{Y|\beta}(y|\boldsymbol{\beta};\boldsymbol{\lambda})dP_B(\boldsymbol{\beta};\boldsymbol{\phi}) \tag{8.5}$$

which contains unknown quantities. However, these parameters can be estimated from the marginal density for Y

$$dP_Y(y;\boldsymbol{\lambda},\boldsymbol{\phi}) = \int_{\mathrm{B}} dP_{Y|\beta}(y|\boldsymbol{\beta};\boldsymbol{\lambda})dP_B(\boldsymbol{\beta};\boldsymbol{\phi}) \tag{8.6}$$

yielding a compound distribution and the corresponding posterior likelihood function for $(\boldsymbol{\lambda},\boldsymbol{\phi})$.

If $\boldsymbol{\beta}$ is not of interest, we simply have a means of deriving a suitable statistical model for the data generating mechanism in the presence of unexplained variability. If it is of interest, an estimated posterior density can be obtained from Bayes' formula.

Example
Consider count data, assumed to follow a Poisson distribution,

$$P_{Y|\mu}(y|\mu) = \frac{e^{-\mu}\mu^y}{y!}$$

where, however, there is extra variability among the observations. One possible prior distribution for the mean of this distribution is the conjugate prior, a gamma distribution, from Equations (2.78) and (2.79), with dominating measure $\nu(d\mu)$,

$$dP_M(\mu;\zeta,\gamma) = \frac{\gamma^{\zeta+1}\mu^{\zeta}e^{-\gamma\mu}}{\Gamma(\zeta+1)}d\mu \tag{8.7}$$

as shown in Section 2.6. As we have seen, the posterior distribution is given by

$$dP_{M|y}(\mu|y;\zeta,\gamma) = \frac{(\gamma+1)^{\zeta+y+1}\mu^{\zeta+y}e^{-(\gamma+1)\mu}}{\Gamma(\zeta+y+1)}d\mu \tag{8.8}$$

and the resulting compound distribution is negative binomial,

$$P_Y(y;\zeta,\gamma) = \frac{\Gamma(\zeta+y+1)}{\Gamma(\zeta+1)y!}\frac{\gamma^{\zeta+1}}{(\gamma+1)^{\zeta+y+1}} \tag{8.9}$$

□

8.2.3 KALMAN FILTER

Because the Bayesian approach is specifically designed to incorporate prior knowledge or beliefs into the decision-making process, it is especially important when this knowledge is accumulating over time.

Suppose that we have a random variable being observed over time, with *measurement equation*

$$\mathrm{E}[Y_t] = \boldsymbol{\beta}_t^{\mathrm{T}}x_t \tag{8.10}$$

The observations will not be independent and the parameter vector can also be changing in time. Thus, we can take this parameter to have a distribution with *state transition equation*

$$\mathrm{E}[\boldsymbol{\beta}_t] = \mathbf{T}_t\boldsymbol{\beta}_{t-1} \tag{8.11}$$

so that successive values are linked by a square transition matrix, $\mathbf{T}_t$. This is called a *dynamic linear model.*

Example
If we have $\mathrm{E}[Y_t] = \mu + \beta_t$ with

$$\mathrm{E}[\beta_t] = \rho\beta_{t-1} \tag{8.12}$$

we have a first-order autoregression model and if

$$\mathrm{E}[\beta_t] = 0 \tag{8.13}$$

a random effects model. □

If the distributions are taken to be normal, successive predictions, as more observations become available, can be obtained from Bayes' formula by a procedure known as the *Kalman filter.* This approach can be generalized to members of the exponential family, yielding *dynamic generalized linear models* (West *et al.*, 1985).

8.3 Types of prior distributions

Generally, in estimation problems, continuous prior distributions are used so that only sets of parameter values have positive probabilities. Any point parameter value can have positive density but will have zero prior probability. On the other hand, as we shall see, in testing hypotheses, models with point estimates can be assigned positive prior probabilities.

In contrast to random variables, where the dominating measure for empirical observations is always clearly defined by the unit of measurement, the procedure for making a choice for the prior distribution of parameters is not. This adds an additional element of discussion among Bayesians.

Examples
1. The conjugate prior for an exponential family, with dominating measure $\nu(d\boldsymbol{\theta})$, based on the canonical parameter, is

$$dP_\Theta = \mathrm{e}^{\boldsymbol{\theta}^{\mathrm{T}}\boldsymbol{\zeta}-\gamma\kappa(\boldsymbol{\theta})}d\boldsymbol{\theta} \tag{8.14}$$

However, if the dominating measure is $\nu(d\boldsymbol{\mu})$, based on the mean value parameter, it will be

$$\begin{aligned} dP_M &= \mathrm{e}^{\boldsymbol{\theta}^{\mathrm{T}}\boldsymbol{\zeta}-\gamma\kappa(\boldsymbol{\theta})}d\boldsymbol{\mu} \\ &= \frac{d\boldsymbol{\mu}}{d\boldsymbol{\theta}}\mathrm{e}^{\boldsymbol{\theta}^{\mathrm{T}}\boldsymbol{\zeta}-\gamma\kappa(\boldsymbol{\theta})}d\boldsymbol{\theta} \end{aligned} \tag{8.15}$$

where the derivative is not generally a constant.

2. For the Poisson distribution, we obtained the conjugate posterior distribution for the mean value dominating measure, $\nu(d\mu)$, in Equation (8.8) above. It yields a mean of $\mathrm{E}[\mu|y] = (y+\zeta+1)/(\gamma+1)$ and mode $(y+\zeta)/(\gamma+1)$. In contrast, with dominating measure $\nu(d\theta)$, the posterior distribution is

$$
\begin{aligned}
dP_{\Theta|y}(\theta|y;\zeta,\gamma) &= \frac{\mathrm{e}^{\theta(y+\zeta)-(\gamma+1)\mathrm{e}^{\theta}}d\theta}{\int_\Theta \mathrm{e}^{\theta(y+\zeta)-(\gamma+1)\mathrm{e}^{\theta}}d\theta} \\
&= \frac{\mu^{\zeta+y-1}\mathrm{e}^{-(\gamma+1)\mu}d\mu}{\int_M \mu^{\zeta+y-1}\mathrm{e}^{-(\gamma+1)\mu}d\mu} \\
&= \frac{(\gamma+1)^{\zeta+y}\mu^{\zeta+y-1}\mathrm{e}^{-(\gamma+1)\mu}}{\Gamma(\zeta+y)}d\mu \qquad (8.16)
\end{aligned}
$$

which has mean $\mathrm{E}[\mu|y] = (y+\zeta)/(\gamma+1)$ and mode $(y+\zeta-1)/(\gamma+1)$. □

Thus, care must be taken to select parameter values in the prior distribution which are appropriate to the measure chosen.

Generally, if the parametrization within a given model is changed, decisions made through Bayes' formula will be modified, because the Jacobian of the prior distribution, taking the change into account, modifies the posterior distribution. The prior weights given to the various parameter values will have changed. The relative height of the density at two points is not invariant under monotone non-linear transformations of the parameter. This can lead to extreme contradictions.

Example

Suppose that the observations are generated by a location–shifted unit exponential distribution,

$$dP_{Y|\psi}(y|\psi) = \mathrm{e}^{-(y-\psi)}dy \qquad 0<\psi<y \qquad (8.17)$$

a special case of Equation (7.72), and that the prior is a Cauchy distribution,

$$dP_\Psi(\psi) = \frac{d\psi}{\pi(1+\psi^2)} \qquad (8.18)$$

from Equation (2.94) with $\mu = 0$, $\sigma = 1$. The posterior is, then,

$$dP_{\Psi|y}(\psi|y) \propto \frac{\mathrm{e}^{\psi}}{1+\psi^2}d\psi \qquad 0<\psi<y \qquad (8.19)$$

which is an increasing function, so that values of ψ close to y have high posterior probability.

Now let us take, instead, $\alpha = e^{\psi}$ as the parameter, with Jacobian $1/\alpha$, so that the new posterior distribution is

$$dP_{A|y}(\alpha|y) \propto \frac{d\alpha}{1 + [\log(\alpha)]^2} \qquad 1 < \alpha < e^{y} \tag{8.20}$$

which is decreasing on $(1, e^{y})$, so that values of α close to 1, or of ψ close to 0, have high posterior probability. The transformed prior, which still should represent the same knowledge or belief, has heavily down-weighted the larger parameter values. (Berger, 1985b, p. 144) □

For regression models, one possible way to avoid such problems is to construct the prior in terms of the systematic component, $h(\mathbf{x}_i, \boldsymbol{\beta})$ in Equation (1.7), instead of the individual parameters, $\boldsymbol{\beta}$. Although parametrization invariance results, the prior now depends on the study design, i.e. on $\mathbf{X}$.

In most situations, the presence of such contradictions would appear to outweigh the possible advantages of coherency. The reader may also like to consider some of the implications of the paradoxes outlined in Section B.3.7. The choice of an appropriate prior distribution to represent knowledge or belief is one of the most difficult aspects of Bayesian decision-making; Berger (1990) gives a useful discussion.

8.3.1 NON-INFORMATIVE PRIORS

As we have seen, for the rational justification of Bayesian decision-making to be complete, measures for ignorance or non-informativeness must be available. In circumstances where there is a natural parametrization, with an accompanying measure, transformations will not be pertinent and a distribution representing ignorance will be available.

Example

In the location–scale family, the parameters correspond to changes of the location and scale of the unit of measurement of the observations. Thus, they have clearly defined measures themselves. A flat prior is still flat after a linear transformation. □

Thus, we may argue that, if $y - \psi$ is the pivotal quantity for a location parameter, ψ, then a reasonable non-informative prior distribution would leave it as a pivotal quantity under the posterior:

$$dP_{\Psi|y}(y - \psi|y) \propto dP_{\Psi}(\psi) dP_{Y|\psi}(y - \psi|\psi) \tag{8.21}$$

implying that the prior is uniform on Ψ. A similar argument can be used for the pivotal quantity of a scale parameter.

A non-informative distribution will usually be an *improper prior density* which is one which is not integrable ($P_{\psi} \notin \mathcal{L}^1$):

$$\int_{\Psi} dP_{\Psi}(\psi) = \infty \tag{8.22}$$

This does not necessarily affect formal theory, because the posterior distribution may still be integrable. However, paradoxes can arise, especially when integrating to obtain marginal distributions of parameters (Dawid and Stone, 1972; Dawid *et al.*, 1973; Seidenfeld, 1979). When an (improper) uniform prior is used, the posterior distribution is sometimes called an *integrated likelihood.*

Examples

1. For a location parameter, all intervals $(a, a + \Delta)$ might be thought to have equal prior probability, for any given Δ and all a. Then, the parameter has a uniform prior density, but it is an improper density, because the integral over the real line is infinite.

2. For a scale parameter, all intervals $(a, a\Delta)$ might be thought to have equal prior probability, for any given Δ and all a. Then, the parameter has a prior density proportional to its reciprocal, so that the probability of the interval $(a, a\Delta)$ is proportional to the logarithm of the parameter. Again, we obtain an improper density. □

An important theme when the Bayesian approach is used in scientific research is that the likelihood function should dominate the prior. This can be accomplished by only defining the (non-informative) prior over that part of the parameter space where the likelihood function is important. This also eliminates the problem of the prior density being improper. Of course, this is no longer a true prior.

One way of developing an informative prior is to argue that current knowledge or beliefs are equivalent to the results of a hypothetical study of a certain size, with no other previous information. (As we shall see below, this is close to an interpretation of conjugate priors.) If ignorance can be formulated as the appropriate prior distribution for such hypothetical circumstances, priors expressing knowledge are immediately available in this way, by combining the ignorance prior with the hypothetical study (Novick and Hall, 1965).

Bayesian and frequentist probabilities In many situations, the use of an improper uniform prior leads to the same results as would the frequentist decision-making of Chapter 7. This is surprising, because frequentist statements are about long run probabilities for the random variable, or a statistic derived from it, while Bayesian statements are about probabilities of parameter values. They are not operating on the same measure space.

The conditions for agreement between frequentist and Bayesian results, for a scalar parameter ψ with sample size n, fixed before the study, are

(1) it must be possible to write the density $f(y;\psi)$ in the form $f[z(y) - g(\psi)]$, where $z(\cdot)$ and $g(\cdot)$ are monotonic functions with $-\infty < z, g < \infty$;

(2) the prior for $g(\psi)$ must be uniform over $\mathbf{R}$ (Bartholomew, 1965).

Because frequentist decision-making depends on the sample space, agreement can also be obtained in certain other conditions, such as sequential sampling. On the other hand, because the influence of the prior distribution diminishes as $n \to \infty$, the two approaches will tend to agree for large samples. This convergence occurs most rapidly for Jeffreys' prior of Section 8.3.3 below (Welch and Peers, 1963).

8.3.2 VAGUE PRIORS

A *vague prior* is a very diffuse proper distribution.

Example

If the prior is a normal distribution, then it would be vague if it had a very large variance. □

Often, the posterior distribution corresponding to a non-informative prior can be obtained as the limit of the posterior for a vague prior.

8.3.3 INVARIANT PRIORS

As another approach to prior ignorance, one might argue that the appropriate prior is one which makes the expected information constant. In Section 2.4.2, for the one-parameter exponential family, this was called the transformation giving stable variance:

$$g(\theta) = \int_{\theta_0}^{\theta} \sqrt{\mathrm{I}_\theta(u)}du \tag{8.23}$$

In the same way, in Section 3.5.2, we saw, more generally, how measures proportional to the square root of the expected information matrix are parametrization invariant. This implies that a useful non-informative prior can be found on this basis. It is known as *Jeffreys' prior* (Jeffreys, 1946).

Example

For the Poisson distribution, the expected information is proportional to $1/\mu$, so that Jeffreys' prior is proportional to $1/\sqrt{\mu}$. □

However, this prior violates the sufficiency principle, because it is obtained by averaging over the sample space, so that different posterior distributions can result from proportional likelihood functions.

Example
For the binomial distribution, we find that Jeffreys' prior is proportional to $1/\sqrt{\pi(1-\pi)}$, while that for the negative binomial is proportional to $1/\sqrt{\pi}(1-\pi)$, although their likelihood functions are proportional. □

Another problem is that 'in higher dimensions, non-informative priors, such as Jeffreys prior, can have hidden features that have a dramatic, and unrecognized, effect on the answer' (Berger and Bernardo, 1992).

The distribution of the observations, i.e. the likelihood function, multiplied by Jeffreys' prior, is closely related to the approximation to the conditional likelihood in Equation (5.56), upon which the modified profile likelihood is based. If the observed information is used instead of the expected, the relationship is even closer.

Among non-informative prior distributions, Jeffreys' prior, which is proportional to the square root of the expected information, has the special property that it is parametrization invariant. However, this is not the only prior with this invariance property. For parameter transformation $g^* \in \mathcal{G}^*$, a group of transformations defined as in Section 7.3.3, the *right invariant Haar prior* is defined by

$$\int_{\Psi g^*} dP_\Psi^r(\psi) = \int_\Psi dP_\Psi^r(\psi) \tag{8.24}$$

and the *left invariant Haar prior* by

$$\int_{g^*\Psi} dP_\Psi^l(\psi) = \int_\Psi dP_\Psi^l(\psi) \tag{8.25}$$

Jeffreys' prior, based on the expected information, is left invariant.

Example
For the normal distribution, Jeffreys' prior for the mean and variance is proportional to $1/\sigma^2$, while the right invariant Haar prior is proportional to $1/\sigma$. This is generally true for members of the location–scale family. □

Thus, one must be careful. No one has yet been able to, nor will anyone ever likely be able to, define unambiguously, in an absolute sense, a non-informative prior.

8.3.4 REFERENCE PRIORS

The goal in constructing a reference prior is that it should have a minimal influence on the Bayesian analysis when compared to that provided by the data (Berger and Bernardo, 1992).

A non-informative prior is one which injects no information into the posterior. As an extension of this idea, we should be able to calculate what prior, in fact, provides minimal information, after the fact, in a given situation. To do this, we require a suitable definition of information. Jeffreys' prior used a second-order approximation, based on the expected information. We shall now consider the Shannon expected information about ψ, when the prior distribution is dP_ψ. This is defined to be

$$\begin{aligned}\mathrm{I}_S(dP_\psi) &= \int_{\mathcal{Y}}\int_{\Psi} \log\left(\frac{P_{\Psi|y}(\psi|y)}{P_{\Psi}(\psi)}\right) dP_{\Psi|y}(\psi|y)dP_Y(y) \qquad (8.26)\\ &= \int_{\Psi} \log\left(\frac{\mathrm{e}^{\int_{\mathcal{Y}} \log[dP_{\Psi|y}(\psi|y)]dP_{Y|\psi}(y|\psi)}}{dP_{\Psi}(\psi)}\right) dP_{\Psi}(\psi)\end{aligned}$$

which is the expected value, with respect to the marginal distribution of Y, of the Kullback–Leibler distance between the prior and posterior. This quantity will increase with the number of observations. For an infinite number of observations, we would know the true value of ψ (if the model function is correct!). Thus, the limit of this expected information, as $n \to \infty$, measures the amount of information about ψ missing when the prior is dP_Ψ. The reference prior is that which maximizes this limiting quantity, the prior for which we know the least about the parameter. It will be that which is most changed by observing data and reflects the notion that observation of data is carried out because one expects to learn about the parameters. From this, a reference posterior distribution can be obtained.

The second expression in Equation (8.26) is maximized by the reference prior,

$$dP_{\Psi}(\psi) \propto \mathrm{e}^{\int_{\mathcal{Y}} \log[dP^*_{\Psi|y}(\psi|y)]dP_{Y|\psi}(y|\psi)} \qquad (8.27)$$

where $dP^*_{\Psi|y}(\psi|y)$ is the asymptotic posterior density of ψ. As we shall see below, the asymptotic approximations of Chapter 5 will also hold for the posterior distribution. Thus, in regular conditions, this distribution will be normal with mean, the m.l.e., and variance, the inverse of the expected information. Because $\lim_{n\to\infty} \hat{\psi} = \psi$, Equation (8.27) reduces to

$$dP_{\Psi}(\psi) \propto \sqrt{\mathrm{I}(\psi)} \qquad (8.28)$$

Jeffreys' prior. This is not surprising because that prior uses a second-order measure of information and we have now used a second-order asymptotic approximation.

However, if a reference prior is required for only some of the parameters, the others being nuisance parameters, the result will not generally be Jeffreys' prior. Indeed, in the multidimensional case, Jeffreys' prior will dominate the likelihood function for some transformations of the parameter of interest. To solve this problem, first the conditional reference prior,

$dP_{\Phi|\lambda}(\phi|\lambda)$, must be found for the nuisance parameters, given the parameters of interest. Then, the reference prior, $dP_{\Lambda}(\lambda)$, for the parameters of interest is derived for the marginal model,

$$dP_{Y|\lambda}(y|\lambda) = \int_{\Phi} dP_{Y|\lambda,\phi}(y|\lambda, \phi) dP_{\Phi|\lambda}(\phi|\lambda) \tag{8.29}$$

formed by integrating out the nuisance parameters. Thus, the resulting reference prior is

$$dP_{\Lambda\Phi}(\lambda, \phi) = dP_{\Phi|\lambda}(\phi|\lambda) dP_{\Lambda}(\lambda) \tag{8.30}$$

(Bernardo, 1979). A number of different ways of doing this have been proposed. In the extreme case, all parameter elements could be ordered and a stepwise procedure of the same order as the dimension of the parameter vector used.

As with Jeffreys' prior, this more general reference prior approach violates the sufficiency principle, because it is obtained by use of expectations over the sample space. In addition, it should be noted that any attempt to construct non-informative priors for a particular parameter of interest entails an increase in informativeness about the other parameters (Clarke and Wasserman, 1993).

8.3.5 CONJUGATE PRIORS

If closure under sampling is important, then a conjugate distribution to the distribution of the random variable being observed must be found. Such a condition severely constrains the form of possible prior distributions. We saw in Section 2.6 that such distributions do exist for members of the exponential family, as well as for the mean parameters of the exponential dispersion family. Thus, we know that for n i.i.d. observations from the exponential family,

$$dP_{Y|\theta}(\mathbf{y}|\boldsymbol{\theta}) = \exp\left[\sum \boldsymbol{\theta}^{\mathrm{T}}\mathbf{t}(y_i) - n\kappa(\boldsymbol{\theta}) - \sum t_0^*(y_i)\right] d\mathbf{y} \tag{8.31}$$

a prior density of the form

$$dP_{\Theta}(\boldsymbol{\theta}; \boldsymbol{\zeta}, \gamma) \propto \exp\left[\boldsymbol{\theta}^{\mathrm{T}}\boldsymbol{\zeta} - \gamma\kappa(\boldsymbol{\theta})\right] \tag{8.32}$$

where $(\boldsymbol{\zeta}, \gamma)$ have known values, will yield a posterior density for $\boldsymbol{\theta}$ of the same form:

$$dP_{\Theta|y}\left(\boldsymbol{\theta}|\mathbf{y};\ \boldsymbol{\zeta} + \sum \mathbf{t}(y_i),\ \gamma + n\right) \tag{8.33}$$

Examples

1. For the Poisson distribution with conjugate prior, the posterior distribution, from Equations (2.81) and (8.8), is

$$dP_{M|y}(\mu|\mathbf{y};\zeta,\gamma) = \frac{(\gamma+n)^{\zeta+y_\bullet+1}\mu^{\zeta+y_\bullet}\mathrm{e}^{-(\gamma+n)\mu}}{\Gamma(\zeta+y_\bullet+1)}d\mu \tag{8.34}$$

The prior has contributed information equivalent to γ observations so that it is as if we had, in fact, $n+\gamma$ observations, with total sum of observations, $y_\bullet+\zeta$.

2. For a normal distribution with mean zero and unknown variance,

$$dP_{Y|\sigma^2}(\mathbf{y};\sigma^2) = \exp\left[-\frac{\sum y_i^2}{2\sigma^2} - \frac{n}{2}\log(2\pi\sigma^2)\right] d\mathbf{y}$$

so that the conjugate prior is

$$dP_\Sigma(\sigma^2;\zeta,\gamma) \propto \exp\left[-\frac{\zeta}{\sigma^2} - \gamma\log(\sigma^2)\right] \tag{8.35}$$

which is a gamma distribution for σ^{-2}.

The posterior distribution for the variance will be

$$dP_{\Sigma|y}(\sigma^2|\mathbf{y};\zeta,\gamma) \propto \exp\left[-\frac{\zeta+\sum y_i^2}{\sigma^2} - (\gamma+n)\log(\sigma^2)\right] \tag{8.36}$$

Again, it is as if we had, in fact, $n+\gamma$ observations, here with total sum of squares, $\sum y_i^2+\zeta$. □

The posterior obtained from a conjugate prior resembles closely the formula that we obtained in Section 3.6.4 by combining likelihood functions containing prior information. Indeed, as a function of the unknown parameter, the two formulae are proportional: compare Equations (3.83) and (8.34). However, here the prior 'information' is not empirical, but founded on beliefs embedded in a model of human behaviour. It is nevertheless given the same standing, within the posterior distribution, as the empirical information. The two are weighted by their 'sample sizes', respectively n and γ.

In general, when we use a conjugate distribution in the exponential family, it is as if we had actually $n+\gamma$ observations with each sufficient statistic increased by ζ_j over that observed. Thus, according to the fixed parameter values chosen in the prior distribution, the likelihood function actually observed is being correspondingly modified. Care must obviously be taken to use only properly chosen prior parameter values, representing previous knowledge or belief. The use of arbitrary values makes *any* conclusion possible.

Improper priors, for the non-informative situation, can usually be obtained from the conjugate by choosing (limiting) special values of the parameters (ζ, γ). For the exponential family, Jeffreys' prior is a special case of the corresponding conjugate distribution, usually with $\gamma \to 0$, i.e. as the 'prior sample size' becomes negligible.

Example
For the Poisson distribution, the conjugate gamma distribution for prior ignorance can be taken with $\zeta = \frac{1}{2}$ and $\gamma = 0$, which yields Jeffreys' prior given above. Because $\gamma = 0$, the sample size, n, is not being increased when such a non-informative prior is used. □

We saw in Section 2.6 that finite mixtures of conjugate priors can also be used with the exponential family. *Any* prior density for this family, not just a conjugate one, can be approximated arbitrarily closely by such a mixture (Dalal and Hall, 1983).

8.3.6 HIERARCHICAL PRIORS

Suppose that we have a series of independent random variables, Y_i, each having a distribution containing a parameter ψ_i. In considering the prior knowledge or beliefs about these parameters, it may be reasonable to assume that their distributions are *exchangeable* (Lindley and Smith, 1972; de Finetti, 1974a, II, pp. 211–224). In other words, the knowledge is the same about all of the parameters so that it would not be changed by any permutation of the indices. Then, a prior distribution may be obtained by introducing a further parameter, which, itself, has a distribution:

$$dP_{\Psi}(\psi) = \int \prod dP_{\Theta|\mu}(\theta_i|\mu) dP_M(\mu) \tag{8.37}$$

This is a compound distribution. We now have a three-stage model: the likelihood function, the prior for its parameters, and the prior for the prior's parameters. This is called a *hierarchical Bayes* analysis.

Example
Suppose that the random variables have normal distributions with unknown means μ_i and known variance σ^2. The maximum likelihood estimates are $\hat{\mu}_i = y_i$. Now suppose that the means also have a normal distribution, with unknown mean μ_μ and known variance σ^2_μ. If the mean of the prior distribution is uniform over the real line, an improper prior, the posterior distribution of μ_i has mean

$$\mathrm{E}[M_i|\mathbf{y}] = \frac{\frac{y_i}{\sigma^2} + \frac{\bar{y}_\bullet}{\sigma^2_\mu}}{\frac{1}{\sigma^2} + \frac{1}{\sigma^2_\mu}} \tag{8.38}$$

This estimate is a weighted average of the m.l.e., y_i, and the overall mean, $\bar{y}_\bullet$, with weights inversely proportional to the known variances. The m.l.e. is pulled towards the central value, $\bar{y}_\bullet$, with extreme values having the greatest displacement. This result is closely related to the James–Stein estimator (Section 7.4.2) and the same comments apply.

This can be interpreted in another way, based on a random effect model (Sections 6.4.1 and 8.2.2) if μ_μ is simply taken as unknown, without a uniform prior distribution. Then, the log likelihood function is

$$\mathrm{l}(\boldsymbol{\mu}, \mu) = -\frac{1}{2}\left[\frac{\sum(y_i - \mu_i)^2}{\sigma^2} + \frac{\sum(\mu_i - \mu)^2}{\sigma_\mu^2}\right] \tag{8.39}$$

which yields maximum likelihood estimates identical to the expected value just given. □

Hierarchical priors provide a Bayesian decision-making approach to empirical Bayes problems.

8.3.7 ROBUST PRIORS

Classically, robustness has referred to how sensitive inferences are to the particular statistical model describing the phenomenon under study (Section 4.1.2). For Bayesians, this has been of minor concern, perhaps because all possible statistical models must have positive prior probability density. Instead, Bayesian robustness primarily concerns sensitivity of inference to perturbations of the prior distribution.

Various sets of priors, say $\{R_\Psi\}$, have been proposed as appropriate when one is uncertain about the exact choice. These include

- ε-contaminated priors (Huber, 1964; Berger and Berliner, 1986),

$$\{R_\Psi : R_\Psi = (1-\varepsilon)P_\Psi + \varepsilon P_\Psi^*, P_\Psi^* \in \mathcal{P}_\Psi\} \tag{8.40}$$

- density ratio priors (DeRobertis and Hartigan, 1981),

$$\left\{R_\Psi : \frac{dR_\Psi(\psi_1)}{dR_\Psi(\psi_2)} \leq \frac{a(\psi_1)}{b(\psi_2)}\right\} \tag{8.41}$$

- density bounded priors (Lavine, 1991),

$$\left\{R_\Psi : b(\psi) \leq \frac{dR_\psi}{d\nu} \leq a(\psi)\right\} \tag{8.42}$$

- total variation metric priors (Wasserman and Kadane, 1992),

$$\{R_\Psi : \sup|R_\Psi - P_\Psi| \leq \varepsilon\} \tag{8.43}$$

Here, $\varepsilon \in [0, 1]$ is a suitable measure of uncertainty about the chosen prior, P_Ψ, and $a(\cdot)$ and $b(\cdot)$ are bounded non-negative functions with $a(\cdot) \geq b(\cdot)$ almost everywhere. Such expressions will usually be difficult to compute for high dimensional parameters.

Bayesian methods generally lack robustness when the likelihood is concentrated in a tail of the prior, i.e. when a poor guess has been made or a rare event has occurred.

8.4 Point estimation

8.4.1 POSTERIOR ESTIMATES

Posterior mean If a point estimate is required, the mean of the posterior distribution, called the *posterior mean*,

$$\mathrm{E}[\psi|y] \stackrel{\text{def}}{=} \frac{\int_\Psi \psi dP_{Y|\psi}(y|\psi)dP_\Psi(\psi)}{\int_\Psi dP_{Y|\psi}(y|\psi)dP_\Psi(\psi)} \tag{8.44}$$

may be used.

Example
Consider n i.i.d. Poisson random variables, with a conjugate gamma prior distribution and dominating measure $\nu(d\mu)$, as in Equation (8.7). The expected value of a gamma random variable is

$$\mathrm{E}[\mu] = \frac{\zeta + 1}{\gamma} \tag{8.45}$$

which is our prior estimate of the value of the mean. The m.l.e. will be

$$\hat{\mu} = \frac{\sum y_i}{n} \tag{8.46}$$

which is what the data tell us about the mean. Then, our posterior estimate of the mean will be

$$\mathrm{E}[\mu|\mathbf{y}] = \frac{\sum y_i + \zeta + 1}{n + \gamma} \tag{8.47}$$

showing how the prior belief and the observed data are combined together.□

Obviously, in such point estimation, one will want to have $\mathrm{E}[\mu]$ as accurate as possible. Then, the size of γ, the precision, indicates one's confidence in this prior value.

If Laplace's method is used to approximate the two integrals in the posterior mean, we obtain

$$\mathrm{E}[\psi|\mathbf{y}] \doteq \sqrt{\frac{|\mathrm{I}_p(\check{\psi})|}{|\mathrm{I}_{p\psi}(\tilde{\psi})|}}\frac{\mathrm{L}_{p\psi}(\tilde{\psi})}{\mathrm{L}_p(\check{\psi})} \tag{8.48}$$

where L_p is the posterior likelihood, with corresponding 'm.l.e.', and I_p is the corresponding information (Tierney and Kadane, 1986). This can be extended to general functions of the parameter and to elements of multi-dimensional parameter vectors. The close relationship to the modified and adjusted profile likelihood is evident.

We may also note that the modified profile likelihood for a future observation (Section 3.6.5) is Laplace's approximation to the posterior predictive distribution using a uniform prior.

Maximum posterior estimate A second possible point estimate is the mode. Because this is at the point of highest probability, it is called the *maximum posterior estimate*. This is also sometimes known as a method of penalized maximum likelihood estimation, because the prior provides a penalty against parameter values with low prior probability.

Example
For n i.i.d. Poisson random variables, with conjugate prior, the maximum prior estimate is ζ/γ and the maximum posterior estimate is

$$\tilde{\mu} = \frac{\sum y_i + \zeta}{n + \gamma} \tag{8.49}$$

which, as would be expected for a skewed distribution, is smaller than the posterior mean estimate, but again illustrates the combination of prior belief with the observations. □

Generally, in Bayesian decision-making, the likelihood function comes to dominate the prior distribution as the sample size increases. Thus, the asymptotic results of Chapter 5 generally also hold here. With a smooth prior, as the sample size increases, the likelihood function will become negligible except near the maximum likelihood estimate, so that, for example, the normal approximation of Section 5.3.3 will hold. Within this region near the m.l.e., the prior will be essentially flat, so that the posterior is approximately normal with mean, the m.l.e., and variance, the inverse of the observed information. (For conditions in which this will not hold, see Gelman *et al.*, 1995, pp. 101–104.) However, if, as is usual, the density is used as the likelihood function for continuous variables in Equation (3.5), estimates will be inconsistent if the precision of measurement is not infinite.

Both of these estimates reduce to the corresponding prior estimate when $n \to 0$. On the other hand, as n increases, the prior belief plays a continuously decreasing role; the maximum posterior estimate, then, approaches the m.l.e. The 'prior sample size', γ, reflects confidence in the prior expected parameter value. Generally, for point estimation, it is preferable to understate, rather than overstate, this size.

Prediction If several models are being entertained as tools for prediction, then, in a decision-making context, a combined model, weighted by the posterior probabilities of the individual models (Section 8.6.2) may be used. Such an approach would not generally be contemplated in a scientific context.

Example
The two possible models are

$$\mu_i = \beta_0 + \beta_1 x_i$$

and

$$\mu_i = \alpha$$

with respective posterior probabilities, p_1 and p_2. Then, the combined model for prediction would be

$$\mu_i = p_1\beta_0 + p_2\alpha + p_1\beta_1 x_i \tag{8.50}$$

Notice how the slope is reduced or shrunk ($p_1 < 1$). □

8.4.2 BAYES RISK

Because point estimation is almost always a decision problem, a loss function will be involved. The decision theoretic *Bayes risk* is defined as the expected value, with respect to the prior, of the frequentist risk in Equation (7.3):

$$\begin{aligned}\text{risk}_B(dP_\psi, \delta) &\stackrel{\text{def}}{=} \text{E}_{P_\Psi}\{\text{E}_\Psi[\text{loss}(\psi, \delta)]\} \\ &= \int_\Psi \text{E}_\Psi[\text{loss}(\psi, \delta)]dP_\Psi \qquad (8.51)\\ &= \int_{\mathcal{Y}}\int_\Psi \text{loss}(\psi, \delta)dP_{\Psi|y}dP_Y\end{aligned}$$

The Bayes expected loss of a decision, d, is defined to be

$$\text{loss}_B(\psi, d) \stackrel{\text{def}}{=} \int_\Psi \text{loss}(\psi, d)dP_\Psi \tag{8.52}$$

The *Bayes decision rule* is the δ which minimizes the Bayes risk

$$\text{risk}_{DR}(dP_\Psi) \stackrel{\text{def}}{=} \inf_\delta \int_\Psi \mathrm{E}_\Psi[\text{loss}(\boldsymbol{\psi},\delta)]dP_\Psi \tag{8.53}$$

In this way, with a subjective prior, the frequentist collective decision model is changed into an individual one.

A point estimate, $t(\mathbf{y})$, is a *generalized Bayes estimate* for $g(\boldsymbol{\psi})$ under mean squared error loss, Equation (7.59), if

$$t(\mathbf{y}) = \frac{\int_\Psi g(\boldsymbol{\psi})dP_{Y|\psi}(\mathbf{y}|\boldsymbol{\psi})dP_\Psi(\boldsymbol{\psi})}{\int_\Psi dP_{Y|\psi}(\mathbf{y}|\boldsymbol{\psi})dP_\Psi(\boldsymbol{\psi})} \tag{8.54}$$

with the corresponding Bayes risk,

$$\text{risk}_B(dP_\Psi) = \int_\Psi \mathrm{E}_\Psi[\{\delta - g(\boldsymbol{\psi})\}^2]dP_\Psi$$

The posterior mean is such a Bayes estimate. In a similar way as with minimum variance unbiased estimators, we see that this criterion is particularly useful when the (posterior) distribution is normal with constant variance (Section B.2.1). It may be quite unsuitable when higher cumulants are non-zero, because it ignores the asymmetry.

Examples

1. For the canonical parameter in the exponential family, the generalized Bayes estimate is

$$\begin{aligned}\mathbf{t}(\mathbf{y}) &= \frac{\int \boldsymbol{\theta}\mathrm{e}^{\boldsymbol{\theta}^{\mathrm{T}}\mathbf{y}-\kappa(\boldsymbol{\theta})}dP_\theta}{\int \mathrm{e}^{\boldsymbol{\theta}^{\mathrm{T}}\mathbf{y}-\kappa(\boldsymbol{\theta})}dP_\theta} \\ &= \frac{\partial}{\partial\mathbf{y}}\log\int \mathrm{e}^{\boldsymbol{\theta}^{\mathrm{T}}\mathbf{y}-\kappa(\boldsymbol{\theta})}dP_\theta\end{aligned} \tag{8.55}$$

2. The generalized Bayes estimate of the mean value parameter in the exponential family, under a conjugate prior, having dominating measure $\nu(d\boldsymbol{\theta})$, with mean squared error loss, is linear in $\mathbf{y}$:

$$\begin{aligned}\mathbf{t}(\mathbf{y}) &= \frac{\int_\Theta \boldsymbol{\mu}(\boldsymbol{\theta})\mathrm{e}^{\boldsymbol{\theta}^{\mathrm{T}}\mathbf{y}-\kappa(\boldsymbol{\theta})}dP_\theta}{\int_\Theta \mathrm{e}^{\boldsymbol{\theta}^{\mathrm{T}}\mathbf{y}-\kappa(\boldsymbol{\theta})}dP_\theta} \\ &= \mathbf{y} + \frac{\int_\Theta \frac{\partial}{\partial\boldsymbol{\theta}} f_\Theta(\boldsymbol{\theta};\boldsymbol{\zeta},\gamma)\mathrm{e}^{\boldsymbol{\theta}^{\mathrm{T}}\mathbf{y}-\kappa(\boldsymbol{\theta})}d\boldsymbol{\theta}}{\int_\Theta \mathrm{e}^{\boldsymbol{\theta}^{\mathrm{T}}\mathbf{y}-\kappa(\boldsymbol{\theta})}dP_\theta} \\ &= \frac{\mathbf{y}+\boldsymbol{\zeta}}{n+\gamma}\end{aligned} \tag{8.56}$$

where $f_\Theta(\boldsymbol{\theta};\boldsymbol{\zeta},\gamma)d\boldsymbol{\theta} = dP_\Theta$. This is not exactly the same result for the posterior mean as we obtained above for the Poisson distribution, in Equation (8.47), because the dominating measure has changed. □

The maximum posterior estimate is a generalized Bayes estimate under zero–one loss, while a posterior median estimate corresponds to absolute value loss. For the robust Bayesian, the use of utilities, upon which a loss function is based, introduces a second layer of sensitivity concerns.

8.5 Interval estimation

8.5.1 INTERVALS BY INTEGRATION

In Bayesian decision-making, interval estimates are simply obtained by integration of the posterior distribution over the appropriate interval or region, giving the probability that a parameter lies in it. This is called a *credibility interval* or *region.*

Example
For the Poisson distribution, with conjugate prior for the mean, a credibility interval, (μ_1, μ_2), within which the mean lies with given posterior probability, $1 - \alpha$, is obtained by integrating:

$$\int_{\mu_1}^{\mu_2} \frac{(\gamma + n)^{\zeta + y_\bullet + 1} \mu^{\zeta + y_\bullet} \mathrm{e}^{-(\gamma+n)\mu}}{\Gamma(\zeta + y_\bullet + 1)} d\mu = 1 - \alpha \tag{8.57}$$

For this to be usable, the parameters ζ and γ must have known values. □

As with confidence intervals, a credibility interval is not unique. When an interval about some value, such as the posterior mean, is required, this may be obtained in a number of ways. For example, we can take equal tail probabilities on either side or the *region of highest posterior density.*

Example
For the Poisson distribution, suppose that the conjugate prior has $\zeta = \gamma = 1$. Then,

$$2(n+1)\mu \sim \chi^2_{2y_\bullet + 2} \tag{8.58}$$

from which credibility intervals with equal tail probabilities can be obtained.

A region of highest posterior density is given by choosing c such that

$$\int_{\{\mu | f_{M|y}(\mu|y) \geq c\}} \frac{(\gamma + n)^{\zeta + y_\bullet + 1} \mu^{\zeta + y_\bullet} \mathrm{e}^{-(\gamma+n)\mu} d\mu}{\Gamma(\zeta + y_\bullet + 1)} = 1 - \alpha \tag{8.59}$$

□

The latter is the region with minimal Lebesgue measure of all regions with the same posterior probability:

The region of highest posterior density gives the shortest credibility interval.

proof: For given posterior probability, α, suppose that the region of highest posterior density is $[\psi_1, \psi_2]$ containing the mode, so that

$$\int_{\psi_1}^{\psi_2} dP_{\Psi|y}(\psi|y) = 1 - \alpha \tag{8.60}$$

with $f_{\Psi|y}(\psi_1|y) = f_{\Psi|y}(\psi_2|y) > 0$. Let $[\psi_1^*, \psi_2^*]$ be any other interval with $\psi_2^* - \psi_1^* < \psi_2 - \psi_1$. Let us consider first $\psi_1 \leq \psi_1^*$.

If $\psi_2^* \leq \psi_1$, then

$$\begin{aligned}\int_{\psi_1^*}^{\psi_2^*} f_{\Psi|y}(u|y)du &\leq f_{\Psi|y}(\psi_2^*)(\psi_2^* - \psi_1^*)\\ &\leq f_{\Psi|y}(\psi_1)(\psi_2 - \psi_1)\\ &\leq \int_{\psi_1}^{\psi_2} f_{\Psi|y}(u|y)du\end{aligned}$$

If $\psi_2^* > \psi_1$, then

$$\begin{aligned}\int_{\psi_1^*}^{\psi_2^*} f_{\Psi|y}(u|y)du &= \int_{\psi_1}^{\psi_2} f_{\Psi|y}(u|y)du + \int_{\psi_1^*}^{\psi_1} f_{\Psi|y}(u|y)du\\ &\quad - \int_{\psi_2^*}^{\psi_2} f_{\Psi|y}(u|y)du\\ &\leq \int_{\psi_1}^{\psi_2} f_{\Psi|y}(u|y)du + f_{\Psi|y}(\psi_1)(\psi_1 - \psi_1^*)\\ &\quad - f_{\Psi|y}(\psi_2)(\psi_2 - \psi_2^*)\\ &= \int_{\psi_1}^{\psi_2} f_{\Psi|y}(u|y)du + f_{\Psi|y}(\psi_1)[(\psi_2^* - \psi_1^*) - (\psi_2 - \psi_1)]\\ &\leq \int_{\psi_1}^{\psi_2} f_{\Psi|y}(u|y)du\end{aligned}$$

The demonstration is similar for $\psi_1 > \psi_1^*$. □

The region of highest posterior density need not necessarily contain the posterior mean if the posterior distribution is not symmetric or is multimodal; it will contain the posterior mode, the maximum posterior estimate.

In addition, such regions are not parametrization invariant, i.e. the shortest credibility interval for ψ cannot simply be transformed to yield the shortest one for, say, $1/\psi$, although it will still have the same posterior probability level. This can easily lead to contradictions.

Example
In the example of the location–shifted exponential distribution near the beginning of Section 8.3, the region of highest posterior density for ψ is of the form (c_1, y), but, when calculated through $\alpha = e^{\psi}$, it is $(0, \log c_2)$. These regions may even be disjoint. □

Thus, Bayesian conclusions are generally highly dependent on the parametrization of the statistical model, for a given prior.

8.5.2 LIKELIHOOD REGIONS

Likelihood intervals, and, more generally, regions, as studied in Chapter 3, have at least two interpretations in the Bayesian context. When a uniform prior, perhaps defined only over the region where the likelihood is important, is used, to yield an integrated likelihood, regions of highest posterior density will be identical to likelihood regions. These have the property of always having a posterior probability greater than or equal to their prior probability. However, from this Bayesian perspective, such a region is not parametrization invariant, unless the prior is changed when the parameter is transformed. Thus, norming a likelihood by integrating does not automatically ensure that the resulting posterior has a reasonable probability interpretation.

In the robust Bayesian perspective, the likelihood region is robust in the sense that the resulting posterior probability is least sensitive to changes in the prior with respect to the class of ε-contaminated priors. Here, we have a class of priors so that the prior is not fixed; the posterior credibility region is defined directly in terms of the normed likelihood function. A likelihood ratio, then, has the robust interpretation of being the posterior odds comparing two models under prior ignorance. Thus, in contrast to the region of highest posterior density with uniform prior, here this robust credibility region is parametrization invariant (Wasserman, 1989).

8.5.3 NUISANCE PARAMETERS

For multidimensional parameters, the specification of the prior distribution becomes very difficult because it must be multivariate. Most often, a product of independent univariate distributions is used (Lavine *et al.*, 1991). However, as we have seen in Chapters 3 and 6, in almost all cases, the likelihood function does not factor, implying that no empirical information can be obtained independently for the separate elements of the parameter vector. Thus, assuming independent prior distributions would be unreasonable, even contradictory. We *know* that the parameters are dependent.

Example
Suppose that we have two populations with response variables following normal distributions having, respectively, means μ_1 and μ_2 and variances σ_1^2 and σ_2^2. We are interested in the difference in means. If we take independent priors, $dP_{M_1} dP_{M_2} dP_{\Sigma_1} dP_{\Sigma_2}$, this would imply that, even if we know the value of σ_1^2, this would provide no information about σ_2^2. However, for comparison of the means to make sense, we must usually know that the variances are similar or at least that there is some relationship between them. □

As well, with ill-defined dependence, there will be a danger of dilation of the posterior probabilities (Section B.3.5).

Factoring the posterior distribution Of course, if the likelihood function factors, and independent prior distributions can be assumed, the problem of nuisance parameters is solved. Thus, the marginal and conditional methods of Chapter 6, applied to the likelihood function, are also useful in a Bayesian context. If the likelihood function, and the prior distribution, factor, separating nuisance parameters from those of interest, then the posterior will also factor. One is relieved of the obligation to develop a prior distribution for the nuisance parameters, which may be of quite high dimension, and of performing the corresponding high dimensional integration on the multivariate posterior distribution. This will be true even in the approximate situations studied in that chapter.

Another possibility, in certain cases, is that the nuisance parameters can be transformed to orthogonality with respect to the parameter of interest within the posterior distribution using procedures similar to those for information orthogonality in Section 6.1.3 (Tibshirani and Wasserman, 1994).

Integration When the posterior distribution does not factor, the simplest approach to nuisance parameters is to integrate them out, as we did in Equation (8.29). In this way, we obtain the marginal distribution of the parameter of interest.

Example
Suppose that we have two Poisson distributed random variables, with means μ_i, and that we are interested in the ratio of these means, $\lambda = \mu_1/\mu_2$. If we take independent conjugate gamma distributions for the two means, the posterior joint density of the means is the product of independent gamma distributions

$$dP_{M_1M_2|y_1y_2}(\mu_1,\mu_2|y_1,y_2) \propto \prod_{i=1}^{2} \mu_i^{\zeta_i+y_i} \mathrm{e}^{-(\gamma_i+1)\mu_i} d\mu_i \tag{8.61}$$

Then, the marginal posterior density of λ is obtained by change of variables and integration in the same way as for a ratio of Chi-squared distributions, so that

$$\frac{\gamma_2(\gamma_1+1)(\zeta_2+1)(y_2+\zeta_2+1)}{\gamma_1(\gamma_2+1)(\zeta_1+1)(y_1+\zeta_1+1)}\lambda \sim \mathrm{F}_{2y_1+2\zeta_1+2,2y_2+2\zeta_2+2} \tag{8.62}$$

an F distribution. (Cox and Hinkley, 1974, p. 371) □

However, one may question if such marginal posterior distributions are any more appropriate for decision-making than is modelling only marginal distributions of response variables which are interdependent, instead of developing a full multivariate model.

Maximization An alternative method, especially if a maximum posterior estimate is being used, is to maximize the posterior distribution over the nuisance parameters to yield a *profile posterior distribution*. This is numerically simpler than integration, and often provides a good approximation to it, as can be shown by applying a Laplace approximation. However, the same objection can be made to it as to a profile likelihood: nuisance parameters are assumed to take their most (posterior) probable value, giving intervals which are too narrow.

On the other hand, integration means taking a weighted *average* value of those nuisance parameters. Here, the nuisance parameters are effectively assumed to be equal to this average, giving intervals for a parameter of interest which may be narrower than the profile posterior distribution, whereas wider intervals are required.

Conditioning Thus, extreme caution should be taken in integrating out nuisance parameters. The posterior density of the parameter of interest can be written

$$\begin{aligned} f_\Lambda(\boldsymbol{\lambda}|\mathbf{y})d\boldsymbol{\lambda} &= \int_\Phi dP_{\Lambda\Phi}(\boldsymbol{\lambda},\boldsymbol{\phi}|\mathbf{y}) \\ &= \int_\Phi dP_{\Lambda|\phi}(\boldsymbol{\lambda}|\boldsymbol{\phi},\mathbf{y})dP_\Phi(\boldsymbol{\phi}|\mathbf{y}) \end{aligned} \tag{8.63}$$

The first factor in the integrand, $dP_{\Lambda|\phi}(\boldsymbol{\lambda}|\boldsymbol{\phi},\mathbf{y})$, indicates how the distribution of the parameter of interest changes for different values of the nuisance parameter, while the second factor is a weighting in the average.

If the density, $f_\Phi(\boldsymbol{\phi}|\mathbf{y})$, is narrow, then the integration is almost the same as assuming the nuisance parameter to take its posterior modal value.

The narrow posterior reflects relative certainty about the value of this parameter. On the other hand, if this density is rather flat, the way in which $dP_{\Lambda|\phi}(\boldsymbol{\lambda}|\boldsymbol{\phi}, \mathbf{y})$ varies with $\boldsymbol{\phi}$ will be important to any conclusions drawn.

Example
Suppose that we have n i.i.d. normal observations. Let μ be the parameter of interest and σ^2 the nuisance parameter. With a flat non-informative prior, the conditional posterior, $f_{M|\sigma}(\mu|\sigma, \mathbf{y})$, is $\mathrm{N}(\hat{\mu}, \sigma^2/n)$ and the marginal posterior, weighting, distribution for σ is related to a Chi-squared distribution with $n-1$ d.f. This yields a marginal posterior for μ which is a Student t distribution with $n-1$ d.f., centred at $\hat{\mu}$ and scaled by $\hat{\sigma}/\sqrt{n}$.

Thus, the integrated marginal posterior for μ resembles the conditional normal distribution at $\sigma^2 = \hat{\sigma}^2$, but with the thicker tails of a Student t distribution. In contrast, the conditional posterior for μ, although retaining the same maximum, has greater or smaller dispersion depending on the fixed value of the nuisance parameter, σ^2. (Box and Tiao, 1973, p. 98) □

In other examples, the range of (posterior) probable values of the parameter of interest will vary depending on the (conditional) value of the nuisance parameter.

More generally, additional care must be taken if decisions are to be made marginally about both $\boldsymbol{\lambda}$ and $\boldsymbol{\phi}$. Values of each may be found which are individually probable but jointly improbable. Thus, such decisions are generally not compatible. The marginal probability of, say, $\boldsymbol{\lambda}$ applies irrespective of the value of $\boldsymbol{\phi}$, whereas the joint probability is for specific values of $\boldsymbol{\phi}$.

Thus, we conclude, as in Chapter 6, that, if the posterior distribution does not factor, individual parameters cannot usually be interpreted in isolation from each other. They must be studied simultaneously. One possibility, where feasible, is to look at the conditional posterior distribution, $dP_{\Lambda|\phi}(\boldsymbol{\lambda}|\boldsymbol{\phi}, \mathbf{y})$, for various values of $\boldsymbol{\phi}$, rather than the marginal distribution.

8.6 Testing

8.6.1 POSTERIOR ODDS AND BAYES FACTORS

If the hypothesis is made that $\boldsymbol{\psi} \in \Psi_0 \subset \Psi$, the posterior probability of this hypothesis can be obtained by integration. This can be used as a test without the need to compare it to an alternative hypothesis (although, of course, it is being implicitly compared to its complement, $\boldsymbol{\psi} \in \Psi \setminus \Psi_0$).

If two completely specified models, indexed by $\boldsymbol{\psi}_1$ and $\boldsymbol{\psi}_2$, are to be compared, the *posterior odds* can be used:

$$\frac{P_{\Psi|y}(\psi_1|y)}{P_{\Psi|y}(\psi_2|y)} \stackrel{\text{def}}{=} \frac{P_{Y|\psi}(y|\psi_1)}{P_{Y|\psi}(y|\psi_2)} \times \frac{P_{\Psi}(\psi_1)}{P_{\Psi}(\psi_2)} \tag{8.64}$$

This formulation shows clearly how the likelihood ratio, now called the *Bayes factor* (Good, 1958), is weighted by the prior odds.

The posterior odds has two major advantages over the usual Bayes' formula of Equation (8.2):

(1) Because the norming factor in the denominator has cancelled, it contains no reference to any exhaustive totality of possible models.
(2) The differential element has also disappeared, so that it is invariant under one-to-one parameter transformations (Fisher, 1912; Barnard, 1949).

If the models compared have equal prior probability, the posterior odds are just a likelihood ratio. Thus, the latter, including the normed likelihood, have a Bayesian interpretation in terms of posterior odds, given that the prior odds are even.

Indeed, the ratio

$$\frac{\mathrm{L}(\psi_1)}{\mathrm{L}(\psi_1)+\mathrm{L}(\psi_2)} \tag{8.65}$$

also has an interpretation as a conditional significance level (Berger *et al.*, 1994). Thus, this is another case where the direct likelihood, frequentist, and Bayesian conclusions can agree.

More complex sets of contiguous models can also be compared by ratios of integrated posterior distributions. However, a large parameter region has, inherently, more chance of dominating the comparison than a small region, so that such ratios should be interpreted with care (see Section 3.1.3).

8.6.2 SHARP HYPOTHESES

The simple null hypothesis, $\psi = \psi_0$, of the frequentist, is here called a *sharp hypothesis.* As we have seen, one particular situation where this is important is model selection, with the hypothesis that certain parameters take point values, simplifying out of the model. Then, we have

$$\frac{P_{\Psi|y}(\psi_0|y)}{P_{\Psi|y}(\psi \neq \psi_0|y)} = \frac{p_0 P_{Y|\psi}(y|\psi_0)}{(1-p_0)\int_{\psi\neq\psi_0} P_{Y|\psi}(y|\psi_0)dP_{\Psi|\psi\neq\psi_0}(\psi)} \tag{8.66}$$

where p_0 is the prior probability that $\psi = \psi_0$, with the probability for other possible parameter values given by $(1-p_0)dP_{\Psi|\psi\neq\psi_0}(\psi)$. This is an extension of previous ideas in this chapter, because we are now giving positive prior probabilities to point hypotheses, not just to sets of model parameters. Here, we have $P(H_0) = p_0$ and $P(H_A) = 1 - p_0$.

We are giving a finite prior probability, p_0, to the sharp hypothesis, H_0, or to the *single* model, ψ_0, and another finite probability, $1 - p_0$, to the composite hypothesis, H_A, or to a usually infinite set of other models, over which it must be spread. This contrasts to estimation with a continuous prior, where any point parameter value has zero prior probability. In other words, we have counting measure on the model of H_0 and Lebesgue measure on those of H_A. A uniform prior would have each equal to one-half. Some Bayesians have even argued that this is an objective prior, because it gives equal weight to the two hypotheses, although they contain radically different numbers of models.

Example
Consider n observations from a normal distribution with known variance, σ^2, and H_0: $\mu_0 = 0$. Let the conditional prior distribution for the mean be $\mathrm{N}(0, \sigma_\mu^2)$. Then,

$$\frac{P_{M|y}(\mu_0|y)}{P_{M|y}(\mu \neq \mu_0|y)} = \frac{p_0}{1-p_0} \times \frac{\exp\left[-\frac{\sum y_i^2}{2\sigma^2}\right]}{\frac{1}{\sqrt{2\pi\sigma_\mu^2}} \int \exp\left[-\frac{\sum (y_i-\mu)^2}{2\sigma^2}\right] \exp\left[-\frac{\mu^2}{2\sigma_\mu^2}\right] d\mu}$$

$$= \frac{p_0}{1-p_0} \times \left[1 + \frac{n\sigma_\mu^2}{\sigma^2}\right]^{\frac{1}{2}} \exp\left[-\frac{n\bar{y}_\bullet^2}{2\sigma^2\left(1 + \frac{\sigma^2}{n\sigma_\mu^2}\right)}\right] \qquad (8.67)$$

Thus, for example, if $\bar{y}_\bullet = 0$, the posterior odds in favour of $\mu = \mu_0$ increase with n, as would be expected. □

In contrast to the pure significance test, which only concerns one model, and the frequentist hypothesis test, which is asymmetric in comparing models, here we have the asymmetric hypotheses, one a point and the other an infinite collection of models, but compared in a symmetric manner. A frequentist likelihood ratio test compares the null or sharp hypothesis model to the most likely model for the data, an extreme contrast, while the Bayes test compares it to a weighted average of all other possible models, some of which may be more likely or probable and others less than the null. Thus, the two tests have quite different interpretations and the latter will generally give less extreme results than the former.

Suppose now that we wish to compare two composite hypotheses, H_0 and H_A, where all of the parameter values in the models are not completely specified. We require priors, $dP(\psi)$, for the parameters, as well as those, p_i, for the two models. Then, the unknown parameters are removed by integration, before calculating the posterior odds for the two models:

$$\frac{P(H_0|y)}{P(H_A|y)} = \frac{p_0}{p_A} \times \frac{\int_\Psi P_{Y|\Psi}(y|\psi_0)dP_\Psi(\psi_0)}{\int_\Psi P_{Y|\Psi}(y|\psi_A)dP_\Psi(\psi_A)} \tag{8.68}$$

where $p_A = 1 - p_0$. The ratio of integrated or 'prior average' likelihoods, again known as a Bayes factor, is weighted by the two hypothesis priors. In contrast to Bayesian point estimates, these Bayes factors tend to be sensitive to the choice of priors for the parameters. As is clear from Equation (8.68), the marginal density for Y must be proper for each hypothesis.

In a robust Bayes context, we would be interested in the lower bound on the posterior probability of the null hypothesis, $P(H_0|y)$, over all possible prior distributions, dP_Ψ. This is given when the Bayes factor is equal to the normed likelihood of the null model (Berger and Sellke, 1987):

$$P(H_0|y) \geq \frac{p_0 P_{Y|\psi}(y|\psi_0)}{p_0 P_{Y|\psi}(y|\psi_0) + p_A P_{Y|\psi}(y|\hat{\psi})} \tag{8.69}$$

i.e. the 'prior', dP_Ψ, has a point mass only at the maximum likelihood estimate. Because of the averaging over less likely models, all other priors yield less extreme results than the normed likelihood. (Recall that the latter is the basis of frequentist likelihood ratio tests.)

8.6.3 JEFFREYS–LINDLEY PARADOX

Let us now investigate in more detail the implications of testing sharp hypotheses in the Bayesian approach. This can best be done by means of examples.

Examples

1. In Equation (8.67), $\lim_{\sigma_\mu^2 \to \infty} P_{M|y}(\mu_0|y) = 1$ for all p_0 and $\bar{y}_\bullet$. As prior belief about μ becomes more diffuse, μ_0 is chosen with probability one, no matter what data are observed or what non-zero prior weight was given to that hypothesis.

2. Consider again the normal distribution with known variance, but let us now compare the model H_0 with $\mu_0 = 0$ to H_A with μ_A unspecified. We shall here take a proper uniform prior for μ_A: $dP(\mu_A) = 1/(2c)$ on $(-c, c)$ for c large. Then, the Bayes factor for comparing the two hypotheses is

$$\frac{2c\sqrt{\frac{n}{\sigma^2}}\exp\left[-\frac{n\bar{y}_\bullet^2}{2\sigma^2}\right]}{\int_{\sqrt{n}(\bar{y}_\bullet - c)/\sigma}^{\sqrt{n}(\bar{y}_\bullet + c)/\sigma} \exp[-u^2]du}$$

For a given frequentist significance level, we would fix $\sqrt{n}\bar{y}_\bullet/\sigma$. The denominator rapidly approaches unity as c increases. Thus, this Bayes factor becomes arbitrarily large as $c \to \infty$ or $n \to \infty$, for any fixed value of the significance level, increasingly favouring H_0 over H_A.

Table 8.1. Bayes factors and likelihood ratios for the model $\pi_0 = 0.5$ for four Bernoulli experiments, from Edwards *et al.* (1963).

n	50	100	400	10 000
y	32	60	220	5098
$\hat{\pi}$	0.64	0.60	0.55	0.51
Bayes factor	0.82	1.09	2.17	11.69
Normed likelihood	0.14	0.13	0.13	0.15

Table 8.2. Bayes factors and likelihood ratios for the model $\pi_0 = 0.5$ for six Bernoulli experiments.

n	25	100	600	1500	5000	8000
y	13	52	312	780	2600	4160
$\hat{\pi}$	0.52	0.52	0.52	0.52	0.52	0.52
Bayes factor	4.03	7.43	12.12	9.32	1.03	0.12
Normed likelihood	0.98	0.92	0.62	0.30	0.02	0.0016

Any other fixed proper prior shows the same effect as $n \to \infty$. This results because the prior assigns increasing weight to values of μ which have very small likelihood. (Aitkin, 1991)

3. Let us look, finally, at a very influential example, from Edwards *et al.* (1963). Here, n Bernoulli trials are undertaken, with true probability π, and a sharp hypothesis of $\pi_0 = 0.5$. For a beta prior on π, from Equation (C.20), the Bayes factor is

$$(n+1)\binom{n}{y}\pi_0^y(1-\pi_0)^{n-y} \tag{8.70}$$

This is given, along with the likelihood ratio, for four data sets in Table 8.1. Although the data alone in all four cases provide about the same evidence against the sharp hypothesis, as compared to the best model for those data, the Bayes factor tends increasingly to favour it.

A second table shows that this Bayes factor is not a monotone function with n. Consider the values in Table 8.2, all with the same m.l.e., and with decreasing plausibility for the model $\pi_0 = 0.5$, according to the normed likelihood. The Bayes factor favours this model less for small samples than for 600 observations, but finally going against it only for more than 5000 observations. □

These are examples of the *Jeffreys–Lindley paradox.* In the light of such results, some Bayesians have argued that sharp hypotheses are unreasonable, and that only interval hypotheses should be used. This conclusion is logical when Lebesgue measure is attributed to a parameter (but recall the problems with coherency). However, it raises the further question if comparing unequal-sized intervals by posterior odds ratios is meaningful.

Example
Suppose that one wants to compare $\psi \in (0,1)$ and $\psi \in [1,5)$. With a uniform prior, the posterior can favour $\psi \in [1,5)$ even though the likelihood function indicates stronger support for values of $\psi \in (0,1)$. (See the discussion in Section 3.1.3.) □

Others have held that the improper prior causes the problem and that Bayesian decision-making should be limited to informative priors. Example 2 above demonstrates that the problem does not lie here. Thus, this paradox appears to imply that, if one already has enough prior information or belief to place a point mass (counting measure) on one hypothesis, but not on other possible individual models, so that the latter set has Lebesgue measure, then empirical data are not necessary. In this sense, the prior probability of a sharp hypothesis should always be zero (Novick, 1969). But this evidently creates problems for model selection.

8.6.4 BAYESIAN INFORMATION CRITERION (BIC)

If a parameter is truly random, with Lebesgue measure, then the probability that it is zero, i.e. that it disappears from the model, is zero. This can pose problems for model selection, unless one is prepared to eliminate a parameter simply because zero lies in a region of high posterior probability.

In a way similar to the asymptotic development leading to the Akaike information criterion (AIC) of Section 7.4.2, Bayesian arguments can be developed to derive a criterion for model selection. However, in this case, the criterion depends on the sample size, n. Thus, an asymptotic approximation to model selection by Equation (8.66) is given by modifying the deviance to $\mathrm{D} + p\log(n)$, where p is the dimension of the unknown parameter vector in a model (Schwarz, 1978). This has come to be called the Bayesian information criterion (BIC). It can be obtained from the calibrated deviance of Equation (3.69) by setting $a = 1/\sqrt{n}$.

As already mentioned there, the BIC has similar characteristics to the Jeffreys–Lindley paradox, tending to indicate the simplest model as n becomes very large. However, in contrast to the AIC, the BIC will consistently select the true model out of two contenders because $P(\chi^2_p > p\log(n))$ goes to zero in the limit.

8.7 Bayesian and frequentist decision-making

In contrast to the direct likelihood approach, which uses models as informative approximations to an underlying data generating mechanism, both the frequentist and the Bayesian assume that models, in some sense, could be true. The central element of frequentist theory is the hypothesis test, while the Bayesian approach turns around parameter estimation in a pre-given family of models. The former ensures the integrity of the conclusions by specifying hypotheses and the latter by specifying the prior, both before observing the data. The Bayesian looks more to the past and the frequentist to the future.

For a frequentist, hypothesis testing and point estimation are complementary, whereas, for a Bayesian, the problems are completely different. In testing, models have a positive prior probability, whereas, in estimation, with continuous priors, any special point value of a model parameter has zero prior probability.

The frequentist must be able to specify exhaustively all observations which might possibly have arisen under the given data generating conditions – the sample space. The Bayesian must be able to specify exhaustively all models which might possibly have generated the observed data – the model space. For direct likelihood inference, neither complete list is necessary.

The frequentist model of human decision-making can be interpreted as one of collective behaviour, through the long run repetition of observations. It can provide useful criteria for judging the operating characteristics of decision-making procedures. The Bayesian model, especially the coherent subjectivist one, is individualist, based on personal prior knowledge or beliefs. Although this can be overcome by looking at a collection of priors, the Bayesian paradigm then loses its justification in terms of coherency. In contrast to the statistical models for a phenomenon under study, decision-making models are rarely checked, nor is it clear just how they might be.

The frequentist is asking how well the decision-making model works on average, when the situation is replicated many times, always with the same true value of the unknown parameter, but the observations changing randomly according to the given probability model. The Bayesian is asking how well it works, on average, when the situation is replicated many times, always with the same observed data, but the unknown parameter changing randomly according to the posterior distribution (Efron in discussion of Blyth, 1970).

Both frequentist and Bayesian decision-making attack the problem of comparing *sets* of models, something which cannot be done with the likelihood function alone. Whereas the frequentist tends to look at comparisons of extreme parameter values in the set, the Bayesian uses averages over the set. The Bayesian approach has difficulty with fully specified models (sharp hypotheses), as compared to the frequentist, and direct likelihood, because, for Bayesians, parameters generally do not have point values, but Lebesgue measure. In both approaches, examination of likelihood regions can play a central role, whether through the Neyman–Pearson lemma, the integrated likelihood, or robustness arguments. With rare exceptions, neither allows even approximately compatible decision-making among models with differing dimensions of parameters.

The frequentist use of the sample space generally means that checking model goodness of fit is at least implicitly involved in all procedures. In contrast, the Bayesian requirements, that all possible models have positive prior probability (density) and that posterior probabilities are conditional on one of the model functions being correct, can induce a mentality whereby model checking is considered to be of minor importance. (For one possible approach to this, see Draper, 1995.)

In point estimation, Bayes rules outperform frequentist rules when based on reasonable prior belief; if the Bayesian is ignorant or misinformed about the parameters, frequentist rules will be superior (Samaniego and Reneau, 1994).

Reliable prior information and knowledge are necessary for the design, monitoring, and interpretation of any study. If they can be specified in the precise mathematical form of probabilities, then Bayesian decision-making can be used. However, if the prior information is empirical, as we have seen, it can be combined directly as a likelihood, without need for Bayes' formula. On the other hand, if no clearly appropriate prior is available, the formal Bayesian approach should be avoided. In all cases, the likelihood function should be reported, so that anyone can add their own prior distribution. Thus, in scientific work, the likelihood function would be shown in the results section, while subjective priors and/or hypothesis tests could be given in a discussion or interpretation section.

Further reading

The subjective Bayesian point of view owes a debt to de Finetti (1937, 1974a, b). Savage (1954, 1971) and Lindley (1972, 1985) have also made prominent contributions. Other important papers include Edwards *et al.* (1963) and Pratt *et al.* (1964). For the robust and decision theoretic approaches, see Berger (1985b), as well as Berger and Wolpert (1988). Bayesian texts close to some of the ideas presented in this book are those of Box and Tiao (1973) and Gelman *et al.* (1995). The latter centres around random effects models (called hierarchical models) and flat priors, priors which, if used for inferences, yield results identical to those from the simpler direct likelihood approach.

Bayesians have not often discussed evaluation of the goodness of fit of their models or diagnostics; see Box and Tiao (1968), Box (1980), Rubin (1984), and Gelman *et al.* (1995, pp. 161–185).

Gelman *et al.* (1995) provide many examples of the empirical Bayes approach. de Jong (1988) shows how the Kalman filter can be used to construct complex likelihood functions for stochastic process models.

For comparisons of the frequentist and Bayesian methods, including conditions under which they can give the same results, see Lindley (1958), Welch and Peers (1963), Bartholomew (1965, 1971), Pratt (1965), Casella and Berger (1987), Dawid (1991), Severini (1991), Datta and Ghosh (1995), and Gelman *et al.* (1995, pp. 104–111).

The body of Fisher's and Barnard's work emphasizes the contrast between scientific inference and decision-making. For some problems arising from the simplicity of decision-making models, see Blyth (1970, 1972a, b).

8.8 Exercises

(1) (a) Compare the conjugate and non-informative Jeffreys' priors for the mean of the binomial, Poisson, normal, gamma, and inverse Gaussian distributions.
(b) What values of the two parameters in the conjugate prior give Jeffreys' prior in each case?
(c) For what parametrization are these uniform prior distributions?
(Novick, 1969)

(2) Consider n i.i.d. observations from a normal distribution with known variance, $\sigma^2 = 4$. Suppose that we have a uniform prior for the mean, μ.
(a) Plot the likelihood function for the mean, with $n = 10$, for the observed values, $\bar{y}_\bullet = 6, 9$, and 12.
(b) Repeat for the reciprocal of the mean, $\alpha = 1/\mu$.
(c) Plot the corresponding posterior distributions in all cases.
(Box and Tiao, 1973, pp. 28–29)

(3) Two engineers are attempting to obtain more accurate estimates of some physical constant which we shall represent by ψ. The first is familiar with the field and can express prior opinion as a normal distribution with mean 900 and variance 400, while the second, a relative novice, believes it to have a mean of 800 and variance of 640. The method of measurement is known to give results having a normal distribution with variance 160.
(a) Plot the prior distributions.
(b) Write down the posterior distributions.
(c) A single observation is made, with $y = 850$. Plot the likelihood function.
(d) Plot the two posterior distributions and compare them to the priors.
(e) A further 99 observations are made with overall mean, $\bar{y}_\bullet = 870$. Plot the new posterior distributions and compare them to the previous ones.
(Box and Tiao, 1973, pp. 15–18)

(4) Give the conjugate prior distribution for the Bernoulli distribution for measures based on
   (a) the canonical parameter, $\theta = \log\left(\frac{\pi}{1-\pi}\right)$;
   (b) the mean value parameter, $\mathrm{E}[Y] = \pi$;
   (c) the complementary log log transform, $\psi = \log[-\log(\pi)]$;
   (d) the inverse sine transform, $\psi = \sin^{-1}(\sqrt{\pi})$.

(5) Suppose that we use Jeffreys' prior with the binomial distribution. We obtain one success in n trials. How large should n be in order to have two to one odds in favour of a success occurring in the next trial?
(Cox and Hinkley, 1974, p. 382–383)

(6) Consider the three category multinomial distribution,

$$f(y_1, y_2; \pi_1, \pi_2, n) = \binom{n}{y_1\ y_2\ y_3} \pi_1^{y_1} \pi_2^{y_2} (1 - \pi_1 - \pi_2)^{y_3}$$

with π_1 the parameter of interest.
   (a) Derive Jeffreys' prior for (π_1, π_2).
   (b) Find the reference prior when π_2 is a nuisance parameter.
   (c) What is the marginal posterior distribution and expected value of π_1 in each case?
(Bernardo in discussion of Kass, 1989)

(7) We have a large finite population of known size N, with a binary random variable Y. We choose a large sample of size n which is, however, substantially smaller than N. Our sample all contains the same value of Y, say 1. We want to estimate the number of ones in the population, say ψ.
   (a) Write down the general likelihood function and its specific form for the sample obtained.
   (b) Derive a reference prior for this model.
   (c) What is the posterior probability that the whole population contains only ones, i.e. that $N = \psi$?
   (d) Suppose now that the parameter of interest is the indicator function of Equation (B.1),

$$\lambda = I(N = \psi)$$

   with nuisance parameter,

$$\phi = \psi^{I(N \neq \psi)}$$

   Derive the reference prior.
   (e) What now is the posterior probability of $N = \psi$?

(f) Relate these differing results to the Jeffreys–Lindley paradox.

(Bernardo and Smith, 1994, pp. 322–323)

(8) Fungal spores are introduced into the earth surrounding the root of a plant. Some of the spores germinate, with a germ-tube emerging. As this grows it bends around and may point towards the adjacent root. With m spores, r germinate and, of these, s are aligned towards the root.

(a) Write down an appropriate likelihood function if π is the probability of germination and ν the conditional probability of alignment.

(b) Derive Jeffreys' prior for this model.

(c) Derive a suitable stepwise reference prior and compare this to Jeffreys' prior.

(d) Consider the transformations $\lambda = \pi\nu$ and $\phi = \pi(1-\nu)/(1-\pi\nu)$. Repeat the derivation of the two sets of priors and discuss the results.

(Crowder and Sweeting, 1989; Polson and Wasserman, 1990)

(9) (a) What is the conjugate prior for the multinomial distribution?

(b) Consider this distribution applied to a 2×2 contingency table, as in Exercise (6.7). We are interested in the log odds ratio, λ.

i. Derive the posterior probability that $\lambda < 1$.

ii. Compare the result to that obtained from Fisher's exact test.

(Cox and Hinkley, 1974, pp. 371–372, 393–394)

(10) Derive the relationships, Equations (8.56), for the generalized Bayes estimate of the mean value parameter of the exponential family given in the example of Section 8.4.2.

(11) (a) For n i.i.d. observations from a normal distribution with known variance and conjugate prior, derive the region of highest posterior density for the mean.

(b) Repeat for a non-informative uniform prior.

(c) How do these intervals compare to the corresponding confidence interval?

(Casella and Berger, 1987, pp. 424–425)

(12) Suppose that y_1 and y_2 are the numbers of successes in n_1 and n_2 independent Bernoulli trials with probabilities π_1 and π_2. Consider the sharp hypothesis, $H_0 : \pi_1 = \pi_2$.

(a) Use the conjugate prior to obtain the posterior odds of this hypothesis.

(b) Find another way of setting up the same test.

(Cox and Hinkley, 1974, pp. 158–160)

(13) Children are given an intelligence test, the result, Y, being a normal random variable with variance 100 and μ the true IQ level of the child. In the population, IQs have a normal distribution with mean 100 and variance 225.
   (a) What is the marginal (compound) distribution of Y?
   (b) What is the posterior distribution of μ?
   (c) If a child scores 115 on a test, what is the distribution of his or her true IQ?

   (Berger, 1985b, p. 128)

(14) Hemophilia is a hereditary illness of the blood which inhibits it from coagulating owing to lack of cytozyme. The disease is due to a recessive gene carried on a sex chromosome. It is transmitted by women, but is found primarily in men.
   (a) Suppose that a woman, who is not hemophilic, has n sons, of whom y are hemophilic.
      i. Develop a model to describe this phenomenon.
      ii. With only this information, how would you make decisions about the unknown parameter?
   (b) Now suppose that the woman has normal parents but that she has a hemophilic brother.
      i. What is the probability that she inherited the gene?
      ii. What is the probability that her sons inherited it?
      iii. From this information, construct a prior distribution for the parameter in the model.
      iv. What is the posterior distribution?
      v. What is the probability that the woman carries the gene if $y = 0$?
      vi. If $y > 0$?

   (Kalbfleisch, 1985b, p. 322)

(15) You observe normally distributed random variables with mean μ and unit variance. Your prior distribution is $\mu \sim \mathrm{N}(\zeta, \gamma)$ and, for robustness, you believe that $1 \le \zeta \le 2$ and $3 \le \gamma \le 4$. If $y = 0$ is observed, what is the range of posterior probabilities for the credibility interval, $-1 \le \mu < 2$?

   (Berger, 1985b, p. 204)

(16) Suppose that the random variable Y has a normal distribution with unknown mean μ and known variance σ^2. You believe that μ has a prior normal distribution with mean μ_0 and variance σ_0^2, both known.
   (a) What is the posterior expectation of μ, given $Y = y$?
   (b) What happens to this expectation as y becomes increasingly different from the prior expectation, μ_0?

(c) Suppose now that your prior distribution is the Student t (Section C.2). What is the posterior expectation?
(d) How does it change as y becomes extreme?
(e) What conclusions can be drawn about outliers and model diagnostics?

(Dawid, 1973)

(17) We are making observations of counts and wish to compare the Poisson distribution of Equation (1.5) and the negative binomial of Equation (3.29). For $n = 2$, we have $y_1 = y_2 = 0$.
(a) If the only two models under consideration are $\mu = 2$ for the Poisson and $\pi = 2/3$ for the negative binomial (which have the same mean), what is the value of the Bayes factor comparing them?
(b) If the parameters are not fixed, but have, respectively, gamma and beta prior distributions, what is the Bayes factor?
(c) Repeat when $y_1 = y_2 = 2$ is observed.

(Bernardo and Smith, 1994, pp. 392–394)

(18) Scores on an educational test were recorded for all eligible students in each of eight schools. These scores may be assumed to have a normal distribution, but there is considerable heterogeneity in the mean score among schools. Thus, two possible models have a common mean and eight different means for the schools, both with common known variance.
(a) For a uniform prior distribution on the means of each model, construct the Bayes factor to compare them.
(b) What problem arises with this result?
(c) Replace the flat prior by a normal distribution with variance, σ_P^2. What conclusion will be drawn as $\sigma_P^2 \to \infty$ for fixed data and fixed prior odds for the two hypothesized models?
(d) Study how the posterior distribution for the different means changes as the number of schools included in the study increases.
(e) Suggest a more reasonable model for the heterogeneity among schools and show how it solves these problems.

(Gelman *et al.*, 1995, pp. 176–177)

Part V

Examples

9
Poisson regression

To summarize the different approaches to statistical inference and decision-making, an analogy with practice in the medical profession may be useful. This profession has an accumulated knowledge of many diseases – the statistical models. A task of the doctor is to discover which one applies to a particular patient. When a radically new drug is developed, doctors will conduct a randomized study to compare its effect to a control group to determine what possible effects the medication has – direct likelihood inference. However, before the drug is released on the market, the hypothesis that it has no effect, or that it is inferior to existing medication, must be rejected in favour of a positive effect, optimizing a long term criteria in the population – frequentist decision-making. Once a doctor begins to use the drug on patients, information obtained from these studies can be incorporated into the diagnoses to decide what is the optimal treatment – Bayesian decision-making.

To illustrate briefly, and practically, many of the inference and decision-making procedures outlined in previous chapters, we shall look at two simple examples, in this chapter and the next.

9.1 Models

In this chapter, we shall examine data collected to study the relationship between life stresses and illnesses. One randomly chosen member of each randomly chosen household in a sample from Oakland, California, USA, was interviewed. In a list of 41 events, respondents were asked to note which had occurred within the last 18 months. The results for those recalling only one such stressful event are given in Table 9.1. We are interested in modelling how the recall of such events depends on the time since them.

For such frequency data, the Poisson distribution of Equation (1.5)

$$f(y;\mu) = \frac{\mathrm{e}^{-\mu}\mu^{y}}{y!}$$

might be appropriate. This is a member of the one-parameter exponential family. The canonical parameter is

Table 9.1. Frequency of recall of one stressful event in the 18 months preceding an interview. (Haberman, 1978, p. 3)

Month	1	2	3	4	5	6	7	8	9
Respondents	15	11	14	17	5	11	10	4	8
Month	10	11	12	13	14	15	16	17	18
Respondents	10	7	9	11	3	6	1	1	4

$$\theta = \log(\mu)$$

so that the canonical link function is the log. Then, the canonical Poisson regression model is

$$\log(\mu_i) = \beta_0 + \beta_1 x_i \tag{9.1}$$

where μ_i is the average number of respondents in month i and $x_i = i$ is the month.

9.2 Likelihood function

The likelihood function for this Poisson model for these data is

$$\begin{aligned} \mathrm{L}(\beta_0, \beta_1) &= \prod_{i=1}^{18} \mathrm{e}^{-\mathrm{e}^{\beta_0+\beta_1 x_i}} \mathrm{e}^{(\beta_0+\beta_1 x_i) y_i} \\ &= \mathrm{e}^{y_\bullet \beta_0 + \beta_1 \sum x_i y_i} \prod_{i=1}^{18} \mathrm{e}^{-\mathrm{e}^{\beta_0+\beta_1 x_i}} \qquad (9.2) \\ &= \mathrm{e}^{147\beta_0 + 1077\beta_1} \prod_{i=1}^{18} \mathrm{e}^{-\mathrm{e}^{\beta_0+\beta_1 x_i}} \end{aligned}$$

so that the sufficient statistics for β_0 and β_1 are, respectively, $y_\bullet = 147$ and $\sum x_i y_i = 1077$.

The log likelihood function is

$$\mathrm{l}(\beta_0, \beta_1) = 147\beta_0 + 1077\beta_1 - \sum_{i=1}^{18} \mathrm{e}^{\beta_0+\beta_1 x_i} \tag{9.3}$$

The score or estimating equations are

$$\begin{aligned} \mathrm{U}_{\beta_0}(\hat{\beta}_0, \hat{\beta}_1) &= 147 - \sum_{i=1}^{18} \mathrm{e}^{\hat{\beta}_0+\hat{\beta}_1 x_i} \\ &= 0 \qquad (9.4) \end{aligned}$$

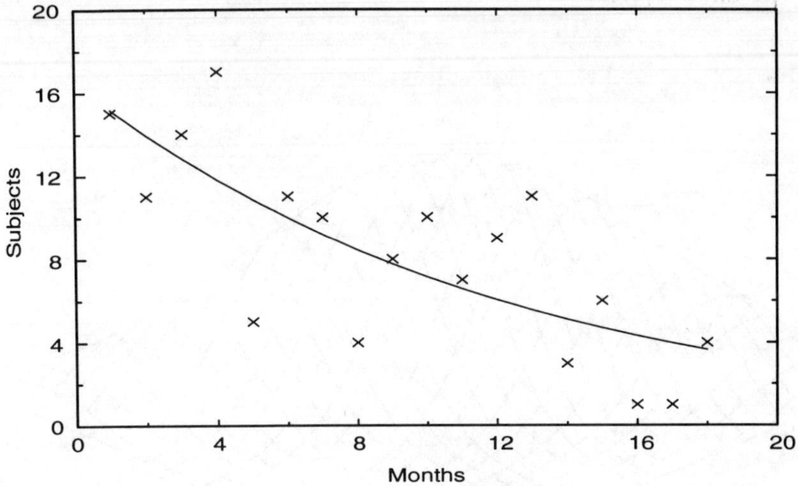

Fig. 9.1. Canonical Poisson regression plot of the recall data of Table 9.1.

$$\begin{aligned}U_{\beta_1}(\hat{\beta}_0, \hat{\beta}_1) &= 1077 - \sum_{i=1}^{18} x_i e^{\hat{\beta}_0 + \hat{\beta}_1 x_i} \\ &= 0\end{aligned}$$

These are non-linear and must be solved simultaneously by some iterative technique such as the Newton–Raphson or scoring method or IWLS. The maximum likelihood estimates of the parameters, so obtained, are $\hat{\beta}_0 = 2.803$ and $\hat{\beta}_1 = -0.0838$. Because the slope is estimated to be negative, the number of subjects recalling events is decreasing with time.

The model can be rewritten

$$\begin{aligned}\hat{\mu}_i &= e^{\hat{\beta}_0 + \hat{\beta}_1 x_i} \\ &= 16.494 e^{-0.083\,77 x_i} \qquad (9.5)\end{aligned}$$

This is a model of exponential decline. The data and the estimated model are plotted in Figure 9.1.

A three-dimensional plot for the normed likelihood is given in Figure 9.2 and, for the deviance, in Figure 9.3. The two parameters are highly interrelated. If one is varied, the m.l.e. of the other changes greatly. We can also plot the normed profile likelihood functions, as in Figures 9.4 and 9.5, as well as the corresponding normed profile deviances, in Figures 9.6 and 9.7. However, the latter do not necessarily take into account the strong inter-relationship between the two parameters. From the normed profile likelihood, 10% intervals are $(2.5, 3.0)$ for β_0 and $(-0.12, -0.05)$ for β_1. By this criterion, a slope of zero is implausible. The graphs are reasonably

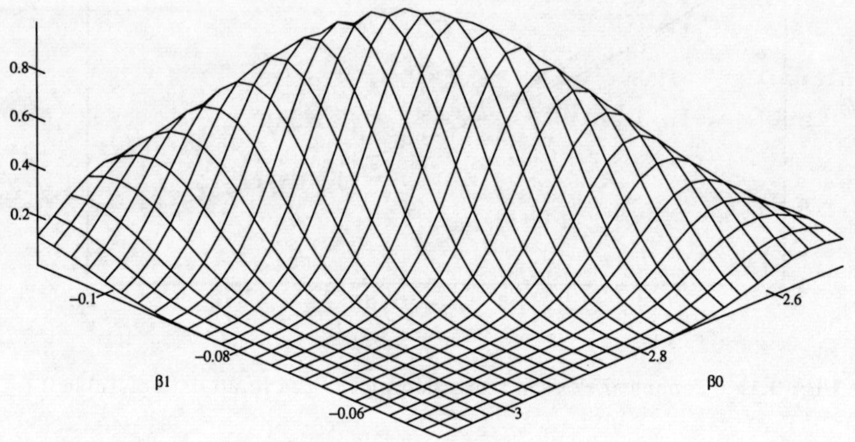

Fig. 9.2. Plot of the likelihood function for the canonical Poisson regression model for the recall data of Table 9.1.

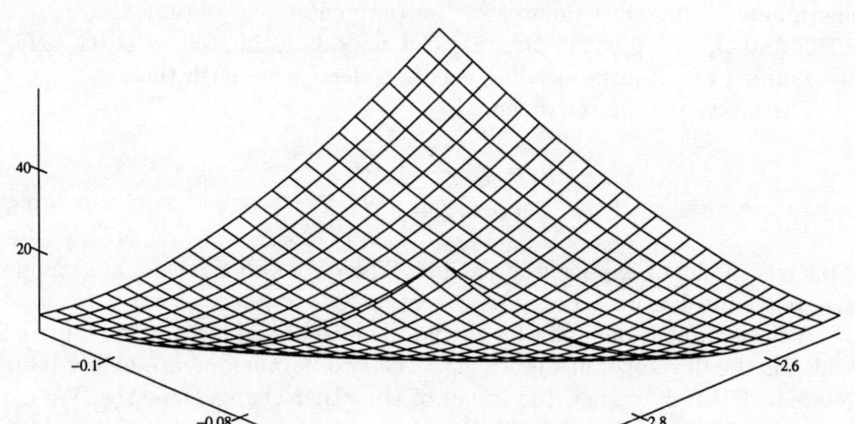

Fig. 9.3. Plot of the deviance function for the canonical Poisson regression model for the recall data of Table 9.1.

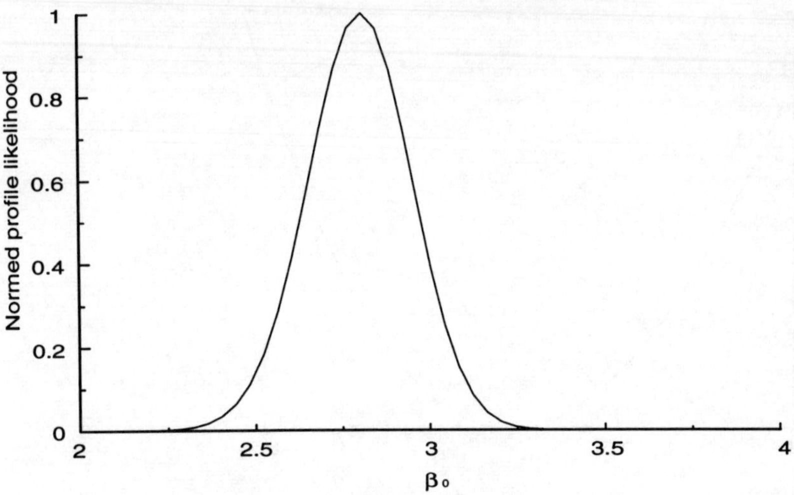

Fig. 9.4. Plot of the profile likelihood function for β_0 in the canonical Poisson regression model for the recall data of Table 9.1.

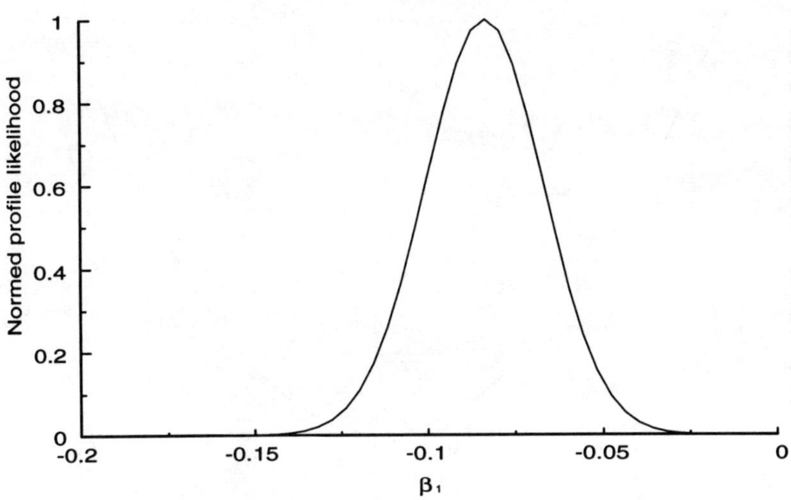

Fig. 9.5. Plot of the profile likelihood function for β_1 in the canonical Poisson regression model for the recall data of Table 9.1.

symmetric, because the parameters are canonical in the exponential family.

The difference in deviance for a model with $\beta_1 = 0$, as compared to the m.l.e., is 26.27, showing that it is implausible. The AIC for the regression model is 28.56, as compared to 52.83 when $\beta_1 = 0$, confirming this conclusion

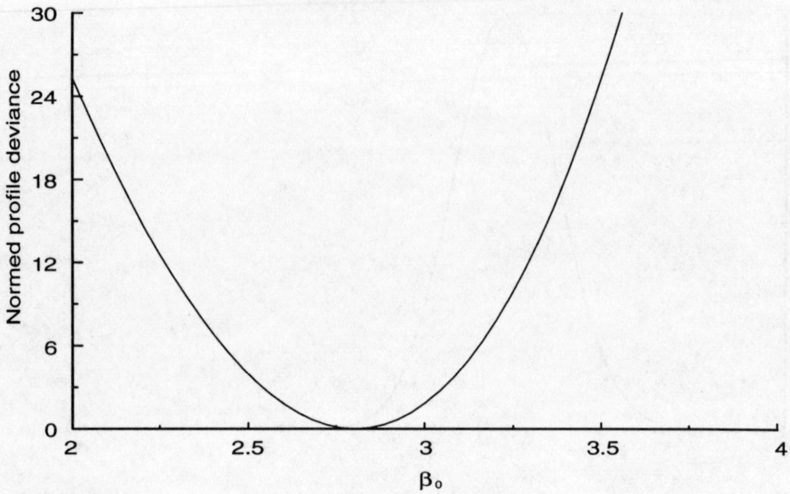

Fig. 9.6. Plot of the profile deviance function for β_0 in the canonical Poisson regression model for the recall data of Table 9.1.

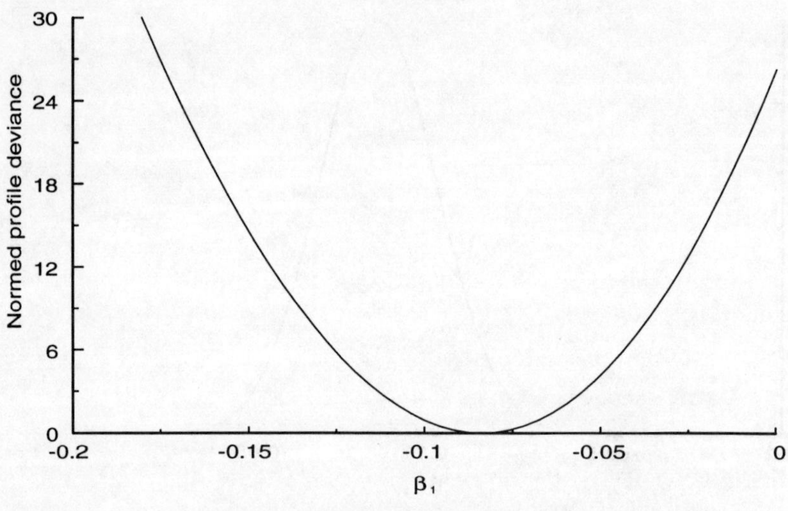

Fig. 9.7. Plot of the profile deviance function for β_1 in the canonical Poisson regression model for the recall data of Table 9.1.

The observed and expected information matrices are both

$$\begin{aligned}\mathbf{J}(\beta_0, \beta_1) &= \mathbf{I}(\beta_0, \beta_1) \\ &= \begin{pmatrix} \sum e^{\beta_0+\beta_1 x_i} & \sum x_i e^{\beta_0+\beta_1 x_i} \\ \sum x_i e^{\beta_0+\beta_1 x_i} & \sum x_i^2 e^{\beta_0+\beta_1 x_i} \end{pmatrix} \end{aligned} \tag{9.6}$$

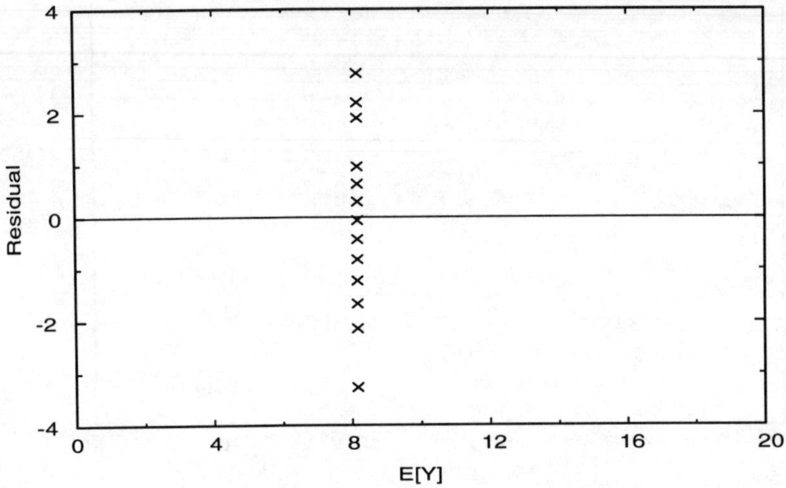

Fig. 9.8. Residual plot for a constant mean in the recall data of Table 9.1.

Their value at the maximum likelihood estimate is

$$\begin{aligned} \mathbf{J}(\hat{\beta}_0, \hat{\beta}_1) &= \mathbf{I}(\hat{\beta}_0, \hat{\beta}_1) \\ &= \begin{pmatrix} 147 & 1077 \\ 1077 & 11\,432 \end{pmatrix} \end{aligned} \tag{9.7}$$

9.3 Goodness of fit

The regression model has a deviance 24.56 larger than the saturated model, but with 16 fewer parameters, showing that it fits the data reasonably well. This deviance is similar in size to that we found above when comparing the regression model to one with $\beta_1 = 0$, but there with only 1 d.f. We see, here, the problem of comparing likelihoods or deviances with different degrees of freedom. Although the two differences in deviance are of similar size, the first has 1 d.f. and the second 16 d.f.

In contrast, the AICs can be compared directly. With zero deviance and 18 parameters, that for the saturated model is 36, considerably larger than the AIC for the regression model, 28.56, given above. Thus, we again conclude that the model fits reasonably well.

Let us compare the residual plots for the regression model with those for the null model, when $\beta_1 = 0$, which has a common mean in all months. The plots of the deviance residuals against the estimated expected values are given in Figures 9.8 and 9.9. For the null model, all observations have the same estimated value, giving the less than useful plot in Figure 9.8. The regression model shows no abnormalities.

The index plots of residuals are given in Figures 9.10 and 9.11. The

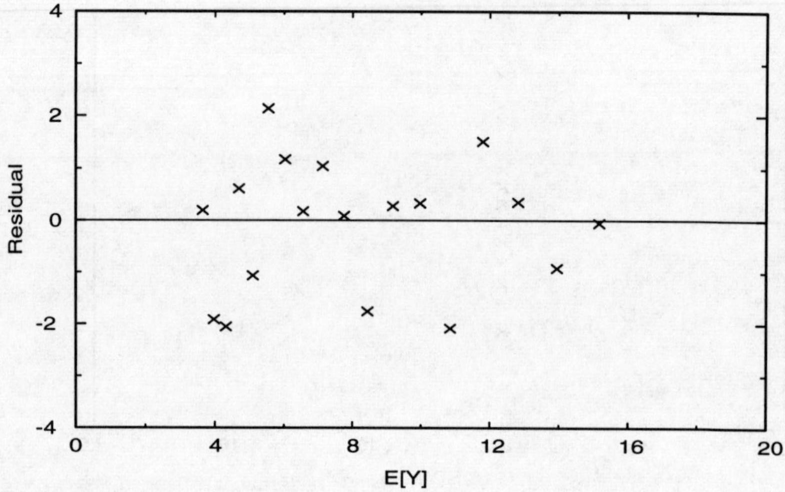

Fig. 9.9. Residual plot for the canonical Poisson regression model for the recall data of Table 9.1.

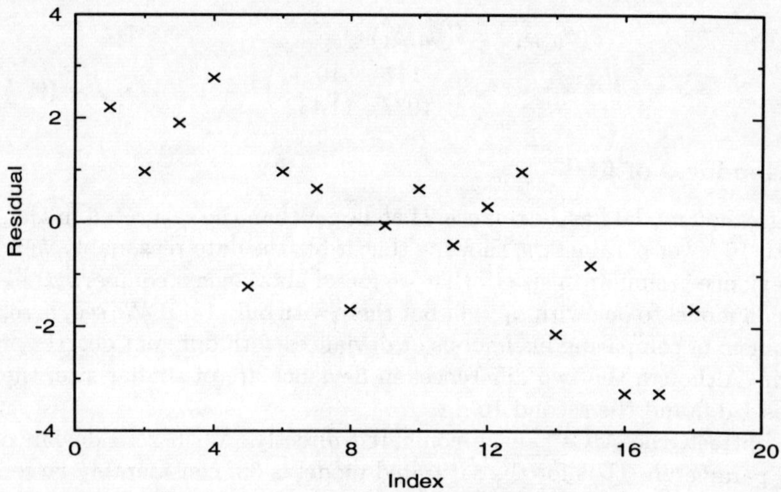

Fig. 9.10. Index plot of residuals for a constant mean in the recall data of Table 9.1.

null model shows positive residuals on the left and negative residuals on the right. With a constant mean, early values are underestimated and later values overestimated. The plot for the regression model shows no such trend. In this example, the residual plot against the explanatory variable, time, is identical to the index plot.

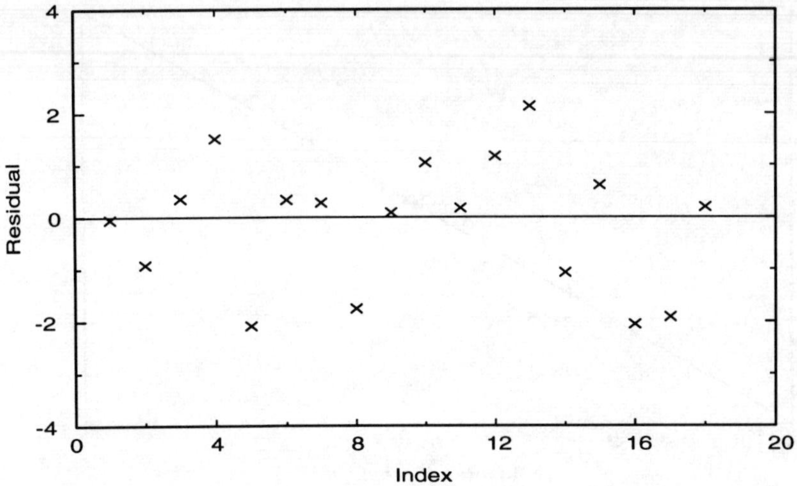

Fig. 9.11. Index plot of residuals for the canonical Poisson regression model for the recall data of Table 9.1.

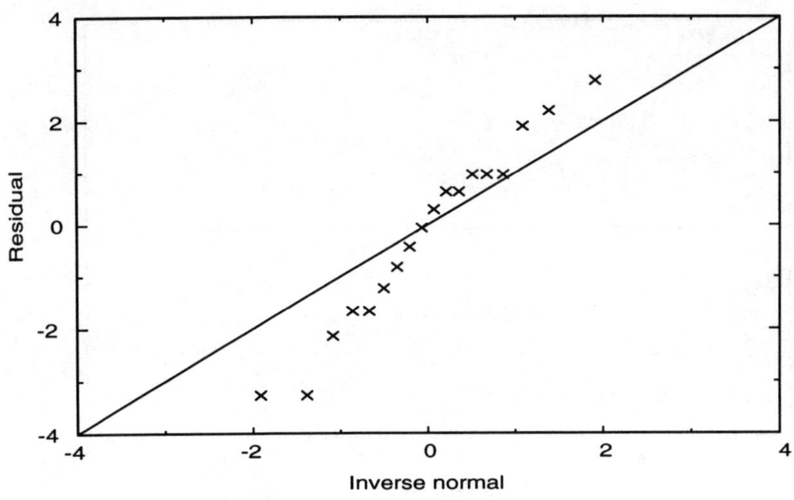

Fig. 9.12. Normal probability plot of residuals for a constant mean in the recall data of Table 9.1.

The normal probability plots are given in Figures 9.12 and 9.13. The departure from a straight line for the null model indicates that it fits poorly. The regression model is much closer to the straight line.

The plots of residuals against leverage are given in Figures 9.14 and 9.15. All points have the same leverage in the null model, again, not a

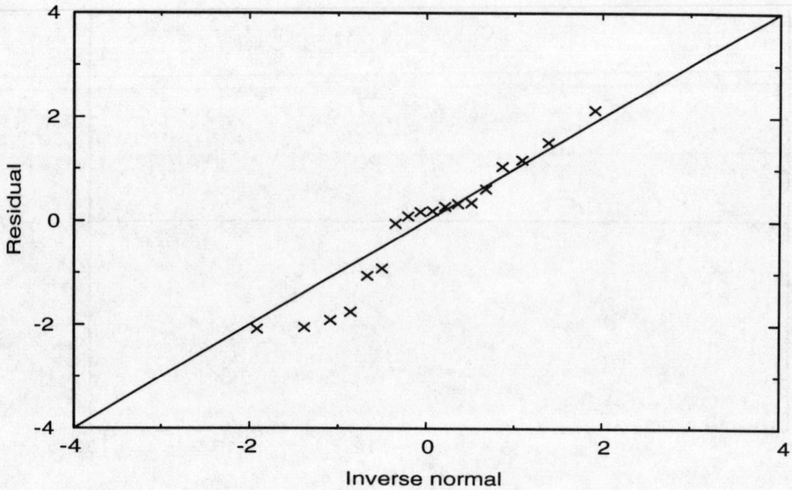

Fig. 9.13. Normal probability plot of residuals for the canonical Poisson regression model for the recall data of Table 9.1.

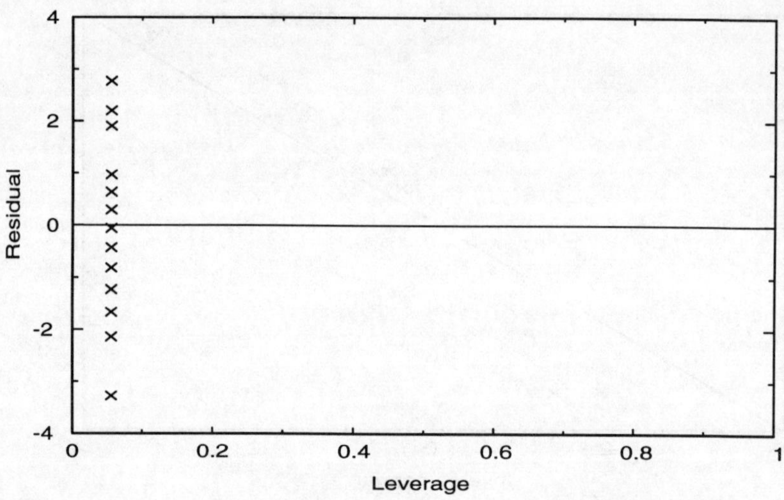

Fig. 9.14. Leverage plot for a constant mean in the recall data of Table 9.1.

very useful plot. In the regression model, only two points show twice the average leverage of 2/18. All of the residual plots seem to point to the regression model fitting well.

The index plots of Cook's distance, for the null and regression models, are given in Figures 9.16 and 9.17. We see that observation 4 influences

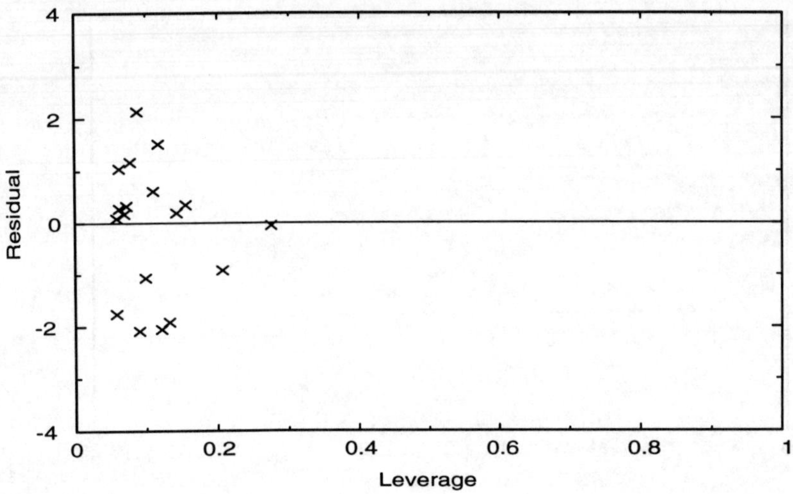

Fig. 9.15. Leverage plot for the canonical Poisson regression model for the recall data of Table 9.1.

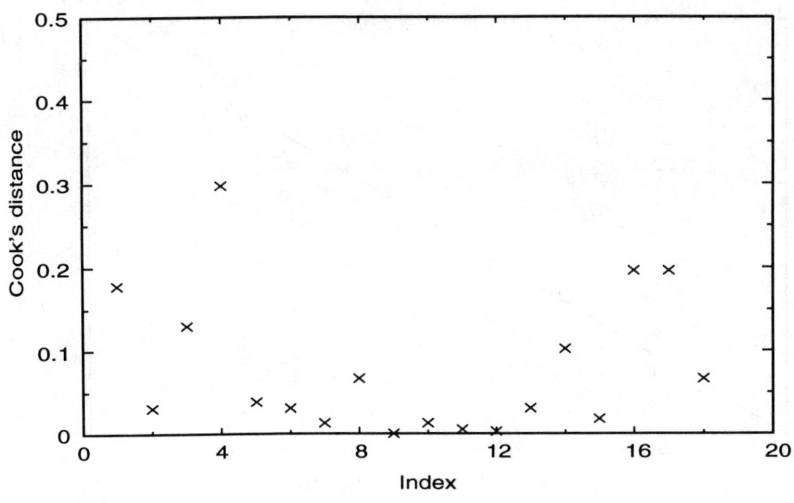

Fig. 9.16. Plot of Cook's distances for a constant mean in the recall data of Table 9.1.

most the parameter estimates for the null model and 13 for the full model, but values less than unity are really not exceptional.

Suppose now that we take these data as representing observations from a distribution of times since the event. If we take this distribution to be exponential, from Equation (1.14), we have $\hat{\mu} = 7.327$. The distribution

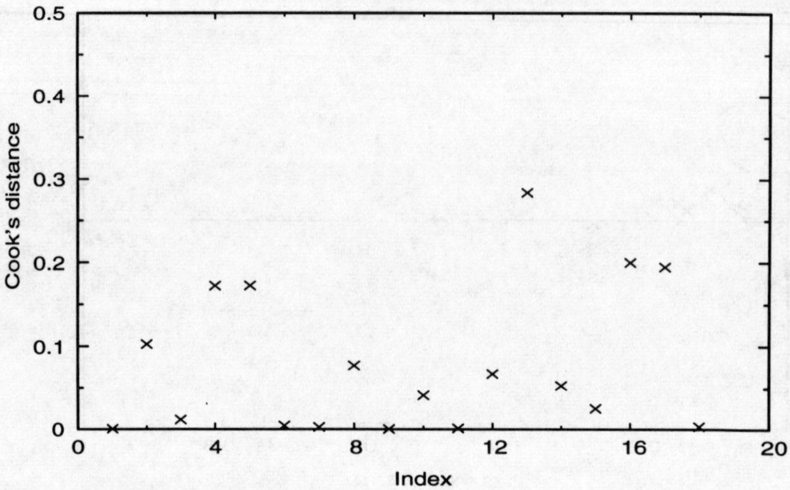

Fig. 9.17. Plot of Cook's distances for the canonical Poisson regression model for the recall data of Table 9.1.

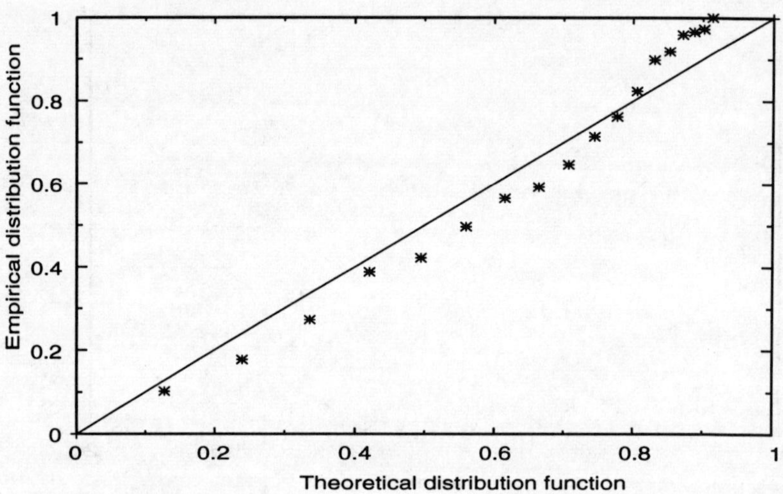

Fig. 9.18. P–P plot for the recall data of Table 9.1.

function is $F(\mu) = 1 - e^{-y/\mu}$, where y is the time since the event. This can be plotted against the observed cumulative relative frequencies, the empirical distribution function. Then, the resulting P–P plot is given in Figure 9.18. This shows that these data are fairly well represented by an exponential distribution.

9.4 Asymptotic inference

The asymptotic covariance matrix of the parameter estimates, the inverse of the information matrix, given above in Equation (9.7), is

$$\begin{pmatrix} 0.021\,95 & -0.002\,07 \\ -0.002\,07 & 0.000\,28 \end{pmatrix}$$

and the correlation matrix is

$$\begin{pmatrix} 1.00 & -0.83 \\ -0.83 & 1.00 \end{pmatrix}$$

The two parameters have highly correlated estimates, as we already saw in looking at the three-dimensional plots of the likelihood and deviance. We can also see that the asymptotic standard errors are 0.148 for $\hat{\beta}_0 = 2.803$ and 0.0168 for $\hat{\beta}_1 = -0.0838$.

The asymptotic Wald tests, respectively 18.91 and −4.99, with P-values both very close to zero, indicate that both parameters are significantly different from zero. We saw above that the change in deviance when β_1 is set to zero is 26.27. This has an asymptotic Chi-squared distribution with one d.f., and an almost zero P-value, again indicating that the hypothesis that this parameter is zero is to be rejected. On the other hand, the P-value for the comparison with the saturated model, with deviance 24.56 on 16 d.f., is 0.078, indicating that the regression model fits reasonably well. However, care must be taken with this. If the time period is allowed to increase with sample size, the degrees of freedom will not remain constant.

9.5 Factoring the likelihood function

Our parameter of interest is the slope, β_1, because it indicates if the number of people recalling events varies with time. The conditional distribution, given the sufficient statistic, $Y_\bullet$, is multinomial, as we saw in Equation (6.29) of Section 6.2.4, and that statistic is a proper cut. From the estimating equation for β_0, we have, for fixed β_1,

$$e^{\hat{\beta}_0} = \frac{y_\bullet}{\sum e^{\beta_1 x_i}}$$

If we substitute this into the likelihood function, to obtain the profile likelihood for β_1, we find that it is the same multinomial distribution. Thus, the profile and conditional likelihoods are identical for β_1 in this model, which is not surprising, given that we have a proper cut.

9.6 Frequentist decision-making

Because we assume that the number of events decreases in time, we can use a one-tailed test for $\beta_1 = 0$ against $\beta_1 < 0$. This UMP likelihood ratio

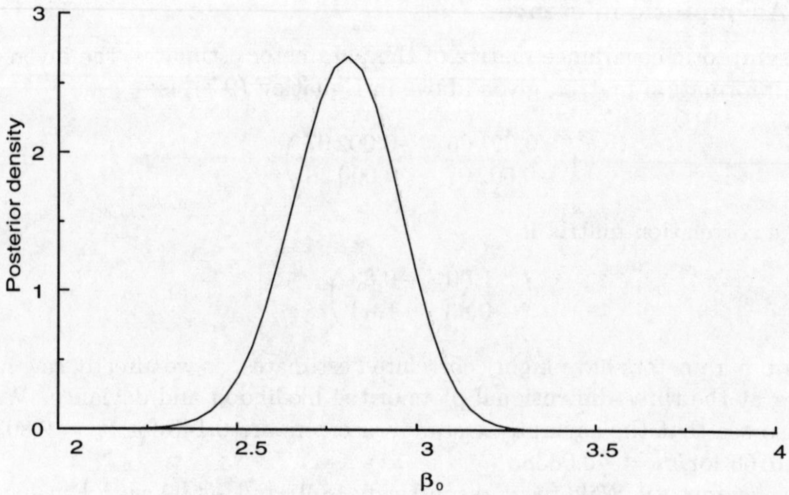

Fig. 9.19. Plot of the marginal posterior density for β_0 in the canonical Poisson regression model for the recall data of Table 9.1.

test is identical to the asymptotic Chi-squared test just given. It rejects the null hypothesis.

Because the parameters are canonical in the exponential family, no minimum variance unbiased point estimator exists.

UMP 95% confidence intervals, obtained from the profile deviances, are $(2.51, 3.09)$ for β_0 and $(-0.117, -0.051)$ for β_1. Approximate 95% confidence intervals can be constructed from the asymptotic standard errors, taking 1.96 standard errors on either side. These are $(2.51, 3.09)$ for β_0 and $(-0.117, -0.051)$ for β_1, which are identical to the UMP critical region. This is not surprising, because the profile likelihood functions were symmetric.

9.7 Bayesian decision-making

Let us use a vague, improper prior. Because the integrals are so complex, we use Laplace's method to obtain a second-order approximation to the integral. The posterior means of the two parameters are $\beta_0 = 2.799$ and $\beta_1 = -0.0841$. This vague prior does not really change the m.l.e.

The two marginal densities are plotted in Figures 9.19 and 9.20. They are very similar to the original likelihood functions.

If we use a normal distribution, centered on zero with variance 0.25, as an informative prior for β_1, with the same flat prior for β_0, this changes the results only very little.

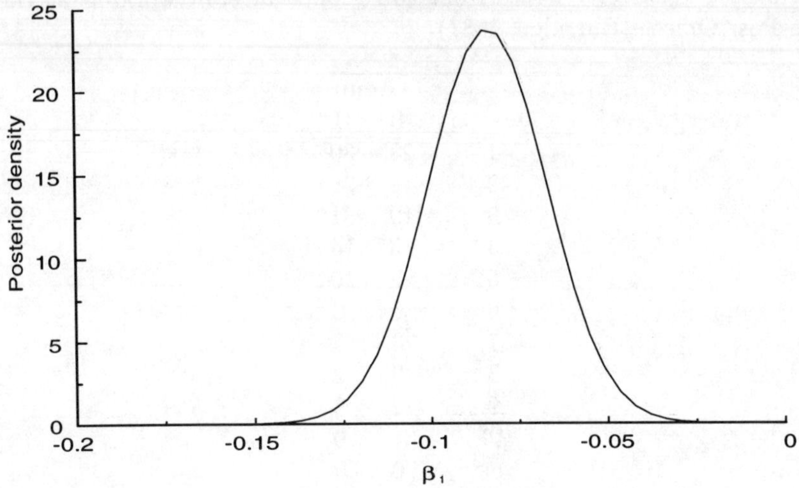

Fig. 9.20. Plot of the marginal posterior density for β_1 in the canonical Poisson regression model for the recall data of Table 9.1.

Table 9.2. Occupants of houses as obtained from a postal survey. (Lindsey and Mersch, 1992)

Occupants	Houses
1	436
2	133
3	19
4	2
5	1
6	0
7	1

9.8 Exercises

(1) In a postal survey, the number of occupants of each house is obtained, with results as in Table 9.2. Obviously, houses without occupants will not answer the survey. A truncated Poisson distribution is obtained by renormalizing the distribution without the zero category. The sum of the series is $1 - e^{-\mu}$, so that the truncated distribution is given by Equation (2.41).

(a) Fit this distribution to the data and draw suitable conclusions.

(b) The logarithm of Equation (2.41) can be written

$$\log[p(y;\mu)] = -\mu - \log(1 - e^{-\mu}) + y\log(\mu) - \log(y!)$$

Table 9.3. Quarterly losses from two groups of staff working at the British Post Office. (Burridge, 1981)

	Group	
Quarter	I	II
1	22	30
2	18	28
3	19	31
4	13	14
5	5	10
6	6	6
7	3	5
8	2	2
9	2	3
10	1	0
11	0	0
12	1	1
13	0	1
14	0	0
15	0	0
16	1	1
17	1	1
18	1	0
19	3	2
20	1	0
21	1	3
22	0	1
23	0	1
24	0	0

$$= \beta_0 + \beta_1 y - \log(y!) \qquad (9.8)$$

which has the form of a Poisson regression. Use the methods of this chapter to reanalyze the data and compare the results with those from the first part of this exercise.

(Lindsey and Mersch, 1992)

(2) Table 9.3 gives the distributions of successive quarterly losses from two groups of staff recruited to the British Post Office in the first quarter of 1973. Two possible models which might be fitted to these data are the log normal and gamma distributions.

(a) In the same way as in the previous exercise for the Poisson distribution, create a Poisson regression model for each of these distributions.

(b) Compare them using the techniques of the present chapter.
(Lindsey and Mersch, 1992)

10
Binomial regression

10.1 Models

A biological assay, or bioassay, is an experiment to study the reaction of a living organism to some material, usually a pharmaceutical preparation. When the only response recorded is presence or absence of the reaction, it is called a quantal response. Usually, the material to be studied is applied at different dose levels in an inert diluent and the response observed. Comparison may be made to a negative control, i.e. inert material, or to a positive control, usually a competing pharmaceutical preparation. In many such assays, the material may cause irreversible change, such as death, so that the subjects can only be used once.

One of the very early quantal response assays, published in 1926, concerned a study of the effect of insulin on mice, as measured by whether or not they had convulsions. The data are given in Table 10.1. We are interested in modelling how the proportion of mice with convulsions varies with the dose of insulin and with the type of preparation. Here, the control

Table 10.1. Numbers of mice with convulsions under various doses of insulin and two preparations. (Finney, 1978, p. 376, from Hemmingsen and Krogh)

Standard preparation			Test preparation		
Dose	With convulsions	Total	Dose	With convulsions	Total
3.4	0	33	6.5	2	40
5.2	5	32	10.0	10	30
7.0	11	38	14.0	18	40
8.5	14	37	21.5	21	35
10.5	18	40	29.0	27	37
13.0	21	37			
18.0	23	31			
21.0	30	37			
28.0	27	30			

is positive.

The idea behind such a model is that each organism reacts only to dose levels greater than some minimum, different for each organism. This is the stochastic data generating mechanism. Then, these threshold levels have a distribution in the population under study, whose cumulative distribution function might be described by $\pi = F(\beta_0 + \beta_1 x)$, where x is the dose and π is the probability of convulsions.

As we have seen, one possible model for such frequency data is based on the binomial distribution of Equation (2.111),

$$f(y;\pi) = \binom{n}{y}\pi^y(1-\pi)^{n-y}$$

The canonical parameter for this distribution is

$$\theta = \log\left(\frac{\pi}{1-\pi}\right)$$

so that the canonical link function is the logit. Thus, the canonical binomial regression model is logistic regression. A simple model for these data allows the logit of the probability of convulsions to depend linearly on dose and on the preparation:

$$\log\left(\frac{\pi_{ij}}{1-\pi_{ij}}\right) = \beta_0 + \beta_1 x_{1i} + \beta_2 x_{2j} \tag{10.1}$$

where π_{ij} is the probability of convulsions at dose x_{1i}, with preparation x_{2j} (coded 0 for standard and 1 for test preparation). This corresponds to a logistic distribution for the threshold levels, with

$$\pi_{ij} = \frac{e^{\beta_0+\beta_1 x_{1i}+\beta_2 x_{2j}}}{1+e^{\beta_0+\beta_1 x_{1i}+\beta_2 x_{2j}}} \tag{10.2}$$

the logistic cumulative distribution function.

Many other models are also possible. For example, Finney (1978) uses a probit link function, the inverse of a normal distribution function, although this will generally give very similar results to the logit. In that case, the normal distribution describes the reaction levels. Note that both the normal and logistic distributions are symmetric.

Another possible data generating mechanism for quantal response data can be described by the multihit model, whereby the organism is thought to be bombarded by the dose, but requires one or more 'hits' to react. Closely related to this is the multistage model in which the organism must pass through several stages before reacting. If the hits are arriving as a Poisson process, i.e. the length of the stages have exponential distributions, then the distribution function will be gamma (Morgan, 1992, pp. 22–25).

10.2 Likelihood function

The likelihood function for our logistic model is

$$
\begin{aligned}
\mathrm{L}(\beta_0, \beta_1, \beta_2) &= \prod_{j=1}^{2} \prod_{i=1}^{m_j} \frac{\mathrm{e}^{(\beta_0+\beta_1 x_{1i}+\beta_2 x_{2j}) y_{ij}}}{(1+\mathrm{e}^{\beta_0+\beta_1 x_{1i}+\beta_2 x_{2j}})^{n_{ij}}} \\
&= \frac{\mathrm{e}^{\beta_0 y_{\bullet\bullet}+\beta_1 \sum_i x_{1i} y_{i\bullet}+\beta_2 \sum_j x_{2j} y_{\bullet j}}}{\prod_j \prod_i (1+\mathrm{e}^{\beta_0+\beta_1 x_{1i}+\beta_2 x_{2j}})^{n_{ij}}} \qquad (10.3) \\
&= \frac{\mathrm{e}^{227\beta_0+4084\beta_1+78\beta_2}}{\prod_j \prod_i (1+\mathrm{e}^{\beta_0+\beta_1 x_{1i}+\beta_2 x_{2j}})^{n_{ij}}}
\end{aligned}
$$

where m_j is the number of dose levels for preparation j, y_{ij} is the number of mice, out of n_{ij}, with convulsions at dose x_{1i} and preparation x_{2j}. Thus, the sufficient statistics for β_0, β_1, and β_2 are, respectively, $y_{\bullet\bullet} = 227$, $\sum_i x_{1i} y_{i\bullet} = 4084$, and $\sum_j x_{2j} y_{\bullet j} = 78$.

The log likelihood function is

$$
\begin{aligned}
\mathrm{l}(\beta_0, \beta_1, \beta_2) &= 227\beta_0 + 4084\beta_1 + 78\beta_2 \\
&\quad - \sum_j \sum_i n_{ij} \log\left(1+\mathrm{e}^{\beta_0+\beta_1 x_{1i}+\beta_2 x_{2j}}\right) \qquad (10.4)
\end{aligned}
$$

The score or estimating equations are

$$
\begin{aligned}
\mathrm{U}_{\beta_0}(\hat\beta_0, \hat\beta_1, \hat\beta_2) &= 227 - \sum_j \sum_i \frac{n_{ij} \mathrm{e}^{\hat\beta_0+\hat\beta_1 x_{1i}+\hat\beta_2 x_{2j}}}{1+\mathrm{e}^{\hat\beta_0+\hat\beta_1 x_{1i}+\hat\beta_2 x_{2j}}} \\
&= 0 \\
\mathrm{U}_{\beta_1}(\hat\beta_0, \hat\beta_1, \hat\beta_2) &= 4084 - \sum_j \sum_i \frac{n_{ij} x_{1i} \mathrm{e}^{\hat\beta_0+\hat\beta_1 x_{1i}+\hat\beta_2 x_{2j}}}{1+\mathrm{e}^{\hat\beta_0+\hat\beta_1 x_{1i}+\hat\beta_2 x_{2j}}} \qquad (10.5) \\
&= 0 \\
\mathrm{U}_{\beta_2}(\hat\beta_0, \hat\beta_1, \hat\beta_2) &= 78 - \sum_j \sum_i \frac{n_{ij} x_{2i} \mathrm{e}^{\hat\beta_0+\hat\beta_1 x_{1i}+\hat\beta_2 x_{2j}}}{1+\mathrm{e}^{\hat\beta_0+\hat\beta_1 x_{1i}+\hat\beta_2 x_{2j}}} \\
&= 0
\end{aligned}
$$

As in the previous chapter, these are non-linear and must be solved simultaneously by some iterative technique. The maximum likelihood estimates of the parameters, so obtained, are $\hat\beta_0 = -2.078$, $\hat\beta_1 = 0.161$, and $\hat\beta_2 = -0.875$. Because the slope for dose is estimated to be positive, the proportion of mice having convulsions is increasing with dose. The coefficient for type of preparation is estimated to be negative, indicating fewer convulsions with the second, test preparation.

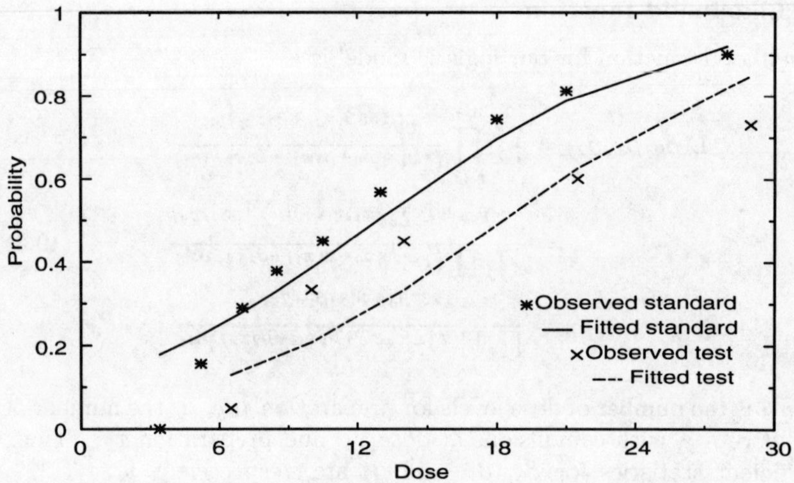

Fig. 10.1. Logistic regression plot of the insulin data of Table 10.1 – dose without interaction.

The model may be rewritten

$$\begin{aligned}\hat{\pi}_i &= \frac{e^{\hat{\beta}_0+\hat{\beta}_1 x_{1i}}}{1+e^{\hat{\beta}_0+\hat{\beta}_1 x_{1i}}}\\ &= \frac{e^{-2.078+0.161 x_{1i}}}{1+e^{-2.078+0.161 x_{1i}}} \qquad \text{Standard preparation} \qquad (10.6)\\ &= \frac{e^{\hat{\beta}_0+\beta_3+\hat{\beta}_1 x_{1i}}}{1+e^{\hat{\beta}_0+\beta_3+\hat{\beta}_1 x_{1i}}}\\ &= \frac{e^{-2.953+0.161 x_{1i}}}{1+e^{-2.953+0.161 x_{1i}}} \qquad \text{Test preparation}\end{aligned}$$

Our model implies the same form of increase in convulsions with dose for both preparations, but at different levels. These are plotted in Figure 10.1. We see that the fitted curves do not follow the observed data very closely. The AIC for our model is 33.10.

The increase in deviance when dose is omitted from the model ($\beta_1 = 0$) is 138.81 and when preparation is removed ($\beta_2 = 0$), 14.91, both with 1 d.f., indicating that both parameters are necessary to the model. The corresponding AICs are 169.91 and 46.01, showing that both are inferior to the full regression model.

One possible defect of the model is that convulsions may depend on dose in a different way for the two preparations. An interaction term, the product of the two explanatory variables, $x_{1i}x_{2j}$, may be added to the model, with a decrease in deviance of 4.23 on 1 d.f. and AIC of 30.87, indicating

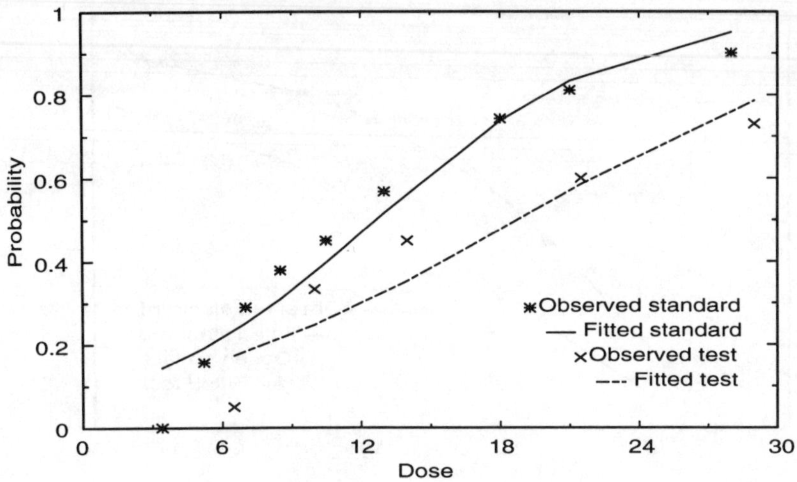

Fig. 10.2. Logistic regression plot of the insulin data of Table 10.1 – dose with interaction.

an improvement. The maximum likelihood estimates of the parameters are now $\hat{\beta}_0 = -2.444$, $\hat{\beta}_1 = 0.193$, $\hat{\beta}_2 = 0.061$, and $\hat{\beta}_3 = -0.066$. Thus, the proportion of convulsions under the test preparation is now higher for very low doses (β_2, the difference in intercept, is positive), but increases less rapidly with dose than for the standard preparation (β_3, the difference in slope, is negative). This model is plotted in Figure 10.2. There is still considerable difference between the fitted curves and the data.

One possibility would be to fit a higher order polynomial in dose. In our case, a simpler approach is to try a transformation of the dose, such as the logarithm. In fact, this is the usual practice in analyzing bioassay data. The new linear regression model will be

$$\log\left(\frac{\pi_{ij}}{1-\pi_{ij}}\right) = \beta_0 + \beta_1 \log(x_{1i}) + \beta_2 x_{2j} \qquad (10.7)$$

This corresponds to a distribution of threshold levels which is log logistic and which is, thus, no longer symmetric.

The maximum likelihood estimates are $\hat{\beta}_0 = -5.553$, $\hat{\beta}_1 = 2.297$, and $\hat{\beta}_2 = -0.929$. This model is plotted in Figure 10.3. The curve is now very close to the observed standard preparation responses, although still somewhat different from those for the test preparation. The AIC is 14.79, substantially better than our previous models. The increases in deviance are 157.12 if log dose is removed from the model and 16.79 for preparation, with corresponding AICs 161.12 and 20.79. The decrease in deviance when the interaction is added is now only 0.36, with AIC 16.44.

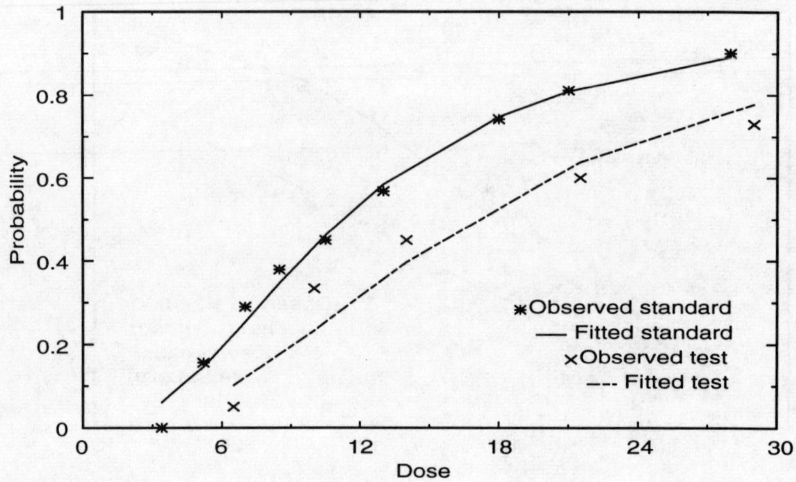

Fig. 10.3. Logistic regression plot of the insulin data of Table 10.1 – log dose without interaction.

We shall look more closely at how well these different models fit the data in the next section. For the moment, we shall only consider the model with log dose and no interaction.

With three or four parameters in a model, it is difficult to plot the likelihood surface for all of them simultaneously. Hence, we shall only look at the normed profile likelihood functions, as shown in Figures 10.4 to 10.6. The corresponding plots of the normed profile deviances are given in Figures 10.7 to 10.9. We see that a model with a zero value of any parameter is very unlikely.

The observed and expected information matrices are both

$$\mathbf{J}(\boldsymbol{\beta}) = \mathbf{I}(\boldsymbol{\beta}) \tag{10.8}$$
$$= \begin{pmatrix} \sum_j \sum_i \frac{n_{ij}C}{(1+C)^2} & \sum_j \sum_i \frac{n_{ij}\log(x_{1i})C}{(1+C)^2} & \sum_j \sum_i \frac{n_{ij}x_{2j}C}{(1+C)^2} \\ \sum_j \sum_i \frac{n_{ij}\log(x_{1i})C}{(1+C)^2} & \sum_j \sum_i \frac{n_{ij}\log^2(x_{1i})C}{(1+C)^2} & \sum_j \sum_i \frac{n_{ij}\log(x_{1i})x_{2j}C}{(1+C)^2} \\ \sum_j \sum_i \frac{n_{ij}x_{2j}C}{(1+C)^2} & \sum_j \sum_i \frac{n_{ij}\log(x_{1i})x_{2j}C}{(1+C)^2} & \sum_j \sum_i \frac{n_{ij}x_{2j}^2 C}{(1+C)^2} \end{pmatrix}$$

where $C = e^{\beta_0+\beta_1 \log(x_{1i})+\beta_2 x_{2j}}$. Their value at the maximum likelihood estimate is

$$\mathbf{J}(\hat{\boldsymbol{\beta}}) = \mathbf{I}(\hat{\boldsymbol{\beta}}) = \begin{pmatrix} 87.89 & 222.01 & 33.05 \\ 222.01 & 584.07 & 90.76 \\ 33.05 & 90.76 & 33.05 \end{pmatrix} \tag{10.9}$$

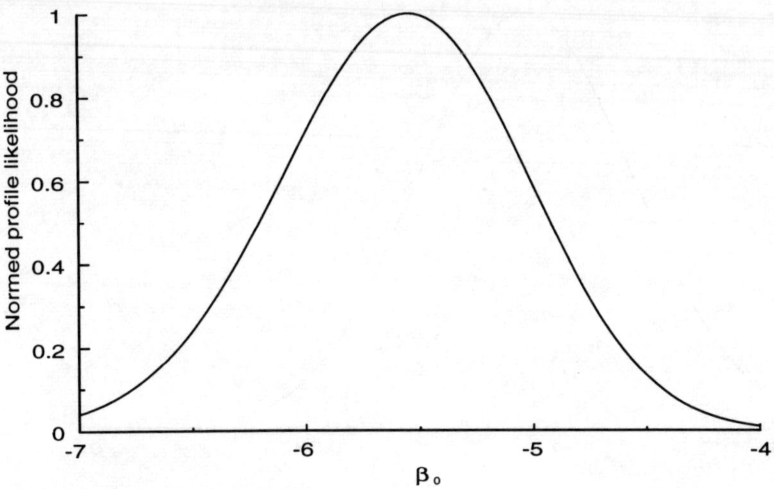

Fig. 10.4. Plot of the profile likelihood function for β_0 in the logistic regression model for the insulin data of Table 10.1.

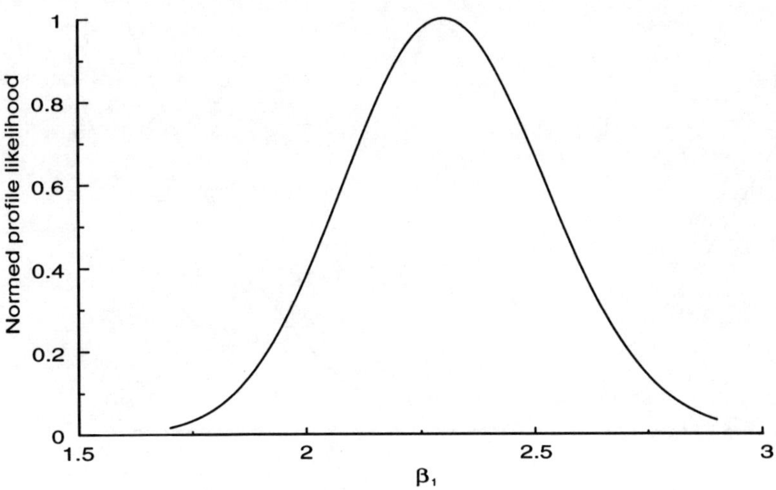

Fig. 10.5. Plot of the profile likelihood function for β_1 in the logistic regression model for the insulin data of Table 10.1.

Often in quantal response models, a quantity of interest is the (log) dose which corresponds to $100p\%$ reaction, called the ED_{100p} (for effective dose). Thus, the ED_{50} is the median of the threshold distribution. For a simple logit model, we have

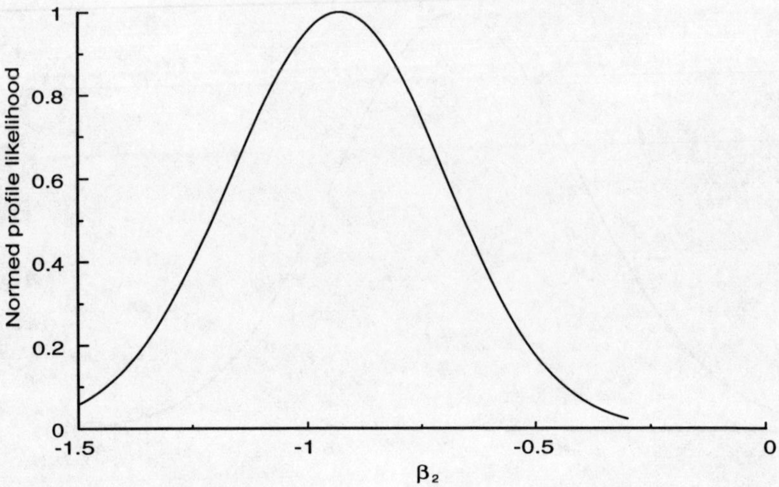

Fig. 10.6. Plot of the profile likelihood function for β_2 in the logistic regression model for the insulin data of Table 10.1.

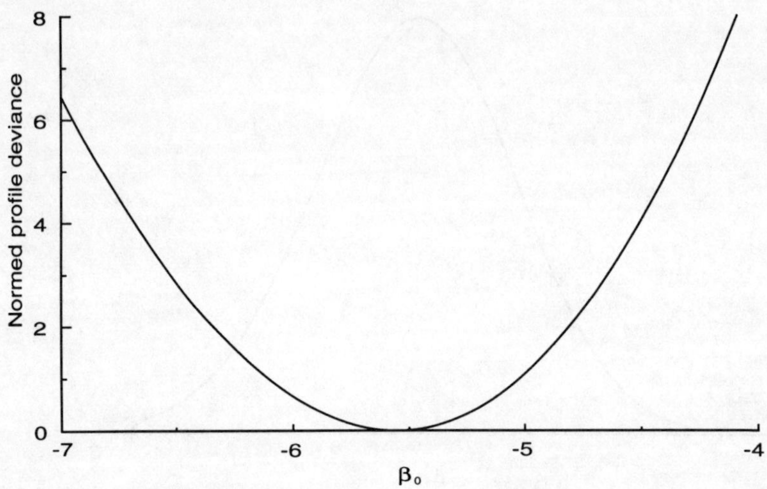

Fig. 10.7. Plot of the profile deviance function for β_0 in the logistic regression model for the insulin data of Table 10.1.

$$\beta_0 + \beta_1 \mathrm{ED}_{100p} = \log\left(\frac{p}{1-p}\right) \qquad (10.10)$$

from which

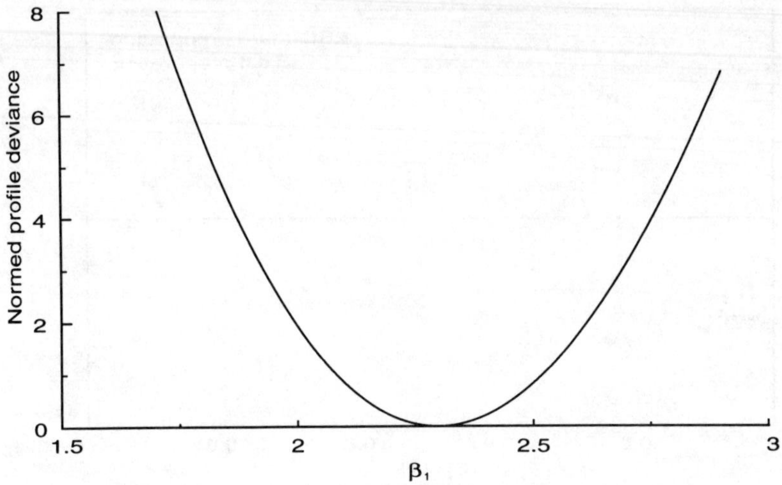

Fig. 10.8. Plot of the profile deviance function for β_1 in the logistic regression model for the insulin data of Table 10.1.

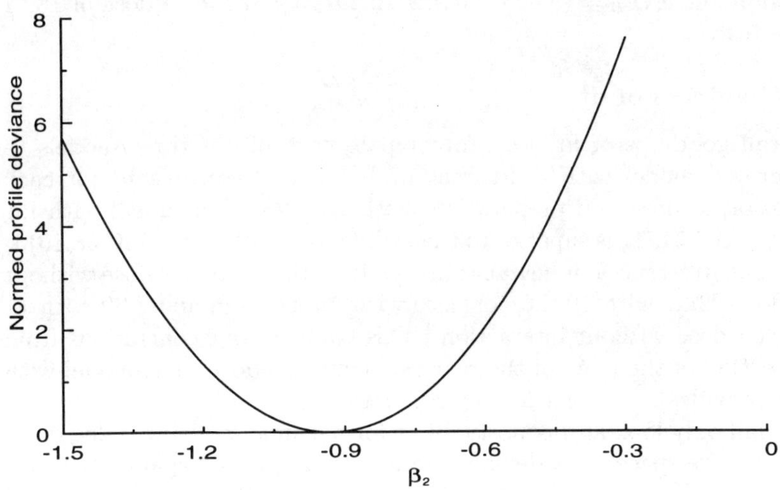

Fig. 10.9. Plot of the profile deviance function for β_2 in the logistic regression model for the insulin data of Table 10.1.

$$\mathrm{ED}_{100p} = \frac{1}{\beta_1}\left[\log\left(\frac{p}{1-p}\right) - \beta_0\right] \tag{10.11}$$

so that $\mathrm{ED}_{50} = -\beta_0/\beta_1$. For our data, the maximum likelihood estimates are $\widehat{\mathrm{ED}}_{50} = 5.553/2.297 = 2.418$ for a dose of 11.2 with the standard

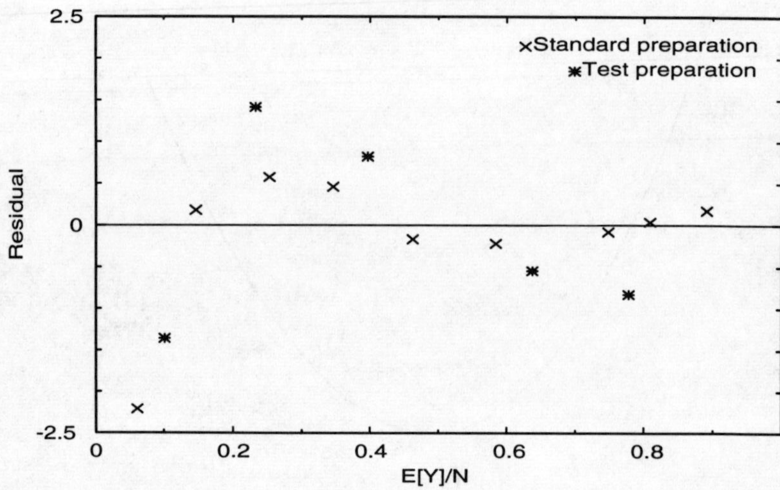

Fig. 10.10. Residual plot against expected proportions of convulsing mice for the insulin data of Table 10.1, for the model with log dose without interaction.

preparation and $\widehat{\mathrm{ED}}_{50} = (5.553 + 0.929)/2.297 = 2.822$ for a dose of 16.81 with the test.

10.3 Goodness of fit

For overall goodness of fit, we can compare each of our three models to the saturated model with a different probability of convulsions for each combination of dose and preparation, with AIC 28. Our model with log dose, with AIC 14.79, is superior to this, while that with dose (AIC 38.10) is poorer. The differences in deviance are 27.10 with 11 d.f. for dose without interaction, 22.87 with 10 d.f. for dose with interaction, and 8.79 with 11 d.f. for log dose without interaction. This confirms our conclusions from the inspection of the plots of the regression curves above. The model with log dose provides a superior fit to these data.

We shall only look at diagnostic plots for our final model with log dose, in order to save space. The deviance residuals are plotted against the estimated expected values in Figure 10.10. We see that the smallest expected values have large negative residuals, indicating overestimation, and that the residuals are generally larger, in absolute value, for the test than for the standard preparation. This was already evident from Figure 10.3. These residuals may also be plotted against the dose (or log dose), as in Figure 10.11. As might be expected, given the close relationship between dose and convulsions, this plot is very similar to the previous one.

The normal probability plot is given in Figure 10.12. Again, we see some systematic departure from the model.

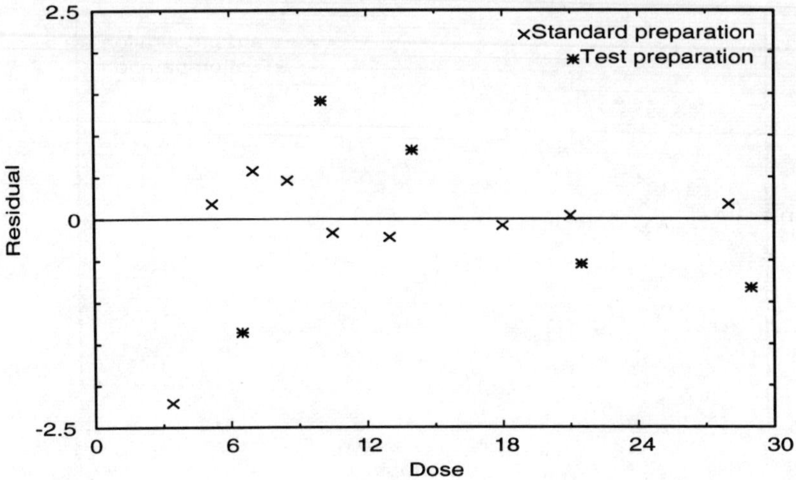

Fig. 10.11. Residual plot against dose for the insulin data of Table 10.1, for the model with log dose without interaction.

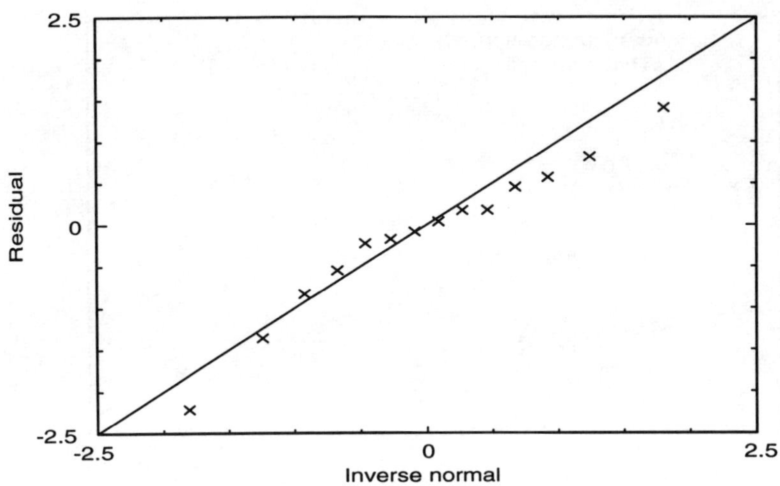

Fig. 10.12. Normal probability plot of residuals for the insulin data of Table 10.1, for the model with log dose without interaction.

The plot of residuals against leverage is given in Figure 10.13. The observations for the test preparation have higher leverage than those for the standard preparation, but none are very large.

The index plot of Cook's distance is given in Figure 10.14. These are consistently larger for the test preparation than for the standard. As well,

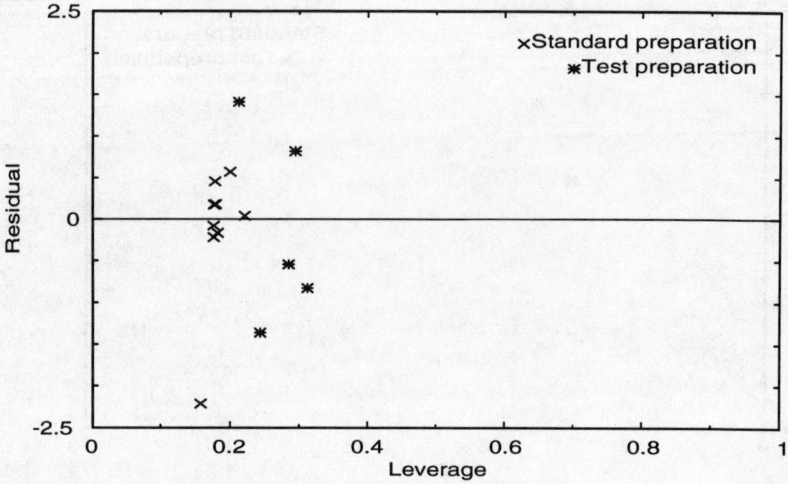

Fig. 10.13. Leverage plot for the insulin data of Table 10.1, for the model with log dose without interaction.

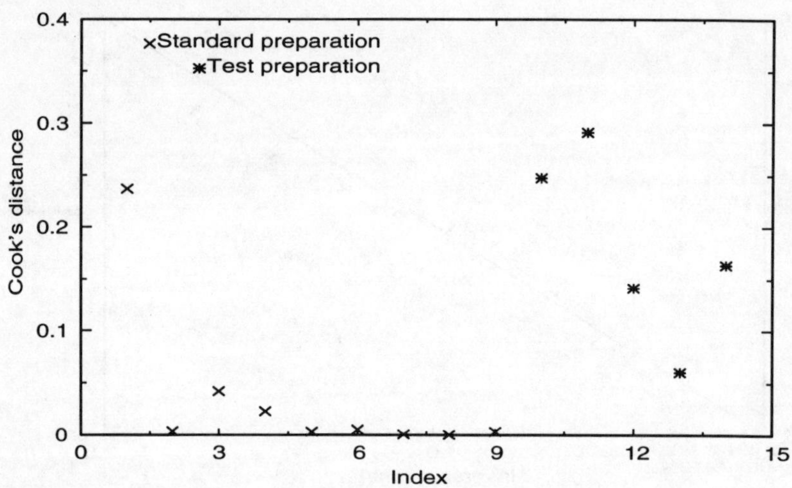

Fig. 10.14. Plot of Cook's distances for the insulin data of Table 10.1, for the model with log dose without interaction.

they are larger for the smallest dose in both preparations.

Thus, we conclude that our model fits the standard preparation best, except for the smallest dose, where it fits neither preparation very well. In spite of the overall good fit, we may be tempted to search for a better model (see Exercise (10.1) below).

10.4 Asymptotic inference

The asymptotic covariance matrix of the parameter estimates, the inverse of the information matrix, given above in Equation (10.9), is

$$\begin{pmatrix} 0.294 & -0.115 & 0.022 \\ -0.115 & 0.048 & -0.017 \\ 0.022 & -0.017 & 0.054 \end{pmatrix}$$

and the corresponding correlation matrix is

$$\begin{pmatrix} 1.00 & -0.97 & 0.18 \\ -0.97 & 1.00 & -0.33 \\ 0.18 & -0.33 & 1.00 \end{pmatrix}$$

The parameters have highly correlated estimates, especially β_0 and β_1. From the first of these matrices, we obtain the asymptotic standard errors of the three parameter estimates, 0.542 for $\hat{\beta}_0 = -5.553$, 0.220 for $\hat{\beta}_1 = 2.297$, and 0.233 for $\hat{\beta}_2 = -0.929$.

The asymptotic Wald tests, respectively -10.24, 10.46, and -3.98, with P-values all very close to zero, indicate that all parameters are significantly different from zero. The change in deviance when β_1 is set to zero is 157.1, and for $\beta_2 = 0$, 16.8, each with an asymptotic Chi-squared distribution with one degree of freedom, and P-values close to zero, again indicating that the hypotheses that these parameters are zero are to be rejected. On the other hand, the P-value for the comparison with the saturated model, with deviance 8.79 on 11 d.f., is 0.64, indicating that the regression model fits well.

The asymptotic distribution of the $\widehat{\mathrm{ED}}_{50}$ will be $\widehat{\mathrm{ED}}_{50} \sim \mathrm{N}(\mathrm{ED}_{50}, \sigma^2_{\mathrm{ED}})$. By the delta method,

$$\hat{\sigma}^2_{\mathrm{ED}} \doteq \frac{1}{\hat{\beta}_1^2}(\mathrm{I}^{11} + 2\widehat{\mathrm{ED}}_{50}\mathrm{I}^{12} + \widehat{\mathrm{ED}}_{50}^{2}\mathrm{I}^{22}) \qquad (10.12)$$

where I^{ij} is the element of the inverse of the information matrix given above. For example, for the standard preparation, with $\widehat{\mathrm{ED}}_{50} = 2.418$, the asymptotic standard error is 0.136.

10.5 Factoring the likelihood function

We saw in Equation (10.9) that the expected information matrix for the model with log dose is estimated as

$$\mathbf{I}(\hat{\beta}) = \begin{pmatrix} 87.89 & 222.01 & 33.05 \\ 222.01 & 584.07 & 90.76 \\ 33.05 & 90.76 & 33.05 \end{pmatrix}$$

The off-diagonal elements are not near zero, showing that the parameters are far from being information orthogonal. Let us study what happens if we make the design matrix closer to design orthogonality. We subtract the mean log dose from all log doses and use the values -1 and 1 for the two preparations. The resulting expected information matrix is

$$\mathbf{I}(\hat{\beta}) = \begin{pmatrix} 87.94 & 6.09 & 21.80 \\ 6.09 & 23.74 & -13.04 \\ 21.80 & -13.04 & 87.96 \end{pmatrix} \tag{10.13}$$

The off-diagonal elements are now considerably reduced, although certainly still not close to zero. If there had been equal numbers of observations for each dose, the result might have been better.

The inverse of the information matrix is

$$\begin{pmatrix} 0.0127 & -0.0054 & -0.0040 \\ -0.0054 & 0.0482 & 0.0085 \\ -0.0040 & 0.0085 & 0.0136 \end{pmatrix}$$

and the corresponding correlation matrix is

$$\begin{pmatrix} 1.00 & -0.22 & -0.30 \\ -0.22 & 1.00 & 0.33 \\ -0.30 & 0.33 & 1.00 \end{pmatrix}$$

We see that the correlation between β_0 and β_1 is substantially reduced from what we found in the previous section.

Equation (4.19) for an exact test for simple logistic regression can be extended to our model:

$$\Pr(t_1, t_2|t_0; \beta_1, \beta_2) = \frac{c(t_0, t_1, t_2)\mathrm{e}^{\beta_1 t_1 + \beta_2 t_2}}{\sum_u \sum_v c(t_0, u, v)\mathrm{e}^{\beta_1 u + \beta_2 v}} \tag{10.14}$$

where $\mathbf{T}^{\mathrm{T}} = (Y_{\bullet\bullet}, \sum\sum \log(x_{1i})Y_{ij}, \sum\sum x_{2j}Y_{ij})$. This is a very complex expression which would take considerable computer time to evaluate, especially because it must be repeated many times to produce a conditional likelihood plot.

One, then, might wish to apply one of the approximations, the modified profile likelihoods, of Section 6.3. However, from the score, Equation (10.5), we see that the slope parameters, which interest us, are a complex function of the intercept parameter. This means that the partial derivatives of Equation (6.40) will be difficult to evaluate for the modified profile likelihood. The adjusted profile likelihood of Equation (6.43) requires transforming the parameters to information orthogonality, again a complex procedure. Here, this will be essential, because, as we have seen, the parameters of our model have substantial correlation.

10.6 Frequentist decision-making

If we have assumed that the proportion of mice with convulsions increases with dose and that the test preparation is no worse than the standard, we can use one-tailed tests for $\beta_1 = 0$ against $\beta_1 > 0$ and $\beta_2 = 0$ against $\beta_2 < 0$. These UMP likelihood ratio tests are identical to the asymptotic Chi-squared tests just given. They reject the null hypotheses.

Because the parameters are canonical in the exponential family, no minimum variance unbiased point estimators exist.

UMP 95% confidence intervals, obtained from the profile deviances, are $(-6.66, -4.51)$ for β_0, $(1.88, 2.74)$ for β_1, and $(-1.39, -0.48)$ for β_2. Approximate 95% confidence intervals can also be constructed from the asymptotic standard errors, taking 1.96 standard errors on either side. These are $(-6.62, -4.49)$ for β_0, $(1.87, 2.73)$ for β_1, and $(-1.39, -0.47)$ for β_2, which are close to the UMP critical regions. This is not surprising, because the profile likelihood functions were reasonably symmetric.

10.7 Bayesian decision-making

Let us use a vague, improper prior for each transformed parameter, $\beta_k^* = e^{\beta_k}$. This ensures that all (new) parameters are positive in the likelihood function. Note that these are not uniform priors for the original parameters, β_k. Again, we use Laplace's method to obtain a second-order approximation to the integral. The posterior means for the three original parameters are $\beta_0 = -5.293$, $\beta_1 = 2.255$, and $\beta_2 = -0.854$. We see that this vague prior for β_k^* moves the m.l.e. of each β_k towards zero.

The three marginal densities are plotted in Figures 10.15 to 10.17. They are very similar to the original likelihood functions, except for the displacement.

10.8 Exercises

(1) (a) Redo the analyses of this chapter using a complementary log log link, either $\log[-\log(\pi_i)]$ or $\log[-\log(1-\pi_i)]$.
(b) Which of these is better?
(c) What improvements and what complications occur?
(Lindsey, 1995b, pp. 68–74)

(2) A quantal response experiment was performed to study the effect of nicotine on the mortality of aphids, giving the results in Table 10.2. Follow the example of this chapter to analyze these data. The ED_{50} value is here of particular interest.
(Morgan, 1992, p. 95)

(3) A serological survey was carried out in Bongono, Zaire, to determine the extent of malarial infection, with results as shown in Table 10.3. The percentage sero-positive is bounded above by a factor reflecting

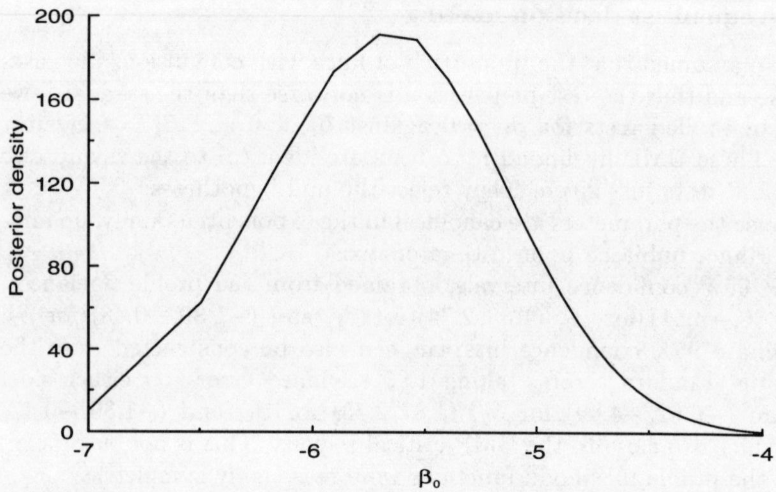

Fig. 10.15. Plot of the marginal posterior density for β_0 for the insulin data of Table 10.1, for the model with log dose without interaction.

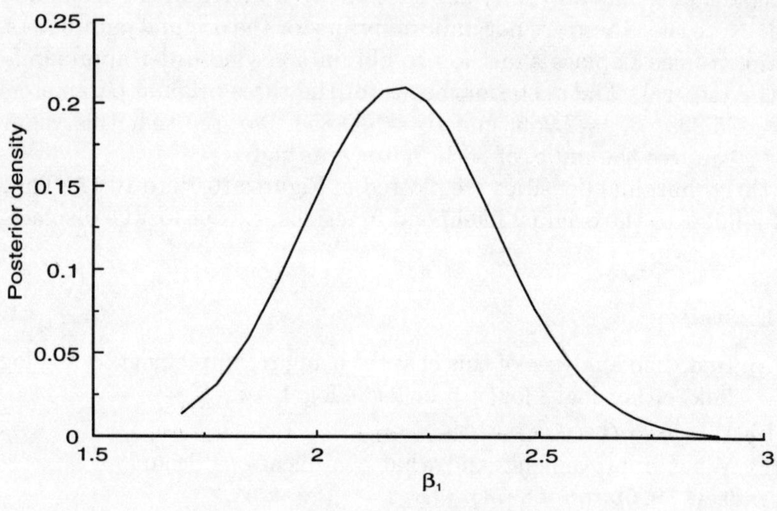

Fig. 10.16. Plot of the marginal posterior density for β_1 for the insulin data of Table 10.1, for the model with log dose without interaction.

the incidence of malaria. Explore how the techniques of this chapter might be used to analyze such data.
(Morgan, 1992, p. 16)

(4) An experiment was performed in a hospital to determine if rocking reduced the probability of babies crying. Each day for 18 days, one

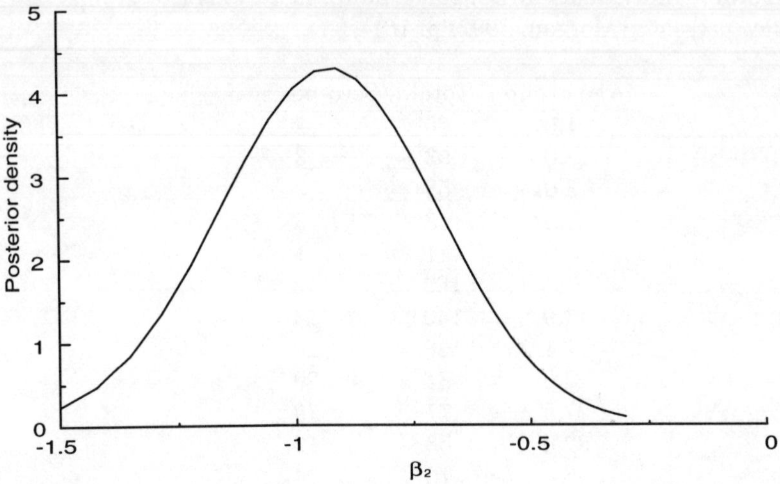

Fig. 10.17. Plot of the marginal posterior density for β_2 for the insulin data of Table 10.1, for the model with log dose without interaction.

Table 10.2. Numbers of aphids dying under various concentrations (%) of nicotine. (Morgan, 1992, p. 95)

Dose	Total	Dead
0.0	45	3
0.0025	50	5
0.005	46	4
0.01	50	3
0.02	46	11
0.03	46	20
0.04	49	31
0.06	50	40
0.08	50	43
0.10	50	48
0.15	50	48
0.20	50	50

baby was chosen at random to be rocked and all of the others present in the ward were not, with results as shown in Table 10.4.

(a) Consider a logistic regression model with no interaction between day and treatment. Plot the profile likelihood for the treatment effect.

(b) Because the day is not of interest, develop a conditional likeli-

Table 10.3. Individuals in Bongono, Zaire, in various age groups, with antibodies present. (Morgan, 1992, p. 16)

Age group	Total	Sero-positive
1.0	60	2
2.0	63	3
3.0	53	3
4.0	48	3
5.0	31	1
7.3	182	18
11.9	140	14
17.1	138	20
22.0	84	20
27.5	77	19
32.0	58	19
36.8	75	24
41.6	30	7
49.7	62	25
60.8	74	44

Table 10.4. Results from an experiment to determine the effect of rocking on babies crying. (Cox, 1966)

Control		Test	
Total	Not crying	Total	Not crying
8	3	1	1
6	2	1	1
5	1	1	1
6	1	1	0
5	4	1	1
9	4	1	1
8	5	1	1
8	4	1	1
5	3	1	1
9	8	1	0
6	5	1	1
9	8	1	1
8	5	1	1
5	4	1	1
6	4	1	1
8	7	1	1
6	4	1	0
8	5	1	1

hood which removes these incidental parameters, again plotting it for treatment effect.

(c) An alternative is to try a random effects model. Plot the likelihood for the treatment effect from this model.

(d) Now perform an appropriate model selection procedure to find what model is suitable for these data.

(e) Plot the likelihood function for treatment, if it is not already one of the above.

(f) Draw the appropriate conclusions.

(Cox, 1966; Lindsey, 1975)

Part VI

Appendices

Appendix A
Elements of measure theory

Probability theory is often considered simply to be a special branch of measure theory. Hence, we briefly review the latter from a non-technical point of view.

A.1 Measure spaces

Consider some space which is a set, Ω, with elements ω, along with a collection, $\mathcal{F}$, of subsets E of Ω, and a corresponding measure, a function ν, defined on $\mathcal{F}$. This measure must obey the following conditions:

(1) $0 = \nu(\emptyset) \leq \nu(E) \leq \nu(\Omega) \quad \forall E \in \mathcal{F}$;
(2) $\nu(E_1 \cup E_2) = \nu(E_1) + \nu(E_2) - \nu(E_1 \cap E_2)$ for $E_i \in \mathcal{F}$;
(3) $\nu\left(\cup_{i=1}^{\infty} E_i\right) = \sum_{i=1}^{\infty} \nu(E_i)$ if $E_i \cap E_j = \emptyset,\ i \neq j$.

We denote this *measure space* by $(\Omega, \mathcal{F}, \nu)$.

If Ω is a countable set of elements, $\mathcal{F}$ may be taken to be all of its subsets. However, if Ω is uncountably infinite, it is not always possible to take $\mathcal{F}$ to be the collection of *all* subsets. We then add the conditions on $\mathcal{F}$:

(1) $\emptyset \in \mathcal{F}, \Omega \in \mathcal{F}$;
(2) $E^c \in \mathcal{F}$ whenever $E \in \mathcal{F}$, where $E^c = \{\omega \in \Omega; \omega \notin E\}$;
(3) $\cup_{i=1}^{\infty} E_i \in \mathcal{F}$ whenever $E_i \in \mathcal{F}, i = 1, 2, \ldots$.

Such a collection, $\mathcal{F}$, is called a σ-algebra.

Suppose that Y is any real-valued function defined on Ω. It is said to be measurable with respect to $\mathcal{F}$ if

$$\{\omega : Y(\omega) \leq y\} \in \mathcal{F} \quad \forall y$$

The σ-algebra generated by Y is defined to be the smallest σ-algebra with respect to which Y is measurable. It consists of exactly those sets, E, which are in every σ-algebra, $\mathcal{F}$, for which Y is $\mathcal{F}$-measurable.

The σ-algebra of *Borel sets*, $\mathcal{B}$, is that generated by the identity function $f(y) = y$ for $y \in (-\infty, \infty)$. In other words, the σ-algebra of Borel sets is the smallest σ-algebra containing every interval $(y_1, y_2]$, where $-\infty \leq y_1 \leq y_2 < \infty$.

In n-dimensional space, Borel sets are defined by the smallest σ-algebra containing all rectangles, $\mathcal{B} = \{(y_1, \ldots, y_n) : a_i < y_i \leq b_i,\ i = 1, \ldots, n\}$. A unique measure can be defined which assigns to each such rectangle its volume, $\nu(\mathcal{B}) = \prod(b_i - a_i)$. When completed by all subsets of measure zero, the sets are called *Lebesgue measurable.* On the other hand, if the space is countable, *counting measure* is defined by the number of elements in any subset, if the number is finite, and infinity otherwise.

The *support* of ν is the minimal closed set $S \in \mathcal{F}$ such that $\nu(S^c) = 0$. If, for each $E \in \mathcal{F}$, $\nu_1(E) = 0 \Rightarrow \nu_2(E) = 0$, then the measure ν_2 is said to be *dominated* by ν_1, written $\nu_2 \ll \nu_1$.

A.2 Integration

For the measure space $(\Omega, \mathcal{F}, \nu)$, if $f(\omega)$ is a ν-measurable function relative to $\mathcal{F}$, the Lebesgue–Stieltjes integral is defined as

$$\nu(fI_E) \stackrel{\text{def}}{=} \int_E f d\nu \tag{A.1}$$

where I_E is the indicator function of Equation (B.1). Note that summation is a special type of integration. Then, the sequence $f_1, f_2, \ldots$ is integrable with respect to the measure, ν, iff $\sum |f_i| < \infty$:

$$\sum f_i = \int_{\mathrm{N}} f d\nu \tag{A.2}$$

where $\mathbf{N}$ is the set of positive integers.

Integration is linear

$$\nu(\alpha f_1 + \beta f_2) = \alpha\nu(f_1) + \beta\nu(f_2) \tag{A.3}$$

where $\alpha, \beta \in \mathbf{R}$.

A measurable mapping, f, for which $\int_\Omega |f|^r d\nu < \infty$, is said to be r-*integrable.* We shall denote this as $f \in \mathcal{L}^r(\Omega, \mathcal{F}, \nu)$.

A.3 Convexity

A *convex set* $\mathcal{C} \subset \mathbf{R}^k$ is such that $c\mathbf{y}_1 + (1 - c)\mathbf{y}_2 \in \mathcal{C}$ for any $\mathbf{y}_1, \mathbf{y}_2 \in \mathcal{C}$ and $0 \leq c \leq 1$. For $k = 1$, a convex set is an interval. Notice how this definition is related to probabilistic ideas in that it represents an averaging of two points. If the set is not convex, the average is not in the set.

A *boundary point* of a set is a point for which all neighbourhoods contain points not in the set. A convex set is open if none of its members are boundary points; it is closed if all boundary points are members. The complement of an open set is closed. A set is bounded if it is strictly contained within some ball of radius r. A compact set is both closed and bounded.

The *convex hull* of an arbitrary set $\mathcal{S} \subset \mathbf{R}^k$ is the set $\mathcal{C} = \{\mathbf{y} : \exists\{\mathbf{y}_i\} \subset \mathcal{S}, \{c_i \in \mathbf{R}^+\}, \sum c_i = 1, \ni \mathbf{y} = \sum c_i \mathbf{y}_i\}$. It is the smallest convex set containing $\mathcal{S}$. If $\mathcal{S} \subset \mathbf{R}^k$ is the support of ν, the closed convex hull of $\mathcal{S}$ is called the *convex support* of ν. The construction of a convex hull guarantees that all averages are in the set. Indeed, *Carathéodory's theorem* states that any member of the convex hull, $\mathcal{C}$, can be represented as an average of at most $k+1$ points of the original set, $\mathcal{S}$. (This does not imply that there is a unique way to do this.)

A *convex function*, $f : \mathbf{R}^k \to \mathbf{R}$, has

$$f[c\mathbf{y}_1 + (1-c)\mathbf{y}_2] \leq cf(\mathbf{y}_1) + (1-c)f(\mathbf{y}_2) \tag{A.4}$$

for all $\mathbf{y}_1, \mathbf{y}_2 \in \mathcal{C}$ and $0 < c < 1$. If f is a convex function, $-f$ is a concave function. The derivatives $\partial f(\mathbf{y})/\partial y_i$ are non-decreasing (if they exist). When the argument is scalar, for any fixed y_0, there exists a constant, d, such that

$$f(y) \geq f(y_0) + d(y - y_0)$$

If all second-order partial derivatives

$$\frac{\partial^2 f(\mathbf{y})}{\partial y_i \partial y_j}$$

exist, f is convex iff the $k \times k$ matrix of such derivatives is non-negative definite for all $\mathbf{y} \in \mathcal{S}$.

Appendix B
Review of probability theory

In this appendix, I present, without proofs, some general results of probability theory with which the reader may be expected to be already familiar. Although concepts from measure theory are used, the reader who is not acquainted with this theory, hopefully, should still be able to follow; a brief outline of basic concepts was given in Appendix A above. Measure theory will only be used sparingly, where absolutely necessary, in the main text. It is introduced here because some knowledge of it is essential to be able to follow many of the important developments in modern theoretical statistics.

B.1 Probability spaces

B.1.1 PROBABILITY

Let Ω be a *sample space* of possible outcomes of some *random experiment*, with each possible outcome or *sample point* denoted by ω, $\mathcal{F}$ be a σ-algebra on Ω called the family of *events*, and P be a set function defined on $\mathcal{F}$.

Call the triplet, $(\Omega, \mathcal{F}, P)$, a *probability space* if P obeys the following conditions:

(1) $P(E) \geq 0 \quad \forall E \in \mathcal{F}$;
(2) $P(\Omega) = 1$;
(3) $P(\bigcup_{i=1}^{\infty} E_i) = \sum_{i=1}^{\infty} P(E_i) \quad \forall E_i \in \mathcal{F}$ and $E_i \bigcap E_j = \emptyset$, $i \neq j$.

P is, then, called a *probability measure* on $(\Omega, \mathcal{F})$. Note that P maps from $\mathcal{F}$ to $[0, 1]$. Only the second point is an additional constraint on general measure theory.

We may interpret such a space in the following way. In an experiment, a point, $\omega \in \Omega$, is chosen 'at random' according to the law P. We shall call this process the underlying *data generating mechanism* (Section 1.1). The chosen point determines the outcome of the experiment, E. For this $E \in \mathcal{F}$, $P(E)$ represents the 'probability' that the point $\omega \in E$ is chosen.

Examples

1. If we toss an unbiased coin, the sample space might be

$$\Omega = \{H, T\}$$

with σ-algebra

$$\mathcal{F} = \text{set of all subsets of } \Omega$$

and probability measure

$$\mathrm{P}(H) = \mathrm{P}(T) = \frac{1}{2}$$

This is an abstract, theoretical description, or *model*, of tossing a coin, ignoring many factors, such as the possibility of the coin landing on its edge.

2. If we toss the coin twice, the sample space is

$$\Omega = \{HH, HT, TH, TT\}$$

with probability measure

$$\mathrm{P}(HH) = \mathrm{P}(HT) = \mathrm{P}(TH) = \mathrm{P}(TT) = \frac{1}{4}$$

A possible event is one head:

$$E \in \{HT, TH\}$$

with $P(E) = \frac{1}{2}$. □

Notice that this definition does not require the selection of a point to be expected to occur repeatedly. It is, rather, a theoretical description of some phenomenon under study. Indeed, empirically, any event is unique (de Finetti, 1974a, I, p. 7); only such a probability model (Section 1.2) as we are constructing by these definitions allows us to ignore differences among unique events as irrelevant to the task at hand and to speak of events as repeatable.

Any event with $P(E) = 0$ is called *null*. A probability space is said to be *complete* if all subsets of null sets are events.

For any event, E, the *indicator function*, $I_E : \Omega \rightarrow \mathbf{R}$, is defined by

$$I_E(\omega) \stackrel{\text{def}}{=} \begin{cases} 1 \text{ if } \omega \in E \\ 0 \text{ otherwise} \end{cases} \tag{B.1}$$

The rest of this appendix is concerned with the bases for describing probabilistic data generating mechanisms by the construction of models.

B.1.2 RANDOM VARIABLES

A *random variable* is a real-valued function $Y : \Omega \to \mathbf{R}$ from Ω to the real line $\mathbf{R}$, such that $Y^{-1} : \mathcal{B} \to \mathcal{F}$, where $\mathcal{B}$ is the Borel σ-algebra. Thus, a random variable is simply a function mapping the space of events into the real line. (With this definition, care must be taken for nominal and ordinal random variables.) If we have

$$\begin{aligned}\{-\infty < Y \le y\} &= \{\omega : -\infty < Y(\omega) \le y\} \\ &\stackrel{\text{def}}{=} Y^{-1}\{(-\infty, y]\} \quad \in \mathcal{F} \quad \forall y \end{aligned} \tag{B.2}$$

$Y \in \mathcal{Y}$ is a measurable mapping from $(\Omega, \mathcal{F})$ to the real line with a Borel σ-algebra $(\mathbf{R}, \mathcal{B})$. Then, Y is called an *$\mathcal{F}$-measurable random variable* or is *measurable* relative to $\mathcal{F}$.

Examples

1. For the unbiased coin, let

$$Y(\omega) \stackrel{\text{def}}{=} \begin{cases} 1 \text{ if } \omega = H \\ 0 \text{ otherwise} \end{cases}$$

2. For two tosses, suppose that the random variable counts the number of heads:

$$Y(\omega) \stackrel{\text{def}}{=} \begin{cases} 0 \text{ if } \omega = TT \\ 1 \text{ if } \omega = HT, TH \\ 2 \text{ if } \omega = HH \end{cases}$$

□

The sum or product of n random variables is a random variable. If the random variable, Y, can take, at most, a countably infinite number of different values, it is said to be *discrete*; otherwise, it is *continuous*. We shall let $y \in \mathcal{Y}$ denote the fixed or *observed value* of the variable for a fixed value of ω.

B.1.3 PROBABILITY LAWS

We can derive from $Y : \Omega \to \mathbf{R}$ and $P : \mathcal{F} \to [0, 1]$ the *probability law*, P_Y, defined on the real line, $(\mathbf{R}, \mathcal{B})$, given by

$$\begin{aligned} P_Y(B) &\stackrel{\text{def}}{=} P(Y^{-1}(B)) \text{ for } B \in \mathcal{B} \\ &= P(Y \in B) \end{aligned} \tag{B.3}$$

or

$$P_Y(y_1 < Y \leq y_2) = P(\omega : y_1 < Y(\omega) \leq y_2) \tag{B.4}$$

This is the probability induced on $\mathbf{R}$ by Y and P. We now have

$$\Omega \stackrel{Y}{\rightarrow} \mathbf{R} \tag{B.5}$$

$$[0,1] \stackrel{P}{\leftarrow} \mathcal{F} \stackrel{Y^{-1}}{\leftarrow} \mathcal{B} \tag{B.6}$$

If $B = (-\infty, y]$, then the function,

$$\begin{aligned} F(y) &= P_Y(Y \leq y) \\ &= P_Y((-\infty, y]) \\ &\stackrel{\text{def}}{=} P(\omega : Y(\omega) \leq y) \end{aligned} \tag{B.7}$$

is called the (cumulative) *distribution function* of Y. The characteristics of this function are

(1) $F : \mathbf{R} \rightarrow [0,1]$
 with $y_1 \leq y_2 \Rightarrow F(y_1) \leq F(y_2)$;
(2) $\lim_{y\rightarrow\infty} F(y) = 1$, $\lim_{y\rightarrow-\infty} F(y) = 0$;
(3) F is right continuous, i.e. $\lim_{h\downarrow\infty} F(y+h) = F(y)$.

When $(\Omega, \mathcal{F}, P) = (\mathbf{R}, \mathcal{B}, P_Y)$, where $P_Y((-\infty, y]) = F(y)$, P_Y is the Lebesgue–Stieltjes measure associated with F or the probability law of Y. Thus, by defining $F(y)$, P_Y is also defined. In cases which concern us, this probability measure will be dominated by a counting or Lebesgue measure called, respectively, discrete and continuous probability laws.

The distribution function of a discrete random variable has jump discontinuities at each observable value and is constant in between. The distribution function of a continuous random variable is continuous. However, a random variable may also be neither discrete nor continuous.

B.1.4 INDEPENDENCE

Sub-σ-algebras $\mathcal{F}_1, \mathcal{F}_2, \ldots$ of $\mathcal{F}$ are *independent* if

$$P(E_{i_1} \bigcap \cdots \bigcap E_{i_n}) = \prod_{j=1}^{n} P(E_{i_j}) \tag{B.8}$$

for $E_{i_k} \in \mathcal{F}_{i_k}$ with $i_1, \ldots, i_n$ distinct, $\forall n$. Random variables Y_i are independent if their σ-algebras are independent. From this, we can deduce that, in such a case,

$$P_Y(Y_1 \leq y_1, \ldots, Y_n \leq y_n) = \prod_{i=1}^{n} P_Y(Y_i \leq y_i) \tag{B.9}$$

Independent and identically distributed (i.i.d.) random variables are independent and all have the same distribution function.

Consider now two independent experiments with probability spaces $(\Omega_1, \mathcal{F}_1, P_1)$ and $(\Omega_2, \mathcal{F}_2, P_2)$. The sample space is the collection of ordered pairs, $\Omega_1 \times \Omega_2 = \{(\omega_1, \omega_2) : \omega_1 \in \Omega_1, \omega_2 \in \Omega_2\}$. Unfortunately, the family of all sets, $\mathcal{F}_1 \times \mathcal{F}_2 = \{E_1 \times E_2 : E_1 \in \mathcal{F}_1, E_2 \in \mathcal{F}_2\}$, is not, in general, a σ-algebra. However, there is a unique smallest σ-algebra of subsets of $\Omega_1 \times \Omega_2$ which contains $\mathcal{F}_1 \times \mathcal{F}_2$. Finally, let $P_{12} : \mathcal{F}_1 \times \mathcal{F}_2 \rightarrow [0, 1]$ be given by $P_{12}(E_1 \times E_2) = P_1(E_1)P_2(E_2)$ for any $E_1 \in \mathcal{F}_1$, $E_2 \in \mathcal{F}_2$. The domain of P_{12} can be extended to the whole σ-algebra. Then, the resulting probability space is called the *product space* of $(\Omega_1, \mathcal{F}_1, P_1)$ and $(\Omega_2, \mathcal{F}_2, P_2)$.

B.2 Integration

B.2.1 EXPECTATION

Expectation is the integral with respect to the probability measure P. Thus, the *mean* or *expected value* of a random variable is the integral of Y with respect to P, given by

$$\mathrm{E}[Y] = \int_\Omega Y dP \tag{B.10}$$

If $\mathrm{E}[|Y|] < \infty$, i.e. $Y \in \mathcal{L}^1$, the random variable is said to be *integrable.*

The term, expected value, for a mean or average is historical, but unfortunate. No one *expects* a random variable to have this value (Hoffmann-Jørgensen, 1994, I, pp. 313–320). Indeed, except for rather rare symmetrical cases, it will not even be the most probable value, and, for counting measure, will generally not even be a possible value.

The *variance* of Y is

$$\begin{aligned}\mathrm{var}[Y] &\stackrel{\mathrm{def}}{=} \mathrm{E}[(Y - \mathrm{E}[Y])^2] \\ &= \mathrm{E}[Y^2] - (\mathrm{E}[Y])^2 \\ &= \int_\Omega (Y - \mathrm{E}[Y])^2 dP\end{aligned} \tag{B.11}$$

If $\mathrm{var}[Y] < \infty$, i.e. $Y \in \mathcal{L}^2$, the random variable is said to be *square integrable.* If $\mathrm{E}[|Y|^r] < \infty$, define the $\mathcal{L}^r$ norm,

$$\|Y\|_r \stackrel{\mathrm{def}}{=} (\mathrm{E}[|Y|^r])^{\frac{1}{r}} \tag{B.12}$$

For $1 \leq p \leq r < \infty$, if $Y \in \mathcal{L}^r$, then $Y \in \mathcal{L}^p$ and $\|Y\|_p \leq \|Y\|_r$.

The appropriateness of the $\mathcal{L}^r$ norm in statistical applications generally depends on stochastic independence and on the variance of each observation being independent of its mean.

B.2.2 RADON–NIKODÝM THEOREM

Suppose that P_0 and P are both σ-finite measures on $(\Omega, \mathcal{F})$ with P absolutely continuous with respect to P_0: $P_0(E) = 0 \Rightarrow P(E) = 0$ $(E \in \mathcal{F})$, i.e. P is dominated by P_0. Then $\exists f \in \mathcal{L}^1(\Omega, \mathcal{F}, P_0)$ called the *density*, such that $P = fP_0$.

$$\begin{aligned} P(E) &= \int_E f dP_0 \\ &= \mathrm{E}[f] \quad \forall E \in \mathcal{F} \end{aligned} \tag{B.13}$$

$$\frac{dP}{dP_0} = f \text{ on } \mathcal{F} \text{ almost surely} \tag{B.14}$$

(See Section B.6.2 below for the meaning of 'almost surely'.) If P_0 and P are both probability measures, f is sometimes called the *likelihood ratio* of P relative to P_0.

B.3 Probability functions

B.3.1 PROBABILITY DENSITY FUNCTION

We have denoted the probability law of Y by $P_Y(B) = P(Y \in B)$, a measure on $(\mathbf{R}, \mathcal{B})$. If h is a Borel measurable function from $\mathbf{R}$ to $\mathbf{R}$, then

$$h(Y) \in \mathcal{L}^1(\Omega, \mathcal{F}, P) \text{ iff } h \in \mathcal{L}^1(\mathbf{R}, \mathcal{B}, P_Y) \tag{B.15}$$

and

$$\mathrm{E}[h(Y)] = \int_{\mathrm{R}} h(u) du \tag{B.16}$$

Y has a *probability density function* (p.d.f.), f_Y, if there exists a Borel function, $f_Y : \mathbf{R} \to [0, \infty]$, such that

$$P(Y \in B) = \int_B f_Y(u) du \quad B \in \mathcal{B} \tag{B.17}$$

This will occur when Y is continuous. A p.d.f. is any such non-negative function with respect to a σ-finite measure. In terms of the Radon–Nikodým theorem, P has density f_Y relative to Lebesgue measure.

Let us define a new random variable

$$Z = F(Y) \tag{B.18}$$

called the *probability integral transform.* Then, Z will have a uniform distribution on $(0, 1)$. Because this transformation applies to any random variable with differentiable distribution function, any continuous distribution can be transformed into any other by such transformations of variables.

The common probability measures are usually specified either by a probability density function, f_Y,

$$P_Y((y_1, y_2]) = \int_{y_1}^{y_2} f_Y(u)du \tag{B.19}$$

for continuous random variables, or a probability mass function, p_Y,

$$P_Y(y_1, \ldots, y_n) = \sum_{i=1}^{n} p_Y(y_i) \tag{B.20}$$

for discrete random variables.

Common examples include the normal, gamma, Weibull, binomial, and Poisson distributions. In Chapter 2, we study three important families of probability measures, the exponential, transformation, and stable families.

When the density contains a parameter, $f(y;\psi)$, a useful identity is

$$\frac{\partial \mathrm{E}[g(Y,\psi)]}{\partial \psi} = \mathrm{E}\left[\frac{\partial g(Y,\psi)}{\partial \psi}\right] + \mathrm{E}\left[g(Y,\psi)\frac{\partial \log f(Y,\psi)}{\partial \psi}\right] \tag{B.21}$$

for some function $g(Y, \psi)$.

B.3.2 JOINT PROBABILITY

The joint probability law of random variables Y and Z is defined by

$$P_{Y,Z}(\Gamma) \stackrel{\text{def}}{=} P[(Y,Z) \in \Gamma] \tag{B.22}$$

which is a map $P_{Y,Z} : \Gamma = \mathcal{B} \times \mathcal{B} \rightarrow [0,1]$. Equivalently, the joint (cumulative) probability distribution is

$$F_{Y,Z}(y,z) \stackrel{\text{def}}{=} P_{Y,Z}(Y \leq y, Z \leq z) \tag{B.23}$$

Y and Z have a *joint probability density function*, $f_{Y,Z}$, if

$$\begin{aligned} P[(Y,Z) \in \Gamma] &= \int_\Gamma f_{Y,Z}(u)du \\ &= \int_{\mathrm{R}}\int_{\mathrm{R}} I_\Gamma(u,v) f_{Y,Z}(u,v)dudv \end{aligned} \tag{B.24}$$

where $I_\Gamma(u,v)$ is the appropriate indicator function. If $Y, Z \in \mathcal{L}^2$, their *covariance* is

$$\mathrm{cov}[Y,Z] \stackrel{\text{def}}{=} \mathrm{E}[(Y - \mathrm{E}[Y])(Z - \mathrm{E}[Z])] \tag{B.25}$$

$$= \mathrm{E}[YZ] - \mathrm{E}[Y]\mathrm{E}[Z]$$

We have already seen, in Equation (B.9), that the joint probability of independent random variables is the product of their individual probabilities. Indeed, if two random variables, Y and Z, are independent,

$$\mathrm{E}[YZ] = \mathrm{E}[Y]\mathrm{E}[Z] \tag{B.26}$$

and

$$\mathrm{var}[Y + Z] = \mathrm{var}[Y] + \mathrm{var}[Z] \tag{B.27}$$

Note, however, that conversely, these do not imply independence.

B.3.3 MARGINAL PROBABILITY

The *marginal distributions* of a joint p.d.f., $f_{Y,Z}$, are given by

$$\begin{aligned} f_Y(y) &= \textstyle\int_{\mathrm{R}} f_{Y,Z}(y, v)dv \\ f_Z(z) &= \textstyle\int_{\mathrm{R}} f_{Y,Z}(u, z)du \end{aligned} \tag{B.28}$$

A marginal distribution for discrete variables is obtained in a similar way, by summation. If Y and Z are independent,

$$F_{Y,Z}(y, z) = F_Y(y)F_Z(z) \tag{B.29}$$

and, if the densities exist,

$$f_{Y,Z}(y, z) = f_Y(y)f_Z(z) \tag{B.30}$$

B.3.4 CONVOLUTIONS

For two independent random variables, Y and Z, the distribution of their sum, $S = Y + Z$, is given by the convolution,

$$\begin{aligned} f_{Y+Z}(y + z) &= \int_{\mathrm{R}} f_Y(s - u)f_Z(u)du \qquad \text{(B.31)} \\ &= \int_{\mathrm{R}} f_Y(u)f_Z(s - u)du \end{aligned}$$

This is often denoted by $f_Y * f_Z$. Convolution is commutative and associative.

B.3.5 CONDITIONAL PROBABILITY

For discrete random variables, Y and Z, with $\mathrm{E}[|Y|] < \infty$, we have the *conditional distribution*

$$P_{Y|Z}(Y=y|Z=z) \stackrel{\text{def}}{=} \frac{P_{Y,Z}(Y=y,Z=z)}{P_Z(Z=z)}, \qquad P_Z(Z=z) \neq 0 \tag{B.32}$$

so that

$$\mathrm{E}[Y|Z=z] = \sum_{y \in \mathcal{Y}} y P_{Y|Z}(Y=y|Z=z) \tag{B.33}$$

Then, $U = E[Y|Z=z]$ is also a random variable defined as

$$\text{if } Z(\omega)=z, \text{ then } U(\omega) \stackrel{\text{def}}{=} \mathrm{E}[Y|Z=z] \tag{B.34}$$

The sample space, Ω, has been partitioned into 'orbits' on which Z is constant. The σ-algebra, say $\mathcal{G}$, so generated consists of all possible unions of these orbits. Then, U is constant, with value u, on Z-orbits or is $\mathcal{G}$-measurable.

Suppose now that Y and Z are continuous random variables with joint p.d.f., $f_{Y,Z}(y,z)$, and marginals f_Y and f_Z. Define

$$f_{Y|Z}(y|z) \stackrel{\text{def}}{=} \begin{cases} \frac{f_{Y,Z}(y,z)}{f_Z(z)} & \text{if } f_Z(z) \neq 0 \\ 0 & \text{otherwise} \end{cases} \tag{B.35}$$

Conditional expectation If h is a Borel function on $\mathbf{R}$ such that

$$\mathrm{E}[|h(Y)|] = \int_{\mathrm{R}} |h(u)| f_Y(u)du < \infty$$

then

$$\mathrm{E}[h(Y)|Z=z] = \int_{\mathrm{R}} h(u) f_{Y|Z}(u|z)du \tag{B.36}$$

is its conditional expectation.

If u is the constant value of $U = \mathrm{E}[Y|Z=z]$ for a given $G = \{Z=z\} \in \mathcal{G}$, then

$$\begin{aligned}\int_G U dP &= u P_Z(Z=z) \\ &= \sum y P_{Y|Z}(Y=y|Z=z) P_Z(Z=z) \\ &= \sum y P_{Y,Z}(Y=y, Z=z) \\ &= \int_G Y dP \quad \forall G \in \mathcal{G}\end{aligned}$$

or

$$\begin{aligned} \mathrm{E}[UI_G] &= \mathrm{E}[\mathrm{E}[YI_G|Z=z]] \\ &= \mathrm{E}[YI_G] \end{aligned} \tag{B.37}$$

where I_G is the appropriate indicator function.

B.3.6 BAYES' FORMULA

For random variables Y and Z with strictly positive joint probability density function on $\mathbf{R}^2$, we have

$$\begin{aligned} f_{Y|Z}(y|z) &= \frac{f_{Y,Z}(y,z)}{f_Z(z)} \\ &= \frac{f_{Z|Y}(z|y)f_Y(y)}{f_Z(z)} \end{aligned} \tag{B.38}$$

Therefore,

$$f_{Y|Z}(y|z) \propto f_Y(y) f_{Z|Y}(z|y) \tag{B.39}$$

where the constant of proportionality, depending on y, is determined by

$$\int_{\mathrm{R}} f_{Y|Z}(u|z)du = 1 \tag{B.40}$$

B.3.7 PARADOXES

Marginal probabilities Several paradoxes can arise when only a set of marginal probabilities, and not their joint probabilities, is considered.

- $P(Y_1 > Y_2)$ can be arbitrarily close to unity although $P(Y_1 \leq c) > P(Y_2 \leq c)$ for all c.
  Example
  Let Y_1 and Y_2 have discrete uniform distributions on $(0, 1]$, such that

$$\begin{aligned} P\left(Y_1 = \frac{i+1}{n}, Y_2 = \frac{i}{n}\right) &= \frac{1}{n} - \frac{\epsilon}{n-1}, \quad i = 1, 2, \ldots, n-1 \\ P\left(Y_1 = \frac{1}{n}, Y_2 = 1\right) &= \frac{1}{n} + \epsilon \end{aligned} \tag{B.41}$$

  where $0 < \epsilon < (n-1)/n$, but

$$P(Y_1 > Y_2) = 1 - \frac{1}{n} - \epsilon \tag{B.42}$$

□

- $P(Y_1 < Y_2)$, $P(Y_2 < Y_3)$, and $P(Y_3 < Y_1)$ may all be greater than 0.5.
  Examples
  1. For dependent random variables,

$$\begin{aligned} P(Y_1 = 1, Y_2 = 2, Y_3 = 3) &= 0.3 \\ P(Y_1 = 3, Y_2 = 1, Y_3 = 2) &= 0.3 \\ P(Y_1 = 2, Y_2 = 3, Y_3 = 1) &= 0.4 \end{aligned} \tag{B.43}$$

  2. For independent random variables,

$$\begin{aligned} &P(Y_1 = 3) = 1 \\ &P(Y_2 = 1) = 0.4,\ P(Y_2 = 4) = 0.6 \\ &P(Y_3 = 2) = 0.6,\ P(Y_3 = 5) = 0.4 \end{aligned} \tag{B.44}$$

□

- With $P(Y_1 < Y_2) > 0.5$, $P(Y_1 < Y_3) > 0.5$, and $P(Y_2 < Y_3) > 0.5$, we may have $P(Y_i = \inf Y_j)$ largest for $i = 3$ and smallest for $i = 1$.
  Examples
  1. For dependent random variables,

$$\begin{aligned} P(Y_1 = 1, Y_2 = 2, Y_3 = 3) &= 0.25 \\ P(Y_1 = 2, Y_2 = 1, Y_3 = 3) &= 0.35 \\ P(Y_1 = 2, Y_2 = 3, Y_3 = 1) &= 0.40 \end{aligned} \tag{B.45}$$

  2. For independent random variables,

$$\begin{aligned} &P(Y_1 = 4) = 1 \\ &P(Y_2 = 1) = 0.22,\ P(Y_2 = 3) = 0.22,\ P(Y_2 = 5) = 0.56 \\ &P(Y_3 = 2) = 0.49,\ P(Y_3 = 6) = 0.51 \end{aligned} \tag{B.46}$$

□

Simpson's paradox It is possible to have

$$P(E_1|E_2) < P(E_1|E_3) \tag{B.47}$$

while, at the same time, both

$$\begin{aligned} P(E_1|E_2 \cup E_4) \geq P(E_1|E_3 \cup E_4) \\ P(E_1|E_2 \cup E_5) \geq P(E_1|E_3 \cup E_5) \end{aligned} \tag{B.48}$$

(Simpson, 1951)

Conditional densities Great care must be taken in the use of conditional densities, as in Equation (B.35), and conditional expectations, as in Equation (B.36), if Z is continuous.

Example
Consider two independent random variables, Y_1 and Y_2, which each have a uniform distribution on $(-1, 1)$. Then, their joint density is given by

$$f_{Y_1Y_2}(y_1, y_2) = \frac{1}{4} \qquad -1 < y_1, y_2 < 1 \tag{B.49}$$

The conditional density at $Y_2 = 0$ is

$$f_{Y_1|y_2}(y_1|Y_2 = 0) = \frac{1}{2} \tag{B.50}$$

Define new variables by the one-to-one transformation

$$\begin{aligned} Z_1 &= Y_1 \\ Z_2 &= \tfrac{Y_2}{1+Y_1^2} \end{aligned} \tag{B.51}$$

with Jacobian $1 + Z_1^2$. The new joint density is

$$f_{Z_1Z_2}(z_1, z_2) = \frac{1+z_1^2}{4} \qquad -1 < z_1 < 1,\ \frac{-1}{1+z_1^2} < z_2 < \frac{1}{1+z_1^2} \tag{B.52}$$

The equivalent conditional statement to that given above has $Z_2 = 0$, which has density

$$f_{Z_1|z_2}(z_1|Z_2 = 0) = K(1 + z_1^2) \tag{B.53}$$

for a constant, K. This is different from the previous result, although they both should have the same logical meaning. (Barnard *et al.*, 1962; see also de Finetti, 1974a, I, pp. 275–276) □

In this example, the two conditional densities impose different units of measurement (Sections 1.2, 2.1.1, and 3.1.1) on the original random variables Y_1 and Y_2.

Dilation of probabilities Conditional probabilities can act in strange and unexpected ways, as a further example will show.

Example
Suppose that a fair coin is tossed twice, yielding the random variables $Y_i, i = 1, 2$, where the flips may not be independent. Our data generating

mechanism can be described by $P_{Y_i}(Y_i = H) = P_{Y_i}(Y_i = T) = \frac{1}{2}$, with $0 \leq P_{Y_1,Y_2}(Y_1 = H, Y_2 = H) \leq \frac{1}{2}$. We toss the coin once. Then,

$$\begin{aligned} 0 &= \inf P_{Y_2|y_1}(Y_2 = H|H) \\ &< \inf P_{Y_2}(Y_2 = H) \\ &= \frac{1}{2} \\ &= \sup P_{Y_2}(Y_2 = H) \\ &< \sup P_{Y_2|y_1}(Y_2 = H|H) \\ &= 1 \end{aligned}$$

with the same relationships if a tail is observed. Merely learning about the first toss, whatever the result, changes our knowledge from a precise statement to complete ignorance about the second toss. This is known as dilation of a set of probabilities. (Seidenfeld and Wasserman, 1993) □

Probabilities of probabilities Let Y be an observable proposition. Now, $P(Y = y) = \pi$ is an unobservable proposition. What is the meaning of the proposition $P[P(Y = y) = \pi] = \pi^*$, which is a meta-probability? For this can lead to statements such as

The proposition asserted in this sentence has probability zero.

This is logically certain, but has probability zero. It is an example of Russell's paradox (Barnard, 1949).

B.3.8 STATISTICS

Suppose that we have n i.i.d. random variables, $Y_1, \ldots, Y_n$, with joint probability law, $P_{Y_1,\ldots,Y_n}(y_1, \ldots, y_n)$, corresponding to the probability measure, P. Let $\mathbf{T}(Y_1, \ldots, Y_n)$ be a real-valued measurable function of the n random variables, called a *statistic*, with corresponding probability law, $P_T(\mathbf{t})$. It defines a partition of the sample space into equivalence classes which are mutually exclusive and collectively exhaustive.

Now suppose that $P_{0Y_1,\ldots,Y_n}(y_1, \ldots, y_n)$ and $P_{0T}(\mathbf{t})$ are a second pair of probability laws corresponding to a different arbitrary measure, P_0, on the same space. The following identity is often useful:

$$\frac{dP_T(\mathbf{t})}{dP_{0T}(\mathbf{t})} = \mathrm{E}\left[\left. \frac{dP_{Y_1,\ldots,Y_n}(y_1, \ldots, y_n)}{dP_{0Y_1,\ldots,Y_n}(y_1, \ldots, y_n)} \right| \mathbf{t} \right] \tag{B.54}$$

if $dP_{Y_1,\ldots,Y_n}(y_1, \ldots, y_n) > 0$ for all possible measures P (including P_0) (Barndorff-Nielsen, 1983, p. 350).

B.4 Martingales

B.4.1 FILTRATIONS

Intuitively, a sequence of random variables, $\{M_t\}$, is said to be a *martingale* with respect to a sequence of random variables, $\{Z_t\}$, the history, if

$$\mathrm{E}[M_t|Z_0, \ldots, Z_{t-1}] = M_{t-1} \tag{B.55}$$

where, of course, M_t must be a function of $Z_0, \ldots, Z_t$. When we work with such sequences, instead of having a product of probabilities of independent random variables, we have a product of conditional probabilities, as in Equation (1.12). Let us now look at these ideas more rigorously.

$(\Omega, \mathcal{F}, \{\mathcal{F}_t\}, P)$ is a *filtered* space if $\{\mathcal{F}_t : t \geq 0\}$, the *filtration*, is an increasing family of sub-σ-algebras of $\mathcal{F}$:

$$\mathcal{F}_0 \subseteq \mathcal{F}_1 \subseteq \cdots \subseteq \mathcal{F} \tag{B.56}$$

Here, the sequence usually refers to time. Then, the information available about ω at time t consists of the values of $Z(\omega)$ for all $\mathcal{F}$-measurable functions Z. For the *natural filtration* of a process, $Z = \{Z_t : t \geq 0\}$,

$$\mathcal{F}_t = \sigma(Z_0, Z_1, \ldots, Z_t) \tag{B.57}$$

the information available at time t is

$$Z_0(\omega), Z_1(\omega), \ldots, Z_t(\omega) \tag{B.58}$$

It is the history of the process.

For a process, $M = \{M_t : t \geq 0\}$, if M_t is $\mathcal{F}_t$-measurable, the process is *adapted* to the filtration. If M is adapted, $M_t(\omega)$ is known at time t. Usually,

$$M_t = g_t(Z_0, Z_1, \cdots, Z_t) \tag{B.59}$$

for some $\mathcal{B}^{t+1}$-measurable function g_t on $\mathbf{R}^{t+1}$. If the filtration to which the process is adapted is generated by the process itself, it is called *self-exciting*.

B.4.2 MARTINGALES

A process, M, is called a martingale if

(1) M is adapted,
(2) $\mathrm{E}[|M_t|] < \infty \quad \forall t$,
(3) $\mathrm{E}[M_t|\mathcal{F}_{t-1}] = M_{t-1}, \quad t \geq 1$.

A martingale is a kind of pure random noise process. It is *square integrable* if $\sup_t \mathrm{E}[M_t^2] < \infty$.

A *supermartingale* has $\mathrm{E}[M_t|\mathcal{F}_{t-1}] \leq M_{t-1}$, while a *submartingale* has $\mathrm{E}[M_t|\mathcal{F}_{t-1}] \geq M_{t-1}$. A supermartingale decreases on average and a submartingale increases.

Examples
1. Let $\{\mathcal{F}_t\}$ be a filtration and Z a random variable. Set

$$Z_t = \mathrm{E}[Z|\mathcal{F}_t] \tag{B.60}$$

Then

$$\begin{aligned}\mathrm{E}[Z_t|\mathcal{F}_{t-1}] &= \mathrm{E}[\mathrm{E}[Z|\mathcal{F}_t]|\mathcal{F}_{t-1}] \\ &= Z_{t-1}\end{aligned} \tag{B.61}$$

so that $\{Z_t\}$ is a martingale if $\mathrm{E}[Z] < \infty$, a submartingale if $\mathrm{E}[Z \vee 0] < \infty$, and a supermartingale if $\mathrm{E}[Z \wedge 0] > -\infty$. This is called a Lévy martingale.
2. Let $\{Z_k\}$ be a sequence of independent random variables with $\mathrm{E}[Z_k] = \mu_k\ \forall k$. If $M_t = Z_1 + \cdots + Z_t$, with $\mathcal{F}_t$ its σ-algebra, for $t \geq 1$

$$\begin{aligned}\mathrm{E}[M_t|\mathcal{F}_{t-1}] &= \mathrm{E}[M_{t-1}|\mathcal{F}_{t-1}] + \mathrm{E}[Z_t|\mathcal{F}_{t-1}] \\ &= M_{t-1} + \mathrm{E}[Z_t] \\ &= M_{t-1} + \mu_k\end{aligned}$$

so that $\{M_t\}$ is a martingale if $\mu_t = 0\ \forall t$, a submartingale if $\mu_t \geq 0$, and a supermartingale if $\mu_t \leq 0\ \forall t$.
3. Let $\{Z_k\}$ be a sequence of independent non-negative random variables with $\mathrm{E}[Z_k] = \mu_k\ \forall k$. If $M_t = Z_1 \cdots Z_t$, with $\mathcal{F}_t$ its σ-algebra, for $t \geq 1$

$$\begin{aligned}\mathrm{E}[M_t|\mathcal{F}_{t-1}] &= \mathrm{E}[M_{t-1}Z_t|\mathcal{F}_{t-1}] \\ &= M_{t-1}\mathrm{E}[Z_t|\mathcal{F}_{t-1}] \\ &= M_{t-1}\mu_k\end{aligned}$$

so that $\{M_t\}$ is a martingale if $\mu_t = 1$, a submartingale if $\mu_t \geq 1$, and a supermartingale if $\mu_t \leq 1\ \forall t$. This is called a Wald martingale. □

B.4.3 STOPPING TIMES

A map $U : \Omega \to \{0, 1, \ldots, \infty\}$ is a *stopping time* if

$$\{U \leq t\} = \{\omega : U(\omega) \leq t\} \in \mathcal{F}_t \quad \forall t \leq \infty \tag{B.62}$$

A stopping time after t depends only on the history up to and including time t: $\{U = t\} \in \mathcal{F}_t$. If M is a martingale and U a stopping time, the *stopped martingale*, $M^U = \{M_{U\wedge t} : t \geq 0\}$, is also a martingale. A stopping time is *optional* if there exists a constant c such that $\{M_{U\wedge t} : t \geq c\}$ is integrable. Thus, by the *optional stopping theorem*, $\lim_{t\to\infty} M_{U\wedge t} = M_U$, while $\mathrm{E}[M_u] = M_0$ if $P(U < \infty) = 1$ and $\lim_{t\to\infty} \mathrm{E}[M_t I(U \geq t)] = 0$.

B.4.4 TRANSFORMS AND THE DOOB DECOMPOSITION

A process $C = \{C_t : t \in \mathbf{N}\}$ is *predictable* or *previsible* if C_{t+1} is $\mathcal{F}_t$-measurable for $t \geq 1$. In other words, at time t with the present information, we can predict the future value, C_{t+1}, with certainty. Then, the *martingale transform* of M by C is

$$T_t = \sum_{k=1}^{t} C_k(M_k - M_{k-1}) \tag{B.63}$$

the discrete analogue of the stochastic integral, $\int C dM$. Note that $T_0 = 0$ and

$$T_t - T_{t-1} = C_t(M_t - M_{t-1}) \tag{B.64}$$

If C is a bounded predictable process and M is a martingale, the resulting martingale transform is a martingale, null at 0.

If $Z = \{Z_t : t \geq 0\}$ is an adapted process with $Z_t \in \mathcal{L}^1$, then Z has a unique *Doob decomposition*

$$Z_t = M_t + A_t \tag{B.65}$$

where M is a martingale,

$$M_t = Z_0 + \sum_{k=1}^{t}(Z_k - \mathrm{E}[Z_k|\mathcal{F}_{k-1}]) \tag{B.66}$$

called the Doob martingale, and A is a predictable process, null at 0, with

$$A_t = \sum_{k=1}^{t} \mathrm{E}[Z_k - Z_{k-1}|\mathcal{F}_{k-1}] \tag{B.67}$$

A is called the *predictor* or *compensator* of the process Z because $Z - A$ is a martingale. All sub- and supermartingales have compensators.

If $\mathcal{F}_t$ is the filtration of the adapted process, Z, then

$$\begin{aligned} dM_t &= Z_t - \mathrm{E}[Z_t|\mathcal{F}_{t-1}] \\ &= Z_t - Z_{t-1} \end{aligned} \tag{B.68}$$

is called a *martingale difference*, because $\mathrm{E}[dM_t|\mathcal{F}_{t-1}] = 0$. Then, the process $M = \{M_t : t \geq 0\}$ given by

$$M_t = \sum_{k=1}^{t} dM_k \tag{B.69}$$

is a martingale, as is

$$M_t = \sum_{k=1}^{t} dM_k C_k \tag{B.70}$$

where $\{C_t\}$ is predictable. One possibility is $C_t = I(U \geq t)$ where U is a stopping time, so that

$$M_t = \sum_{k=1}^{t \wedge U} dM_k \tag{B.71}$$

Example
Consider i.i.d. discrete failure times, $Y_1, \ldots, Y_n$. The series $N = \{N_t : t \geq 0\}$, with $N_t = \sum_{k=1}^{t} dN_k$, where $dN_k = \sum_{i=1}^{n} I_{Y_i=k}$ is the number of failure times equal to k, is called a *counting process.* The conditional distribution of $dN_k|\mathcal{F}_{k-1}$ is binomial with probability π_k and index $Y_k = n - N_{k-1}$, the number still at risk of failure at time k. Thus, $\mathrm{E}[dN_k|\mathcal{F}_{k-1}] = Y_k\pi_k$, so that $dM_k = dN_k - Y_k\pi_k$ are martingale differences and

$$\begin{aligned} M_t &= \sum_{k=1}^{t} dM_k \\ &= N_t - \sum_{k=1}^{t} Y_k\pi_k \end{aligned} \tag{B.72}$$

is a martingale. □

B.5 Generating functions

B.5.1 PROBABILITY GENERATING FUNCTION

For discrete, non-negative random variables, the *probability generating function* is

$$\begin{aligned} G_Y(u) &\stackrel{\text{def}}{=} \mathrm{E}[u^Y] \\ &= \sum_y u^y P_Y(Y = y) \end{aligned} \tag{B.73}$$

$F(y)$ may be reconstructed from $G_Y(u)$. If Y and Z are independent random variables,

$$G_{Y+Z}(u) = G_Y(u)G_Z(u) \tag{B.74}$$

B.5.2 MOMENT GENERATING FUNCTION

The *moment generating function* of a random variable, Y, is the map $M_Y : \mathbf{R} \to \mathbf{R}$ such that

$$M_Y(u) \stackrel{\text{def}}{=} \mathrm{E}[\mathrm{e}^{uY}] \tag{B.75}$$

It is the Laplace transform of the probability measure and only exists if $Y \in \mathcal{L}^r \; \forall r$, i.e. $Y \in \mathcal{L}^\infty$. The ith derivative of $M_Y(u)$ with respect to u is

$$M_Y^{(i)}(u) = \mathrm{E}[Y^i \mathrm{e}^{uY}] \tag{B.76}$$

Hence

$$\begin{aligned} M_Y^{(i)}(0) &= \mathrm{E}[Y^i] \\ &= \mu_i \end{aligned} \tag{B.77}$$

say, the ith moment of Y. (We usually write μ instead of μ_1.)

By Taylor series expansion about $u = 0$,

$$M_Y(u) = 1 + \mu_1 u + \mu_2 \frac{u^2}{2!} + \cdots + \mu_i \frac{u^i}{i!} + \cdots \tag{B.78}$$

For discrete, non-negative random variables,

$$M_Y(u) = G_Y(\mathrm{e}^u) \tag{B.79}$$

and

$$G_Y(u) = M_Y[\log(u)] \tag{B.80}$$

$F(y)$ may be reconstructed from $M_Y(u)$. If Y and Z are independent random variables,

$$M_{Y+Z}(u) = M_Y(u)M_Z(u) \tag{B.81}$$

B.5.3 CUMULANT GENERATING FUNCTION

The *cumulant generating function* is the logarithm of the moment generating function:

$$K_Y(u) \stackrel{\text{def}}{=} \log[M_Y(u)] \tag{B.82}$$

The ith derivative of $K_Y(u)$ at $u = 0$ is the ith cumulant of Y, usually written κ_i. In particular,

$$\begin{aligned} \kappa_1 &= \mu_1 && \text{(B.83)} \\ &= \mathrm{E}[Y] \\ \kappa_2 &= \mu_2 - \mu_1^2 && \text{(B.84)} \\ &= \mathrm{var}[Y] \end{aligned}$$

B.5.4 CHARACTERISTIC FUNCTION

The *characteristic function* of a random variable, Y, is the map $C_Y : \mathbf{R} \to \mathbf{C}$ such that

$$\begin{aligned} C_Y(u) &\stackrel{\text{def}}{=} \mathrm{E}[\mathrm{e}^{iuY}] && \text{(B.85)} \\ &= \mathrm{E}[\cos(uY)] + i\mathrm{E}[\sin(uY)] \end{aligned}$$

for real u. It is the Fourier–Stieltjes transform of the distribution function, $F(y)$:

$$C_Y(u) \stackrel{\text{def}}{=} \int_{\mathrm{R}} \mathrm{e}^{iuv} dF(v) \tag{B.86}$$

It always exists and is, thus, often used instead of the moment generating function. We have

$$\begin{aligned} C_Y(0) &= 1 && \text{(B.87)} \\ |C_Y(u)| &\leq 1 && \text{(B.88)} \\ C_Y^{(n)}(u) &= \mathrm{E}[(iY)^n \mathrm{e}^{iuY}] && \text{(B.89)} \end{aligned}$$

If $\int_{\mathrm{R}} |C_Y(u)| du < \infty$, $F(y)$ may be reconstructed from $C_Y(u)$ by inversion of the Fourier transform,

$$f_Y(y) = \frac{1}{2\pi} \int_{\mathrm{R}} \mathrm{e}^{-iuy} C_Y(u) du \tag{B.90}$$

if a density exists, or else by

$$F(y_1) - F(y_2) = \frac{1}{2\pi} \int_{-\infty}^{\infty} \frac{\mathrm{e}^{-iuy_1} - \mathrm{e}^{-iuy_2}}{-iu} C_Y(u) du \tag{B.91}$$

If Y and Z are independent random variables,

$$C_{Y+Z}(u) = C_Y(u) C_Z(u) \tag{B.92}$$

B.6 Asymptotic results

B.6.1 USEFUL INEQUALITIES

For any non-negative function, $h(\cdot)$,

$$cP_Y(h(Y) \geq c) \leq \mathrm{E}[h(Y)] \qquad \forall c > 0 \tag{B.93}$$

For *Markov's inequality*, set $h(Y) = |Y|$ to obtain

$$cP_Y(|Y| \geq c) \leq \mathrm{E}[|Y|] \qquad \forall c > 0 \tag{B.94}$$

For $Y \in \mathcal{L}^2$, set $h(Y) = Y^2$ to obtain *Chebyshev's inequality*:

$$c^2 P_Y(|Y| \geq c) \leq \mathrm{E}[Y^2] \qquad \forall c > 0 \tag{B.95}$$

or equivalently

$$c^2 P_Y(|Y - \mathrm{E}[Y]| \geq c) \leq \mathrm{var}[Y]$$

For any martingale, $\{M_t\}$,

$$cP\left(\sup_t(|M_t|) \geq c\right) \leq \sup_t(\mathrm{E}[|M_t|]) \qquad \forall c > 0 \tag{B.96}$$

is called the *maximal inequality*, and, if $M_0 = 0$, from Equation (B.95),

$$P(|M_t| < c; t = 1, 2, \ldots, n) \geq 1 - \frac{\mathrm{E}[M_t^2]}{c^2} \qquad \forall c > 0 \tag{B.97}$$

is *Kolmogorov's inequality*.

Hölder's inequality states that, for $a, b > 1$ and $\frac{1}{a} + \frac{1}{b} = 1$,

$$\mathrm{E}[|YZ|] \leq \mathrm{E}[|Y^a|]^{\frac{1}{a}}\mathrm{E}[|Z^b|]^{\frac{1}{b}} \tag{B.98}$$

The special case, the *Cauchy–Schwarz inequality*, with $a = b = 2$, states that

$$(\mathrm{cov}[Y, Z])^2 \leq \mathrm{var}[Y]\mathrm{var}[Z] \tag{B.99}$$

with equality only if Y and Z are linear functions of each other.

For $h(\cdot)$ a continuous convex function of a single real variable, *Jensen's inequality* states that

$$h(\mathrm{E}[Y]) \leq \mathrm{E}[h(Y)] \tag{B.100}$$

B.6.2 MODES OF CONVERGENCE

Almost sure convergence: $Y_n \to Y$ almost surely, almost everywhere, or *with probability one*, if

$$P_Y(\lim_{n\to\infty} Y_n = Y) = 1 \tag{B.101}$$

$\mathcal{L}^r$ convergence: $Y_n \to Y$ in $\mathcal{L}^r$ if each Y_n is in $\mathcal{L}^r$ and $Y \in \mathcal{L}^r$ and

$$\lim_{n\to\infty} \mathrm{E}[|Y_n - Y|^r] = 0 \tag{B.102}$$

also called *norm* or rth *mean convergence.*

Convergence in probability: $Y_n \to Y$ in probability (or in measure) if, for every $\varepsilon > 0$,

$$\lim_{n\to\infty} P_Y(|Y_n - Y| > \varepsilon) = 0 \tag{B.103}$$

This may also be written

$$\operatorname*{plim}_{n\to\infty} Y_n = Y \tag{B.104}$$

or $Y_n \xrightarrow{p} Y$.

Convergence in distribution: This can be expressed in several equivalent ways:

$$\begin{aligned}\lim_{n\to\infty} P_Y(Y_n \le y) &= P_Y(Y \le y)\\ \lim_{n\to\infty} \mathrm{C}_{Y_n}(u) &= \mathrm{C}_Y(u)\\ \lim_{n\to\infty} \mathrm{E}[h(Y_n)] &= \mathrm{E}[h(Y)]\end{aligned} \tag{B.105}$$

for any scalar-valued, bounded, continuous function, $h(\cdot)$. This is often written $Y_n \xrightarrow{D} Y$.

These modes are related in the following ways:

$$(Y_n \to Y \text{ in } \mathcal{L}^r) \Rightarrow (Y_n \to Y \text{ in } \mathcal{L}^s) \quad r > s \ge 1 \tag{B.106}$$

$$(Y_n \to Y \text{ in } \mathcal{L}^r) \Rightarrow (Y_n \to Y \text{ in probability}) \; r \ge 1 \tag{B.107}$$

$$(Y_n \to Y \text{ almost surely}) \Rightarrow (Y_n \to Y \text{ in probability}) \tag{B.108}$$

$$(Y_n \to Y \text{ in probability}) \Rightarrow (Y_n \to Y \text{ in distribution}) \tag{B.109}$$

and no others. Thus, all other modes of convergence imply convergence in distribution, explaining why the latter is often called *weak convergence.*

B.6.3 ASYMPTOTIC LAWS

Laws of large numbers

Let $Y_1, Y_2, \ldots$ be i.i.d. random variables with $\mathrm{E}[|Y_k|] < \infty$ and $S_n = Y_1 + \cdots + Y_n$. Then, *Kolmogorov's strong law of large numbers* is

$$\lim_{n\to\infty} \frac{S_n}{n} = \mathrm{E}[Y_k] \text{ almost surely} \tag{B.110}$$

If $\mathrm{E}[Y_k^2] < \infty$, convergence is in $\mathcal{L}^2$. If convergence is only in probability, we have *Chebyshev's weak law of large numbers.*

An important special case of this is the *law of long run frequency.* Let $E_1, E_2, \ldots$ be a sequence of independent events with equal probability, $P(E_i) = \pi$, and with $S_n = \sum_{i=1}^{n} I_{E_i}$ counting the number of occurrences of E_i. Then, S_n/n converges almost surely to π.

Law of the iterated logarithm

If $Y_1, Y_2, \ldots$ are i.i.d. random variables with $\mathrm{E}[Y_k] = 0$ and $\mathrm{var}[Y_k] = 1$, and $S_n = Y_1 + \cdots + Y_n$, then

$$P_Y\left(\limsup_{n\to\infty} \frac{S_n}{\sqrt{2n\log[\log(n)]}}\right) = 1 \tag{B.111}$$

Central limit theorem

Let $Y_1, Y_2, \ldots$ be i.i.d. random variables with $\mathrm{E}[Y_k] = 0$ and $0 < \sigma^2 = \mathrm{var}[Y_k] < \infty$, $S_n = Y_1 + \cdots + Y_n$, and

$$S_n^* = \frac{S_n}{\sigma\sqrt{n}} \tag{B.112}$$

Then, for $y \in \mathbf{R}$, as $n \to \infty$, S_n^* converges in distribution to a standard normal distribution with mean 0 and variance 1, denoted by $\mathrm{N}(0,1)$:

$$\begin{aligned} \lim_{n\to\infty} P_Y(S_n^* \le y) &= \Phi(y) \text{ in distribution} \\ &= \frac{1}{\sqrt{2\pi}} \int_{-\infty}^{y} \mathrm{e}^{-\frac{u^2}{2}}\, du \end{aligned} \tag{B.113}$$

where $\Phi(\cdot)$ represents the *standard normal probability integral.*

For the generalization to the case where the variance is infinite, a stable distribution, see Section 2.7.2.

Martingale convergence theorem

Let $\{M_t\}$ be a martingale. The above laws of large numbers also hold, more weakly, for square integrable martingales. By the *martingale convergence theorem*, there exists a random variable, M_∞, to which the martingale converges in $\mathcal{L}^1$ (in mean),

$$\lim_{t\to\infty} \mathrm{E}[|M_t - M_\infty|] = 0 \tag{B.114}$$

and $\mathrm{E}[M_t] = \mathrm{E}[M_\infty]$ $\forall t$. If the martingale is square integrable, it converges both with probability one and in $\mathcal{L}^2$. If a sequence is a sub- or supermartingale, thc convergence is with probability one,

$$P_M(\lim_{t\to\infty} M_t = M_\infty) = 1 \tag{B.115}$$

Martingale central limit theorem
If $\{M_t\}$ is a square integrable martingale and S_t is a sum of martingale differences with zero expectation, such as $S_t = \sum_{i=1}^{t}(M_i - M_\infty)$, then, under mild conditions, S_t converges in distribution to a normal distribution, $\mathrm{N}(0, \sigma^2)$, where $0 < \sigma^2 < \infty$, the *martingale central limit theorem.*

B.6.4 O AND o NOTATION

If

$$\limsup_{t\to L} \left| \frac{f(t)}{g(t)} \right| < \infty \tag{B.116}$$

we say

$$f(t) = \mathrm{O}(g(t)) \text{ as } t \to L \tag{B.117}$$

When this is convergence in probability, we write $\mathrm{O}_p(\cdot)$. If $f_n = \mathrm{O}_p(1)$, it is said to be bounded in probability.

If

$$\lim_{t\to L} \frac{f(t)}{g(t)} = 0 \tag{B.118}$$

we say

$$f(t) = \mathrm{o}(g(t)) \text{ as } t \to L \tag{B.119}$$

When this is convergence in probability, we write $\mathrm{o}_p(\cdot)$.

Further reading
For an excellent introduction to probability theory from the perspective presented here, see Williams (1991). For more details on measure theory in relation to probability, the reader may also like to consult Doob (1994) and Hoffmann-Jørgensen (1994).

Averages are central to statistical modelling. Indeed, the operation of taking an average, unfortunately called an expected value, provides a dual to probability theory. For an illuminating account, see Whittle (1992).

Examples of the modern use of such probability theory in statistics abound in the applications of stochastic processes and in Bayesian theory. Two modern texts on the former are Fleming and Harrington (1991) and Andersen *et al.* (1993). For the latter, see Chapter 8.

Appendix C
Normal distribution statistics

C.1 Chi-squared distribution

If the random variables U_i have independent *standard normal distributions*, $N(0,1)$, with mean 0 and variance 1, then the statistic, U_i^2, has a *Chi-squared distribution*, χ_1^2, with one degree of freedom, and a sum of them,

$$\begin{aligned} Z_p &= \sum_{i=1}^{p} U_i^2 \\ &\sim \chi_p^2 \end{aligned} \tag{C.1}$$

a Chi-squared distribution with p degrees of freedom (Pearson, 1900).

We may proceed (approximately) by standardizing some observable random variables, Y_i, so that we have

$$U_i = \frac{Y_i - \hat{\mu}_i}{\sigma} \tag{C.2}$$

and

$$Z = \frac{\sum (Y_i - \hat{\mu}_i)^2}{\sigma^2} \tag{C.3}$$

where σ^2 may not be known.

The Chi-squared distribution with p degrees of freedom has density

$$f(z) = \frac{z^{\frac{p}{2}-1} e^{-\frac{z}{2}}}{\Gamma\left(\frac{p}{2}\right) 2^{\frac{p}{2}}} \tag{C.4}$$

It is plotted for several degrees of freedom in Figure C.1. This distribution is a special case of the gamma distribution of Equation (1.3), with $\mu/\alpha = 2$ and $\alpha = p/2$. It has $E[Z_p] = p$, $\text{var}[Z_p] = 2p$. For p large, $\chi_p^2 \doteq N(p, 2p)$.

C.2 Student t distribution

Suppose that $U \sim N(0,1)$ and $V \sim \chi_p^2$, with U and V being independent random variables. Then, the statistic

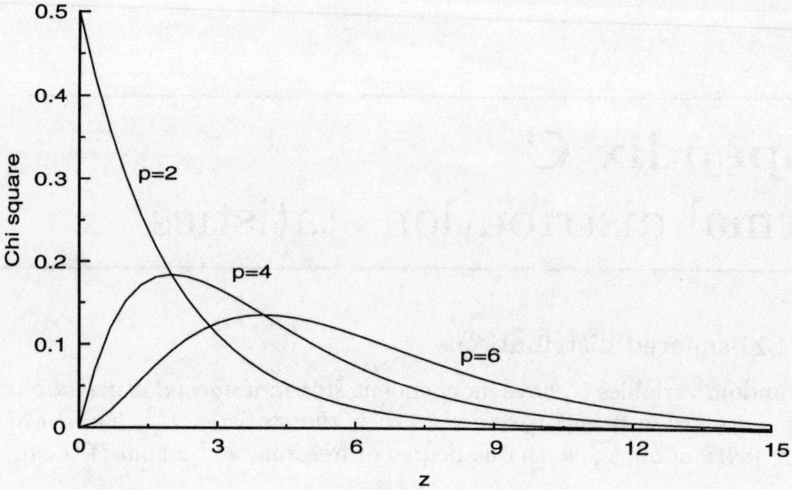

Fig. C.1. Chi-squared distribution for three different degrees of freedom.

$$Z = \frac{U}{\sqrt{V/p}}$$
$$\sim \mathrm{t}_p \tag{C.5}$$

a *Student t distribution* with p degrees of freedom. (Student, 1908)

Often, we may have

$$U = \frac{\bar{Y} - \mu}{\sigma} \tag{C.6}$$

and

$$V = \frac{\sum(Y_i - \hat{\mu}_i)^2}{\sigma^2} \tag{C.7}$$

so that

$$Z = \frac{\sqrt{p}(\bar{Y} - \mu)}{\sqrt{\sum(Y_i - \hat{\mu}_i)^2}} \tag{C.8}$$

which is not a function of the unknown σ^2. In the same way, the transformed random variable,

$$Z_i = \frac{\sqrt{p}(Y_i - \mu)}{\sqrt{\sum(Y_i - \hat{\mu}_i)^2}} \tag{C.9}$$

is said to be *studentized.*

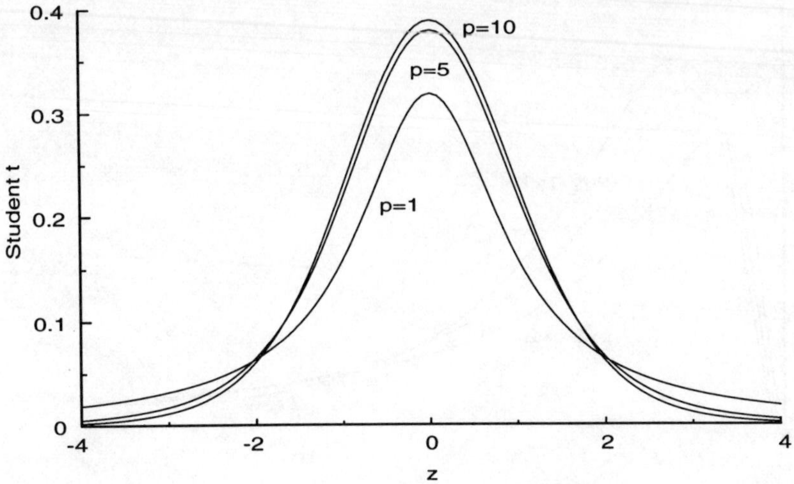

Fig. C.2. Student t distribution plotted for three different degrees of freedom.

The Student t distribution with p degrees of freedom and

$$f(z) = \frac{\Gamma\left(\frac{p+1}{2}\right)}{\sqrt{p\pi}\Gamma\left(\frac{p}{2}\right)}\left(1+\frac{z^2}{p}\right)^{-\frac{p+1}{2}} \tag{C.10}$$

is plotted for several values of p in Figure C.2. We see how it has thicker tails than a normal distribution, but approaches to it as p increases. Asymptotically, we have

$$\lim_{p\to\infty} \mathrm{t}_p = \mathrm{N}(0,1) \tag{C.11}$$

With its heavy tails, the Student t distribution is closely related to the stable family (Section 2.7.2). Thus, one special case of a t distribution, when $p = 1$, is the Cauchy distribution of Equation (2.94).

C.3 F distribution

Suppose that $U \sim \chi^2_p$ and $V \sim \chi^2_q$, with U and V independent random variables. Then, the statistic

$$\begin{aligned} Z &= \frac{U/p}{V/q} \\ &\sim \mathrm{F}_{p,q} \end{aligned} \tag{C.12}$$

an *F distribution* with p and q degrees of freedom (Fisher, 1921, 1924, who used the logarithm of this quantity).

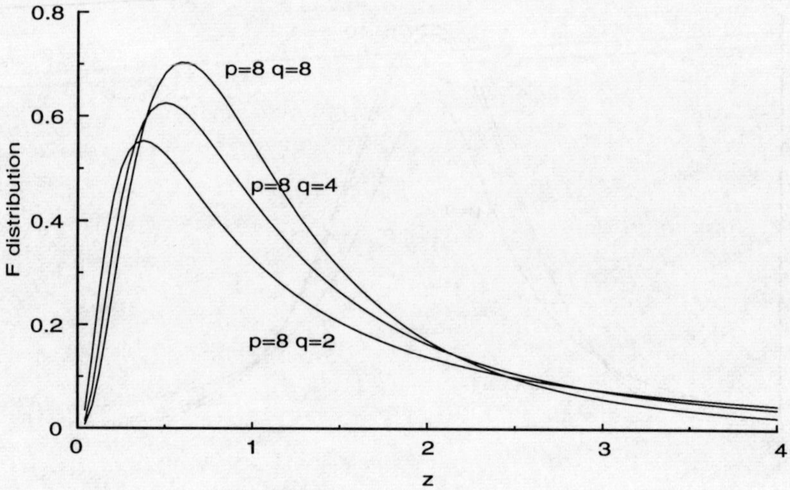

Fig. C.3. F distribution plotted for three different degrees of freedom.

Suppose that

$$U = \frac{\sum(Y_i - \hat{\mu}_{1i})^2}{\sigma^2} \tag{C.13}$$

and

$$V = \frac{\sum(Y_i - \hat{\mu}_{2i})^2}{\sigma^2} \tag{C.14}$$

with $\hat{\mu}_{1i} \neq \hat{\mu}_{2i}$. Then

$$Z = \frac{\sum(Y_i - \hat{\mu}_{1i})^2/p}{\sum(Y_i - \hat{\mu}_{2i})^2/q} \tag{C.15}$$

which is not a function of the unknown σ^2. Note that $\mathrm{t}_q^2 = \mathrm{F}_{1,q}$.

This density is written

$$f(z) = \frac{\left(\frac{p}{q}\right)^{\frac{p}{2}} \Gamma\left(\frac{p+q}{2}\right)}{\Gamma\left(\frac{p}{2}\right)\Gamma\left(\frac{q}{2}\right)} z^{\frac{p}{2}-1} \left(1 + \frac{p}{q}z\right)^{-\frac{p+q}{2}} \tag{C.16}$$

It has been plotted for several degrees of freedom in Figure C.3. The mean and variance are

$$\mathrm{E}[Z] = \frac{q}{q-2} \qquad q > 2 \tag{C.17}$$

$$\mathrm{var}[Z] = \frac{2q^2(p+q-2)}{p(q-2)^2(q-4)} \qquad q > 4 \tag{C.18}$$

Asymptotically, we have

$$\lim_{q\to\infty} \mathrm{F}_{p,q} = \frac{1}{p}\chi_p^2 \tag{C.19}$$

This distribution is a special case of the *beta distribution*, with

$$f(y;\zeta,\gamma) = \frac{y^{\zeta-1}(1-y)^{\gamma-1}}{B(\zeta,\gamma)} \qquad 0 < y < 1 \tag{C.20}$$

where $B(\zeta,\gamma)$ is the beta function and $y^{-1} = 1 + pz/q$, $\gamma = p/2$, and $\zeta = q/2$, with the appropriate Jacobian.

Appendix D
Numerical methods

The two most frequent numerical operations in parametric statistical inference are maximization of a function and averaging or integration. In each case, several fairly general techniques are in wide use and will be presented briefly in this appendix.

D.1 Maximization

The most common use of maximization is to obtain the maximum likelihood estimate. Techniques vary depending on the analytic complexity of the likelihood function. The score equations, $\mathbf{U}(\boldsymbol{\psi}) = \mathbf{0}$, are most often non-linear and cannot be solved explicitly to obtain the maximum likelihood estimates. In such cases, some iterative numerical procedure must be used to obtain the solution.

Parameter transformation (Sections 3.5.2 and 3.6.2 and Chapter 6) to make contours of the likelihood surface more like symmetrical ellipsoids will generally facilitate numerical maximization. We look for

- rapid convergence (few iterations) of non-linear maximization algorithms;
- numerical accuracy at well-determined maxima;
- successful convergence for different starting values.

D.1.1 NEWTON–RAPHSON

A useful general approach is the *Newton–Raphson method*, which uses a quadratic approximation to the function. Start with an initial guess, $\boldsymbol{\psi}_0$, of the parameter values. Then, successive approximations, k, are given by

$$\boldsymbol{\psi}_k = \boldsymbol{\psi}_{k-1} + \mathbf{J}^{-1}(\boldsymbol{\psi}_{k-1})\mathbf{U}(\boldsymbol{\psi}_{k-1}) \tag{D.1}$$

where indices here denote the iterative step and

$$\mathbf{J}(\boldsymbol{\psi}) = -\frac{\partial \mathbf{U}(\boldsymbol{\psi})}{\partial \boldsymbol{\psi}}$$

is the observed information. The procedure is iterated until convergence to the desired degree of accuracy.

If the observed information is replaced by the expected information in the Newton–Raphson method, it is called the *method of scoring.* The Newton–Raphson and scoring methods are identical for the canonical representation in the linear exponential family, because the observed and expected information are equal.

For generalized linear models, the Newton–Raphson equation

$$\boldsymbol{\beta}_k = \boldsymbol{\beta}_{k-1} + \mathbf{J}^{-1}(\boldsymbol{\beta}_{k-1})\mathbf{U}(\boldsymbol{\beta}_{k-1}) \tag{D.2}$$

can be rewritten

$$\mathbf{J}(\boldsymbol{\beta}_{k-1})\boldsymbol{\beta}_k = \mathbf{J}(\boldsymbol{\beta}_{k-1})\boldsymbol{\beta}_{k-1} + \mathbf{U}(\boldsymbol{\beta}_{k-1}) \tag{D.3}$$

or

$$\mathbf{X}^{\mathrm{T}}\mathbf{W}\mathbf{X}\boldsymbol{\beta} = \mathbf{X}^{\mathrm{T}}\mathbf{W}\mathbf{z} \tag{D.4}$$

from Equation (3.96), where

$$z_i = \sum_j x_{ij}\beta_j + [t(y_i) - \mu_i]\left(\frac{\partial \eta_i}{\partial \mu_i}\right) \tag{D.5}$$

This has the form of the score equations for a weighted standard normal linear regression model using least squares, except that $\mathbf{W}$ and $\mathbf{z}$ usually depend on $\boldsymbol{\beta}$ so that it must be solved iteratively. This is known as *iterative weighted least squares* (IWLS).

D.1.2 DERIVATIVE-FREE

By derivative-free, I mean that the derivatives of the function to be maximized need not be supplied analytically.

Finite differences We use the Newton–Raphson method, but calculate the score and information numerically by

$$\mathrm{U}_{\psi_i}(\boldsymbol{\psi}) = \frac{\mathrm{l}(\boldsymbol{\psi} + h_i\mathbf{e}_i) - \mathrm{l}(\boldsymbol{\psi} - h_i\mathbf{e}_i)}{2h_i} \tag{D.6}$$

$$\mathrm{J}_{\psi_i\psi_i}(\boldsymbol{\psi}) = \frac{\mathrm{l}(\boldsymbol{\psi} + h_i\mathbf{e}_i + h_j\mathbf{e}_j) - \mathrm{l}(\boldsymbol{\psi} - h_i\mathbf{e}_i) - \mathrm{l}(\boldsymbol{\psi} - h_j\mathbf{e}_j) + \mathrm{l}(\boldsymbol{\psi})}{h_i h_j}$$

where h_i is the finite difference step-size and $\mathbf{e}_i$ is the ith unit vector. Choices of starting values and step-size are usually critical, especially in non-linear problems.

Nelder–Mead simplex A simplex is a p-dimensional geometrical figure consisting of $p+1$ vertices and all of the line segments connecting them. To start the method, we supply $p+1$ points, defining such a simplex. The algorithm takes a series of steps, usually just moving the point of the simplex where the function is smallest through the opposite face to a higher point. These steps are constructed to preserve the volume of the simplex and continue until some maximum is reached. This algorithm is generally rather slow.

D.1.3 EM ALGORITHM

Many complex statistical models can be set up as if they were describing an incomplete data situation (Section 6.4.2), where a model for the latter would be simpler. The basic idea of this method is very simple:

(1) replace the missing values by estimates (fitted values);
(2) estimate the parameters;
(3) recalculate the missing values in the light of the new parameter estimates;
(4) repeat until convergence.

Let $\mathbf{y}_{obs}$ and $\mathbf{y}_{miss}$ represent, respectively, the observed and unobservable data, so that the hypothetical complete data set would be $(\mathbf{y}_{obs}, \mathbf{y}_{miss})$. The model of interest, then, has likelihood function

$$\begin{aligned} \mathrm{L}(\psi) &= f(\mathbf{y}_{obs}; \psi) \\ &= \int_{\mathcal{Y}_{miss}} f(\mathbf{y}_{obs}, \mathbf{u}; \psi) d\mathbf{u} \end{aligned} \quad \text{(D.7)}$$

However, from Equation (6.68),

$$f(\mathbf{y}_{obs}; \psi) = \frac{f(\mathbf{y}_{obs}, \mathbf{y}_{miss}; \psi)}{f(\mathbf{y}_{miss}|\mathbf{y}_{obs}; \psi)} \quad \text{(D.8)}$$

and the expected value of the log of this with respect to the conditional density, $f(\mathbf{y}_{miss}|\mathbf{y}_{obs}; \psi_0)$, for some fixed value, ψ_0,

$$\begin{aligned} &\int_{\mathcal{Y}_{miss}} \log[f(\mathbf{y}_{obs}, \mathbf{u}; \psi)] f(\mathbf{u}|\mathbf{y}_{obs}; \psi_0) d\mathbf{u} \\ &\quad - \int_{\mathcal{Y}_{miss}} \log[f(\mathbf{u}|\mathbf{y}_{obs}; \psi)] f(\mathbf{u}|\mathbf{y}_{obs}; \psi_0) d\mathbf{u} \end{aligned} \quad \text{(D.9)}$$

often involves relatively simple integrals, especially for members of the exponential family, where the first reduces to the expectation of sufficient statistics.

The EM algorithm consists of the iterative repetition of two steps:

(1) Expectation – compute (D.9) for fixed ψ_0, which, in the linear exponential family, amounts to estimating the sufficient statistics for the complete data.

(2) Maximization – obtain a new ψ_0 by maximizing the first integral of (D.9) over ψ (the change in the second integral will be negative, so that it need not be evaluated). If maximization is difficult, any value which increases this integral can be used.

This procedure has the desirable properties that it can yield the maximum likelihood estimate and that the log likelihood never decreases between two iterations. However, it is usually fairly slow to converge and, with a poor starting value, can converge to a local maximum.

D.1.4 SIMULATED ANNEALING

When the temperature of a metal is lowered slowly in steps, the cooling process is called annealing. At each stage, thermal equilibrium is attained. Molecules lose mobility at lower temperatures and tend to line up in a rigid structure, a state of minimal energy. If temperature is reduced too quickly, an intermediate amorphous state may instead be reached. Function maximization is analogous to this search for the state of minimal energy.

From initial parameter values, and their corresponding function value, $f = f_0$, the following steps are iterated starting with an initial temperature, say $T = T_0$, sufficiently large for any point in the parameter space to be visited with reasonable probability.

(1) Choose a random point in the parameter space and evaluate the function. For example, only one coordinate might be changed to a new random value.

(2) Move to the new point iff the random variable, Z, with uniform distribution on $(0, 1)$, is such that

$$z \leq \exp\left(-\frac{f_i - f_{i+1}}{T}\right) \tag{D.10}$$

where T is the current temperature. In other words, the change in function value is compared to an exponential random variable with mean T. The move is always made if the new point is higher; otherwise there is a finite probability of moving lower.

(3) Iterate the first two points until some criterion of equilibrium is met, counting the number of moves. One possibility is simply to allow a fixed number of iterations, say N, large enough for equilibrium to be reached.

(4) Then, reduce the temperature to ρT $(0 < \rho < 1)$ and repeat until meeting some stopping criterion, such as no moves having been made.

Large N gives a more accurate solution and large ρ more reliability in reaching the maximum. For complex (likelihood) functions with local maxima, this method will be more reliable than those described above.

D.2 Integration

Integration is required in parametric statistical inference to calculate tail areas for the P-values of significance tests and for posterior probabilities, as well as in random effects models. When the integral cannot be solved analytically, the three principal techniques are asymptotic approximations (Section 5.7), quadrature, and sampling. The latter two numerical integration techniques are briefly outlined below.

D.2.1 QUADRATURE

Many numerical techniques approximate the integral by a discrete sum,

$$\int f(u)du \doteq \sum w_i f(y_i) \tag{D.11}$$

where the points, y_i, are called nodes and the w_i, weights.

For example, the Gauss–Hermite integration rule approximates integrals of the form

$$\int_{-\infty}^{\infty} \mathrm{e}^{-u^2} f(u)du \tag{D.12}$$

by taking the normal density as a weight function and the zeros of a Hermite polynomial (see Section 5.7.2) as nodes. Abramowitz and Stegun (1965, p. 924) provide tables of nodes and weights. Liu and Pierce (1994) show how this method is related to the Laplace approximation (Section 5.7.1) and suggest transformations which improve the quadrature approximation.

D.2.2 MONTE CARLO

Rejection sampling Often, the density, $f(y)$, is not available, but only some unnormalized function proportional to it, say $g(y) \propto f(y)$. This can occur, for example, if the denominator of Bayes' formula is complex. Suppose that $h(\cdot)$ is a density from which random numbers can be drawn and that $g(y) \leq ch(y)$ for some constant, c. Then, random numbers can be generated from $f(y)$ by drawing numbers from $h(y)$ and accepting them with probability $g(y)/ch(y)$.

The approximate density, $h(y)$, should be roughly proportional to the target $g(y)$, because otherwise c must be very large and most samples will be rejected. Indeed, the rejection rate is a measure of the efficiency of the procedure.

Importance sampling To approximate the expectation, with respect to $f(y)$, of a continuous function, say $g(y)$, the arithmetic mean, $\sum g(y_i)/n$, can be calculated by direct simulation for n drawings from the corresponding density, $f(y)$. A more efficient approximation can often be obtained by using $\sum g(y_i)f(y_i)/\phi(y_i)$ and drawing from $\phi(y_i)$, which is often taken to be a normal density. However, the variance of such an expectation depends critically on the choice of $\phi(y)$; it should have a shape as similar to $f(y)$ as possible.

If $\phi(y)$ can be chosen such that $g(y)f(y)/\phi(y)$ is fairly constant, then a reasonably precise value of the integral can be obtained. Unfortunately, no reliable methods are available for assessing the accuracy of the approximation produced by this procedure.

Markov chains Suppose that we wish to estimate

$$\mathrm{E}[f] = \sum_{i=0}^{s} f(i)\pi_i \tag{D.13}$$

where $\boldsymbol{\pi}$ is some probability distribution. Let us obtain sample paths from a Markov chain constructed to have equilibrium distribution, $\boldsymbol{\pi}$. Then, choose $\mathbf{P}$, with elements p_{ij}, to be the transition matrix of an irreducible, aperiodic, and not transient Markov chain with $s+1$ states such that

$$P(Y_{t+1} = j|Y_t = i) = p_{ij} \tag{D.14}$$

where Y_t is the state occupied at time t, and such that $\boldsymbol{\pi}$ is its unique stationary distribution, $\boldsymbol{\pi} = \boldsymbol{\pi}\mathbf{P}$. Simulate this Markov chain n times and use the estimate

$$\widehat{\mathrm{E}[f]} = \frac{1}{n}\sum_{i=1}^{n} f(Y_t) \tag{D.15}$$

However, more care must be taken with such Markov chain Monte Carlo (MCMC) methods because observations are dependent, in contrast to the independent values generated by importance sampling. On the other hand, they do have the advantage that they improve at each step whereas importance sampling does not.

In addition, it is usually not possible to start a Markov chain at its equilibrium distribution, which may take some time to reach. The first portion of the sample will usually need to be discarded to reduce the effect of the initial distribution used. Thus, one may make n independent runs of length m, chosen just large enough for the chain to reach equilibrium, and keep the last value of each. At the other extreme, one run of length mn, discarding only a small initial fraction, has the advantages of using much more of the data and being less likely to be affected by the initial values.

Gibbs sampler In physics, the Gibbs distribution describes the equilibrium states of subsystems of very large closed physical systems. It can be written

$$f(\mathbf{y}) = \frac{1}{c}\mathrm{e}^{-U(\mathbf{y})/T} \tag{D.16}$$

where c and T are constants, the former a normalizing constant and the latter the 'temperature'. The smaller is T, the sharper is the mode of the distribution. Note the similarity to Equation (D.10).

An approximation to such a multivariate density, its marginal densities or moments, can be obtained from the corresponding set of p univariate conditional densities in a way closely related to the simulated annealing described above. Starting values of all p variables are selected. Then, a new value is drawn in turn from each conditional density, replacing the corresponding initial value after each step. This is a form of non-stationary Markov chain. The cycle is repeated and, as the iterations proceed, the temperature, T, gradually lowered, again called annealing. We eventually obtain a sample of n multivariate observations which can be smoothed or from which empirical moments can be calculated.

The term, Gibbs sampling, is now generally used for the application of this algorithm to any multivariate distribution, not just the Gibbs.

When the likelihood function is analytically complex, Monte Carlo methods can be applied to generate it as a posterior distribution with a flat prior.

Further reading

Fisher first introduced the Newton–Raphson method for calculating maximum likelihood estimates, now called the Fisher method of scoring, in a number of biological papers in the 1930s.

Hartley (1958) first proposed the EM algorithm, developed by Orchard and Woodbury (1972), and later popularized by Dempster *et al.* (1977); see also Little and Rubin (1987).

Derivative-free optimization techniques are explained in Dennis and Schnabel (1983). For estimation in non-linear models, see Ross (1990).

Brooks and Morgan (1995) provide a good introduction to simulated annealing.

Metropolis and Ulam (1949) introduced Monte Carlo Markov chain methods into statistics. They were further expanded by Hastings (1970) into what is now called the Metropolis–Hastings algorithm. Geman and Geman (1984) developed the special case now called the Gibbs sampler (see also Tierney, 1994).

More generally on numerical techniques, see Press *et al.* (1992). Gelman *et al.* (1995) prove a variety of applications involving many of the techniques discussed in this Appendix.

Bibliography

1. Abramowitz, M. and Stegun, I.A. (1965) *Handbook of Mathematical Functions with Formulas, Graphs, and Mathematical Tables.* New York: Dover.
2. Agresti, A. (1992) A survey of exact inference for contingency tables. *Statistical Science* **7**, 131–177 (with discussion).
3. Aitkin, M. (1991) Posterior Bayes factors. *Journal of the Royal Statistical Society* **B53**, 111–142 (with discussion).
4. Akaike, H. (1973) Information theory and an extension of the maximum likelihood principle. In Petrov, B.N. and Csàki, F., *Second International Symposium on Inference Theory*, Budapest: Akadémiai Kiadó, pp. 267–281.
5. Akaike, H. (1978) A new look at the Bayes procedure. *Biometrika* **65**, 53–59.
6. Akaike, H. (1982) On the fallacy of the likelihood principle. *Statistics and Probability Letters* **1**, 75–78.
7. Altham, P.M.E. (1984) Improving the precision of estimation by fitting a model. *Journal of the Royal Statistical Society* **B46**, 118–119.
8. Altman, D.G. (1991) *Practical Statistics for Medical Research.* London: Chapman and Hall.
9. Andersen, E.B. (1970) Asymptotic properties of conditional maximum-likelihood estimators. *Journal of the Royal Statistical Society* **B32**, 283–301.
10. Andersen, P.K., Borgan, Ø., Gill, R.D., and Keiding, N. (1993) *Statistical Models Based on Counting Processes.* Berlin: Springer Verlag.
11. Anderson-Sprecher, R. (1994) Model comparisons and R^2. *American Statistician* **48**, 113–117.
12. Anscombe, A.J. (1973) Graphs in statistical analysis. *American Statistician* **27**, 17–21.
13. Appleton, D.R. (1995) What do we mean by a statistical model? *Statistics in Medicine* **14**, 185–197.
14. Arnold, B.C. and Strauss, D.J. (1991) Bivariate distributions with conditionals in prescribed exponential families. *Journal of the Royal Statistical Society* **B53**, 365–375.
15. Bar-Lev, S. (1994) A derivation of conditional cumulants in exponential models. *American Statistician* **48**, 126–129.
16. Barnard, G.A. (1946) Sequential tests in industrial statistics. *Journal of the Royal Statistical Society* (Supplement) **8**, 1–26 (with discussion).

17. Barnard, G.A. (1947) A review of "Sequential Analysis" by Abraham Wald. *Journal of the American Statistical Association* **42**, 658–664.
18. Barnard, G.A. (1949) Statistical inference. *Journal of the Royal Statistical Society* **B11**, 115–149 (with discussion).
19. Barnard, G.A. (1954) Sampling inspection and statistical decisions. *Journal of the Royal Statistical Society* **B16**, 151–165 (with discussion).
20. Barnard, G.A. (1965) The use of the likelihood function in statistical practice. *Proceedings of the Fifth Berkeley Symposium*, Volume I, pp. 27–40.
21. Barnard, G.A. (1976) Conditional inference is not inefficient. *Scandinavian Journal of Statistics* **3**, 132–134.
22. Barnard, G.A., Jenkins, G.M., and Winsten, C.B. (1962) Likelihood inference and time series. *Journal of the Royal Statistical Society* **A125**, 321–352 (with discussion).
23. Barndorff-Nielsen, O. (1976) Plausibility inference. *Journal of the Royal Statistical Society* **B38**, 103–131 (with discussion).
24. Barndorff-Nielsen, O. (1978) *Information and Exponential Families in Statistical Theory.* New York: John Wiley.
25. Barndorff-Nielsen, O. (1980) Conditionality resolutions. *Biometrika* **67**, 293–310.
26. Barndorff-Nielsen, O. (1983) On a formula for the distribution of the maximum likelihood estimator. *Biometrika* **70**, 343–365.
27. Barndorff-Nielsen, O. (1986) Inference on full or partial parameters based on the standardized signed log likelihood ratio. *Biometrika* **73**, 307–322.
28. Barndorff-Nielsen, O. (1988) *Parametric Statistical Models and Likelihood.* New York: Springer Verlag.
29. Barndorff-Nielsen, O. (1990) Approximate interval probabilities. *Journal of the Royal Statistical Society* **B52**, 485–496.
30. Barndorff-Nielsen, O. (1994) Adjusted likelihood inference about interest parameters. *Theory of Probability and Applications* **38**, 179–193.
31. Barndorff-Nielsen, O. and Blaesild, P. (1975) S-ancillarity in exponential families. *Sankhyā* **A37**, 354–385.
32. Barndorff-Nielsen, O. and Chamberlin, S.R. (1994) Stable and invariant adjusted directed likelihoods. *Biometrika* **81**, 485–499.
33. Barndorff-Nielsen, O. and Cox, D.R. (1979) Edgeworth and saddle-point approximations with statistical applications. *Journal of the Royal Statistical Society* **B41**, 279–312 (with discussion).
34. Barndorff-Nielsen, O. and Cox, D.R. (1984) Bartlett adjustments to the likelihood ratio statistic and the distribution of the maximum likelihood estimate. *Journal of the Royal Statistical Society* **B46**, 483–495.
35. Barndorff-Nielsen, O. and Cox, D.R. (1989) *Asymptotic Techniques for Use in Statistics.* London: Chapman and Hall.
36. Barndorff-Nielsen, O. and Cox, D.R. (1994) *Inference and Asymptotics.* London: Chapman and Hall.
37. Barndorff-Nielsen, O. and McCullagh, P. (1993) A note on the relation between modified profile likelihood and the Cox-Reid adjusted profile likelihood. *Biometrika* **80**, 321–328.

38. Barnett, V. and Lewis, T. (1984) *Outliers in Statistical Data.* New York: John Wiley.
39. Bartholomew, D.J. (1965) A comparison of some Bayesian and frequentist inferences. *Biometrika* **52**, 19–35.
40. Bartholomew, D.J. (1971) A comparison of Bayesian and frequentist approaches to inferences with prior knowledge. In Godambe, V.P. and Sprott, D.A. (eds.) *Foundations of Statistical Inference.* Toronto: Holt, Rinehart, and Winston, pp. 417–434 (with discussion).
41. Bartholomew, D.J. (1995) What is statistics? *Journal of the Royal Statistical Society* **A158**, 1–20 (with discussion).
42. Bartlett, M.S. (1937) Properties of sufficiency and statistical tests. *Proceedings of the Royal Society, London* **A168**, 268–282.
43. Basawa, I.V. and Scott, D.J. (1983) *Asymptotic Optimal Inference for Nonergodic Models.* New York: Springer Verlag.
44. Basu, D. (1977) On the elimination of nuisance parameters. *Journal of the American Statistical Association* **72**, 355–366, reprinted in Ghosh (1988), pp. 115–141.
45. Bayes, T. (1763) An essay towards solving a problem in the doctrine of chances. *Philosophical Transaction* **53**, 370–418, reprinted in *Biometrika* (1958) **45**, 293–315.
46. Beaton, A.E., Rubin, D.B., and Barone, J.L. (1976) The acceptability of regression solutions: another look at computational accuracy. *Journal of the American Statistical Association* **71**, 158–168.
47. Bennett, J.H. (1990) *Statistical Inference and Analysis: Selected Correspondence of R.A. Fisher.* Oxford: Oxford University Press.
48. Berger, J.O. (1985a) The frequentist viewpoint and conditioning. *Proceedings of the Berkeley Conference in Honor of Jerzy Neyman and Jack Kiefer*, Volume I. Pacific Grove: Wadsworth, pp. 15–44.
49. Berger, J.O. (1985b) *Statistical Decision Theory and Bayesian Analysis.* New York: Springer Verlag.
50. Berger, J.O. (1990) Robust Bayesian analysis: sensitivity to the prior. *Journal of Statistical Planning and Inference* **25**, 303–328.
51. Berger, J.O. and Berliner, L.M. (1986) Robust Bayes and empirical Bayes analysis with ε-contaminated priors. *Annals of Statistics* **14**, 461–486.
52. Berger, J.O. and Bernardo, J.M. (1992) Ordered group reference priors with applications to the multinomial problem. *Biometrika* **79**, 25–37.
53. Berger, J.O., Brown, L.D., and Wolpert, R.L. (1994) A unified conditional frequentist and Bayesian test for fixed and sequential simple hypothesis testing. *Annals of Statistics* **22**, 1787–1807.
54. Berger, J.O. and Sellke, T. (1987) Testing a point null hypothesis: the irreconcilability of P values and evidence. *Journal of the American Statistical Association* **82**, 112–139 (with discussion).
55. Berger, J.O. and Wolpert, R.L. (1988) *The Likelihood Principle: A Review, Generalizations, and Statistical Implications.* Hayward: Institute of Mathematical Statistics.
56. Berkson, J. (1951) Why I prefer logits to probits. *Biometrics* **7**, 327–339.

57. Berman, M. (1981) Inhomogeneous and modulated gamma processes. *Biometrika* **68**, 143–152.

58. Bernardo, J.M. (1979) Reference posterior distributions for Bayesian inference. *Journal of the Royal Statistical Society* **B41**, 113–147.

59. Bernardo, J.M. and Smith, A.F.M. (1994) *Bayesian Theory.* New York: John Wiley.

60. Besag, J. (1974) Spatial interaction and the statistical analysis of lattice systems. *Journal of the Royal Statistical Society* **B36**, 192–236 (with discussion).

61. Birkes, D. (1990) Generalized likelihood ratio tests and uniformly most powerful tests. *American Statistician* **44**, 163–166.

62. Birnbaum, A. (1962) On the foundations of statistical inference. *Journal of the American Statistical Association* **57**, 269–306 (with discussion).

63. Blaesild, P. and Jensen, J.L. (1985) Saddlepoint formulae for reproductive exponential models. *Scandinavian Journal of Statistics* **12**, 193–202.

64. Blyth, C.R. (1970) On the inference and decision models of statistics. *Annals of Mathematical Statistics* **41**, 1034–1053 (with discussion).

65. Blyth, C.R. (1972a) On Simpson's paradox and the sure-thing principle. *Journal of the American Statistical Association* **67**, 364–366 (with discussion).

66. Blyth, C.R. (1972b) Some probability paradoxes in choice from among random alternatives. *Journal of the American Statistical Association* **67**, 366–381 (with discussion).

67. Box, G.E.P. (1953) Non-normality and test on variances. *Biometrika* **40**, 318–335.

68. Box, G.E.P. (1980) Sampling and Bayes' inference in scientific modelling and robustness. *Journal of the Royal Statistical Society* **A143**, 383–430 (with discussion).

69. Box, G.E.P. and Cox, D.R. (1964) An analysis of transformations. *Journal of the Royal Statistical Society* **B26**, 211–252 (with discussion).

70. Box, G.E.P. and Tiao, G.C. (1968) A Bayesian approach to some outlier problems. *Biometrika* **55**, 119–129.

71. Box, G.E.P. and Tiao, G.C. (1973) *Bayesian Inference in Statistical Analysis.* New York: John Wiley.

72. Brooks, S.P. and Morgan, B.J.T. (1995) Optimization using simulated annealing. *Journal of the Royal Statistical Society* **D44**, 241–257.

73. Brown, L.D. (1986) *Fundamentals of Statistical Exponential Families with Applications in Statistical Decision Theory.* Hayward: Institute of Mathematical Statistics.

74. Buckle, D.J. (1995) Bayesian inference for stable distributions. *Journal of the American Statistical Association* **90**, 605–613.

75. Buehler, R.J. (1982) Some ancillary statistics and their properties. *Journal of the American Statistical Association* **77**, 581–594 (with discussion).

76. Burnham, K.P., White, G.C., and Anderson, D.R. (1995) Model selection strategy in the analysis of capture–recapture data. *Biometrics* **51**, 888–898.

77. Burridge, J. (1981) Empirical Bayes analysis of survival time data. *Journal of the Royal Statistical Society* **B43**, 65–75.
78. Buse, A. (1982) The likelihood ratio, Wald, and Lagrange multiplier tests: an expository note. *American Statistician* **36**, 153–157.
79. Butler, R.W. (1986) Predictive likelihood inference with applications. *Journal of the Royal Statistical Society* **B48**, 1–38 (with discussion).
80. Butler, R.W. (1989) Approximate predictive pivots and densities. *Biometrika* **76**, 489–501.
81. Casella, G. and Berger, R.L. (1987) Reconciling Bayesian and frequentist evidence in the one-sided testing problem. *Journal of the American Statistical Association* **82**, 106–111 (with discussion).
82. Chamberlin, S.R. and Sprott, D.A. (1989) Linear systems of pivotals and associated pivotal likelihoods with applications. *Biometrika* **76**, 685–691.
83. Chatfield, C. (1995) Model uncertainty, data mining, and statistical inference. *Journal of the Royal Statistical Society* **A158**, 419–466 (with discussion)
84. Christensen, B.J. and Kiefer, N.M. (1994) Local cuts and separate inference. *Scandinavian Journal of Statistics* **21**, 389–401.
85. Clarke, B. and Wasserman, L. (1993) Noninformative priors and nuisance parameters. *Journal of the American Statistical Association* **88**, 1427–1432.
86. Cochran, W.G. (1937) Problems arising in the analysis of a series of similar experiments. *Journal of the Royal Statistical Society* (Supplement) **4**, 102–118.
87. Consonni, G. and Veronese, P. (1992) Conjugate priors for exponential families having quadratic variance functions. *Journal of the American Statistical Association* **87**, 1123–1127.
88. Cook, R.D. (1977) Detection of influential observations in linear regression. *Technometrics* **19**, 15–18.
89. Cook, R.D. and Weisberg, S. (1982) *Residuals and Influence in Regression.* London: Chapman and Hall.
90. Cox, D.R. (1958) Some problems connected with statistical inference. *Annals of Mathematical Statistics* **29**, 357–372.
91. Cox, D.R. (1966) A simple example of a comparison involving quantal data. *Biometrika* **53**, 215–220.
92. Cox, D.R. (1971) The choice between alternative ancillary statistics. *Journal of the Royal Statistical Society* **B33**, 251–255.
93. Cox, D.R. (1972) Regression models and life tables. *Journal of the Royal Statistical Society* **B34**, 187–220 (with discussion).
94. Cox, D.R. (1975) Partial likelihood. *Biometrika* **62**, 269–276.
95. Cox, D.R. (1978) Foundations of statistical inference: the case for eclecticism. *Australian Journal of Statistics* **20**, 43–59 (with discussion).
96. Cox, D.R. (1988) Some aspects of conditional and asymptotic inference: a review. *Sankhyā* **A50**, 314–337.
97. Cox, D.R. (1990) Role of models in statistical analysis. *Statistical Science* **5**, 169–174.

98. Cox, D.R. and Hinkley, D.V. (1974) *Theoretical Statistics.* London: Chapman and Hall.
99. Cox, D.R. and Reid, N. (1987) Parameter orthogonality and approximate conditional inference. *Journal of the Royal Statistical Society* **B49**, 1–39 (with discussion).
100. Cox, D.R. and Reid, N. (1992) A note on the difference between profile and modified profile likelihood. *Biometrika* **79**, 408–411.
101. Crowder, M. and Sweeting, T. (1989) Bayesian inference for a bivariate binomial distribution. *Biometrika* **76**, 599–603.
102. Dalal, S. and Hall, W.J. (1983) Approximating priors by mixtures of natural conjugate priors. *Journal of the Royal Statistical Society* **B45**, 278–286.
103. Damien, P., Laud, P.W., and Smith, A.F.M. (1995) Approximate random variates generation from infinitely divisible distributions with applications to Bayesian inference. *Journal of the Royal Statistical Society* **B57**, 547–563.
104. Daniels, H.E. (1954) Saddlepoint approximations in statistics. *Annals of Mathematical Statistics* **25**, 631–650.
105. Datta, G.S. and Ghosh, J.K. (1995) On priors providing frequentist validity for Bayesian inference. *Biometrika* **82**, 37–45.
106. David, H.A. (1995) First (?) occurrence of common terms in mathematical statistics. *American Statistician* **49**, 121–133.
107. Davison, A.C. (1986) Approximate predictive likelihood. *Biometrika* **73**, 323–332.
108. Davison, A.C. (1988) Approximate conditional inference in generalized linear models. *Journal of the Royal Statistical Society* **B50**, 445–461.
109. Davison, A.C. and Tsai, C.L. (1992) Regression model diagnostics. *International Statistical Review* **60**, 337–353.
110. Dawid, A.P. (1973) Posterior expectations for large observations. *Biometrika* **60**, 664–667.
111. Dawid, A.P. (1975) On the concepts of sufficiency and ancillarity in the presence of nuisance parameters. *Journal of the Royal Statistical Society* **B37**, 248–258.
112. Dawid, A.P. (1979) Conditional independence in statistical theory. *Journal of the Royal Statistical Society* **B41**, 1–31 (with discussion).
113. Dawid, A.P. (1982) The well-calibrated Bayesian. *Journal of the American Statistical Association* **77**, 605–613 (with discussion).
114. Dawid, A.P. (1991) Fisherian inference and prequential frames of reference. *Journal of the Royal Statistical Society* **B53**, 79–109 (with discussion).
115. Dawid, A.P. and Stone, M. (1972) Expectation consistency of inverse probability distributions. *Biometrika* **59**, 486–489.
116. Dawid, A.P., Stone, M., and Zidek, J.V. (1973) Marginalization paradoxes in Bayesian and structural inference. *Journal of the Royal Statistical Society* **B35**, 189–233 (with discussion).
117. de Finetti, B. (1937) Foresight: its logical laws, its subjective sources. *Annales de l'Institut H. Poincaré* **7**, 1–68.

118. de Finetti, B. (1974a) *Theory of Probability.* 2 Volumes. New York: John Wiley.

119. de Finetti, B. (1974b) Bayesianism: its unifying role for both the foundations and applications of statistics. *International Statistical Review* **42**, 117–130.

120. de Jong, P. (1988) The likelihood for a state space model. *Biometrika* **75**, 165–169.

121. Dempster, A.P., Laird, N.M., and Rubin, D.B. (1977) Maximum likelihood from incomplete data via the EM algorithm. *Journal of the Royal Statistical Society* **B39**, 1–38 (with discussion).

122. Dempster, A.P. and Rubin, D.B. (1983) Rounding error in regression: the appropriateness of Sheppard's correction. *Journal of the Royal Statistical Society* **B45**, 51–59.

123. Dennis, J.E. and Schnabel, R.B. (1983) *Numerical Methods for Unconstrained Optimization and Nonlinear Equations.* Englewood Cliffs: Prentice-Hall.

124. Derman, C., Gleser, L.J., and Olkin, I. (1973) *A Guide to Probability Theory and Application.* New York: Holt, Rinehart, and Winston.

125. DeRobertis, L. and Hartigan, J.A. (1981) Bayesian inference using intervals of measures. *Annals of Statistics* **9**, 235–244.

126. Diaconis, P. and Ylvisaker, D. (1979) Conjugate priors for exponential families. *Annals of Statistics* **7**, 269–281.

127. Diciccio, T. and Efron, B. (1992) More accurate confidence intervals in exponential families. *Biometrika* **79**, 231–245.

128. Dickey, J.M. (1973) Scientific reporting and personal probabilities: Student's hypothesis. *Journal of the Royal Statistical Society* **B35**, 285–305.

129. Dobson, A.J. (1990) *An Introduction to Generalized Linear Models.* London: Chapman and Hall.

130. Doob, J.L. (1994) *Measure Theory.* Berlin: Springer Verlag.

131. Draper, D. (1995) Assessment and propagation of model uncertainty. *Journal of the Royal Statistical Society* **B57**, 45–97 (with discussion).

132. Edelman, D. (1990) A note on uniformly most powerful two-sided tests. *American Statistician* **44**, 219–220.

133. Edwards, A.W.F. (1972) *Likelihood. An Account of the Statistical Concept of* Likelihood *and its Application to Scientific Inference.* Cambridge: Cambridge University Press.

134. Edwards, W., Lindman, H., and Savage, L.J. (1963) Bayesian statistical inference for psychological research. *Psychological Review* **70**, 193–242.

135. Efron, B. (1975) Defining the curvature of a statistical problem (with applications to second order efficiency). *Annals of Statistics* **3**, 1189–1242 (with discussion).

136. Efron, B. (1978) The geometry of exponential families. *Annals of Statistics* **6**, 362–376.

137. Efron, B. (1986a) Why isn't everyone a Bayesian? *American Statistician* **40**, 1–11, 330–331 (with discussion).

138. Efron, B. (1986b) Double exponential families and their use in generalized linear regression. *Journal of the American Statistical Association* **81**, 709–721.

139. Efron, B. and Hinkley, D.V. (1978) Assessing the accuracy of the maximum likelihood estimator: observed versus expected Fisher information. *Biometrika* **65**, 457–487 (with discussion).

140. Efron, B. and Morris, C. (1973) Stein's estimation rule and its competitors – an empirical Bayes approach. *Journal of the American Statistical Association* **68**, 117–130.

141. Ellenberg, J.H. (1976) Testing for a single outlier from a general linear regression model. *Biometrika* **32**, 637–645.

142. Feigin, P.D. (1981) Conditional exponential families and a representation theorem for asymptotic inference. *Annals of Statistics* **9**, 587–603.

143. Feller, W. (1957) *An Introduction to Probability Theory and its Applications.* Volume I. New York: John Wiley.

144. Feller, W. (1966) *An Introduction to Probability Theory and its Applications.* Volume II. New York: John Wiley.

145. Field, C. and Ronchetti, E. (1990) *Small Sample Asymptotics.* Hayward: Institute of Mathematical Statistics.

146. Finney, D.J. (1978) *Statistical Method in Biological Assay.* London: C. Griffin.

147. Fisher, R.A. (1912) On an absolute criterion for fitting frequency curves. *Messenger of Mathematics* **41**, 155–160.

148. Fisher, R.A. (1915) Frequency distribution of the values of the correlation coefficient in samples from an indefinitely large population. *Biometrika* **10**, 507–521.

149. Fisher, R.A. (1920) A mathematical examination of the methods of determining the accuracy of an observation by the mean error, and by the mean square error. *Monthly Notices of the Royal Astronomical Society* **80**, 758–770.

150. Fisher, R.A. (1921) On the 'probable error' of a coefficient of correlation deduced from a small sample. *Metron* **1**, 3–32.

151. Fisher, R.A. (1922a) On the mathematical foundations of theoretical statistics. *Philosophical Transactions of the Royal Society of London* **A222**, 309–368.

152. Fisher, R.A. (1922b) On the interpretation of χ^2 from contingency tables, and the calculation of p. *Journal of the Royal Statistical Society* **85**, 87–94.

153. Fisher, R.A. (1924) On a distribution yielding the error function of several well known statistics. *Proceedings of the International Mathematics Congress*, Toronto, pp. 805–813.

154. Fisher, R.A. (1925) Theory of statistical estimation. *Proceedings of the Cambridge Philosophical Society* **22**, 700–725.

155. Fisher, R.A. (1930) Inverse probability. *Proceedings of the Cambridge Philosophical Society* **26**, 528–535.

156. Fisher, R.A. (1934) Two new properties of mathematical likelihood. *Proceedings of the Royal Society* **A144**, 285–307.

157. Fisher, R.A. (1935a) The logic of inductive inference. *Journal of the Royal Statistical Society* **48**, 39–82 (with discussion).

158. Fisher, R.A. (1935b) The detection of linkage with 'dominant' abnormalities. *Annals of Eugenics* **6**, 187–201.

159. Fisher, R.A. (1941) The asymptotic approach to Behrens's integral, with further tables for the d test of significance. *Annals of Eugenics* **11**, 141–172.

160. Fisher, R.A. (1950) The significance of deviations from expectation in a Poisson series. *Biometrics* **6**, 17–24.

161. Fisher, R.A. (1955) Statistical methods and scientific induction. *Journal of the Royal Statistical Society* **B17**, 69–78.

162. Fisher, R.A. (1958) *Statistical Methods for Research Workers.* Edinburgh: Oliver and Boyd. (First edition, 1925)

163. Fisher, R.A. (1959) *Statistical Methods and Scientific Inference.* Edinburgh: Oliver and Boyd. (First edition, 1956)

164. Fisher, R.A. (1960) *Design of Experiments.* Edinburgh: Oliver and Boyd. (First edition, 1935)

165. Fisz, M. (1962) Infinitely divisible distributions: recent results and applications. *Annals of Mathematical Statistics* **32**, 68–84.

166. Fleming, T.R. and Harrington, D.P. (1991) *Counting Processes and Survival Analysis.* New York: John Wiley.

167. Fraser, D.A.S. (1956) Sufficient statistics with nuisance parameters. *Annals of Mathematical Statistics* **27**, 838–842.

168. Fraser, D.A.S. (1968) *The Structure of Inference.* New York: John Wiley.

169. Fraser, D.A.S. (1990a) Tail probabilities from observed likelihoods. *Biometrika* **77**, 65–76.

170. Fraser, D.A.S. (1990b) Views on conditional and marginal methods of statistical inference. *Statistical Papers* **31**, 83–93.

171. Fraser, D.A.S. and Reid, N. (1988) On conditional inference for a real parameter: a differential approach on the sample space. *Biometrika* **75**, 251–264.

172. Fraser, D.A.S. and Reid, N. (1989) Adjustments to profile likelihood. *Biometrika* **76**, 477–488.

173. Frosini, B.V. (1993) Likelihood versus probability. *Bulletin of the International Statistical Institute* **55**(2), 359–376.

174. Gelman, A., Carlin, J.B., Stern, H.S., and Rubin, D.B. (1995) *Bayesian Data Analysis.* London: Chapman and Hall.

175. Geman, S. and Geman, D. (1984) Stochastic relaxation, Gibbs distributions, and the Bayesian restoration of images. *IEEE Transactions on Pattern Analysis and Machine Intelligence* **PAMI-6**, 721–741.

176. Ghosh, J.K. (1988) *Statistical Information and Likelihood. A Collection of Critical Essays by Dr. D. Basu.* New York: Springer Verlag.

177. Gibbons, J.D. and Pratt, J.W. (1975) P-values: interpretation and methodology. *American Statistician* **29**, 20–25.

178. Girón, F.J., Caro, E., and Domínguez, J.I. (1994) A conjugate family for AR(1) processes with exponential errors. *Communications in Statistics* **A23**, 1771–1784.

179. Goldstein, M. and Howard, J.V. (1991) A likelihood paradox. *Journal of the Royal Statistical Society* **B53**, 619–628 (with discussion).

180. Good, I.J. (1958) Significance tests in parallel and in series. *Journal of the American Statistical Association* **53**, 799–813.

181. Gray, J.B. and Woodall, W.H. (1994) The maximum size of standardized and internally studentized residuals in regression analysis. *American Statistician* **48**, 111–113.

182. Grenander, U. and Miller, M.I. (1994) Representations of knowledge in complex systems. *Journal of the Royal Statistical Society* **B56**, 549–603 (with discussion).

183. Griffiths, D.A. (1980) Interval estimation for the three-parameter lognormal distribution via the likelihood function. *Journal of the Royal Statistical Society* **C29**, 58–68.

184. Haberman, S.J. (1978) *Analysis of Qualitative Data. Volume 1. Introductory Topics.* San Diego: Academic Press.

185. Halmos, P.R. and Savage, L.J. (1949) Application of the Radon-Nikodým theorem to the theory of sufficient statistics. *Annals of Mathematical Statistics* **20**, 225–241.

186. Hand, D.J. (1994) Deconstructing statistical questions. *Journal of the Royal Statistical Society* **A157**, 317–356 (with discussion).

187. Hartley, H.O. (1958) Maximum likelihood estimation from incomplete data. *Biometrics* **14**, 174–194.

188. Hastings, W.K. (1970) Monte Carlo sampling methods using Markov chains and their applications. *Biometrika* **57**, 97–109.

189. Heitjan, D.F. (1989) Inference from grouped continuous data: a review. *Statistical Science* **4**, 164–183 (with discussion).

190. Heitjan, D.F. (1993) Ignorability and coarse data: some biomedical examples. *Biometrics* **49**, 1099–1109.

191. Heitjan, D.F. (1994) Ignorability in general incomplete-data models. *Biometrika* **81**, 701–708.

192. Heitjan, D.F. and Rubin, D.B. (1991) Ignorability and coarse data. *Annals of Statistics* **19**, 2244–2253.

193. Hill, B.M. (1963) The three-parameter lognormal distribution and Bayesian analysis of a point-source epidemic. *Journal of the American Statistical Association* **58**, 72–84.

194. Hill, B.M. (1975) Aberrant behavior of the likelihood function in discrete cases. *Journal of the American Statistical Association* **70**, 717–719.

195. Hill, B.M. (1988) De Finetti's theorem, induction, and $A_{(n)}$ or Bayesian nonparametric predictive inference. *Bayesian Statistics* **3**, 211–241 (with discussion).

196. Hinkley, D.V. (1979) Predictive likelihood. *Annals of Statistics* **7**, 718–728, **8**, 694.

197. Hinkley, D.V. (1980) Likelihood. *Canadian Journal of Statistics* **8**, 151–163.

198. Hinkley, D.V. (1983) Can frequentist inferences be very wrong? A conditional 'yes'. In Box, G.E.P., Leonard, T., and Wu, C.F., *Scientific In-*

ference, Data Analysis and Robustness. San Diego: Academic Press, pp. 85–103.

199. Hinkley, D.V., Reid, N., and Snell, E.J. (1991) *Statistical Theory and Modelling. In Honour of Sir David Cox, FRS.* London: Chapman and Hall.
200. Hoaglin, D.C. and Welsh, R.E. (1978) The hat matrix in regression and ANOVA. *American Statistician* **32**, 17–22.
201. Hoffmann-Jørgensen, J. (1994) *Probability with a View Toward Statistics.* 2 Volumes. London: Chapman and Hall.
202. Hougaard, P. (1982) Parametrizations for non-linear models. *Journal of the Royal Statistical Society* **B44**, 244–252.
203. Hougaard, P. (1986) Survival models for heterogeneous populations derived from stable distributions. *Biometrika* **73**, 387–396.
204. Huber, P.J. (1964) Robust estimation of a location parameter. *Annals of Mathematical Statistics* **35**, 73–101.
205. Hudson, D.J. (1971) Interval estimation from the likelihood function. *Journal of the Royal Statistical Society* **B33**, 256–262.
206. Ibrahim, J.G. and Laud, P.W. (1991) On Bayesian analysis of generalized linear models using Jeffrey's prior. *Journal of the American Statistical Association* **86**, 981–986.
207. Irony, T.Z. (1993) Information in sampling rules. *Journal of Statistical Planning and Inference* **36**, 27–38.
208. James, W. and Stein, C.M. (1961) Estimation with quadratic loss. *Proceedings of the Fourth Berkeley Symposium*, Volume I, pp. 311–319.
209. Janicki, A. and Weron, A. (1994) Can one see α-stable variables and processes? *Statistical Science* **9**, 109–126.
210. Jeffreys, H. (1946) An invariant form for the prior distribution in estimation problems. *Proceedings of the Royal Society* **A186**, 453–461.
211. Jensen, S.T., Johansen, S., and Lauritzen, S.L. (1991) Globally convergent algorithms for maximizing a likelihood function. *Biometrika* **78**, 867–877.
212. Johnson, N.L. and Kotz, S. (1970) *Distributions in Statistics. Continuous Univariate Distributions.* Volume 2. New York: John Wiley.
213. Jørgensen, B. (1986) Some properties of exponential dispersion models. *Scandinavian Journal of Statistics* **13**, 187–197.
214. Jørgensen, B. (1987a) Exponential dispersion models. *Journal of the Royal Statistical Society* **B49**, 127–162 (with discussion).
215. Jørgensen, B. (1987b) Small dispersion asymptotics. *Revista Brasileira de Probabilidade e Estatistica* **1**, 59–90.
216. Jørgensen, B. (1992) Exponential dispersion models and extensions: a review. *International Statistical Review* **60**, 5–20.
217. Jørgensen, B., Martinez, J.R., and Tsao, M. (1994) Asymptotic behaviour of the variance function. *Scandinavian Journal of Statistics* **21**, 223–243.
218. Kalbfleisch, J.D. (1971) Likelihood methods of prediction. In Godambe, V.P. and Sprott, D.A. (eds.) *Foundations of Statistical Inference.* Toronto: Holt, Rinehart, and Winston, pp. 378–392.
219. Kalbfleisch, J.D. (1975) Sufficiency and conditionality. *Biometrika* **62**, 251–268.

220. Kalbfleisch, J.D. and Sprott, D.A. (1969) Applications of likelihoods and fiducial probability to sampling finite populations. In Johnson, N.L. and Smith, H., *New Developments in Survey Sampling.* New York: John Wiley, pp. 358–389.

221. Kalbfleisch, J.D. and Sprott, D.A. (1970) Application of likelihood methods to models involving large numbers of parameters. *Journal of the Royal Statistical Society* **B32**, 175–208 (with discussion).

222. Kalbfleisch, J.G. (1985a) *Probability and Statistical Inference.* Volume 1: *Probability.* New York: Springer Verlag.

223. Kalbfleisch, J.G. (1985b) *Probability and Statistical Inference.* Volume 2: *Statistical Inference.* New York: Springer Verlag.

224. Kass, R.E. (1989) The geometry of asymptotic inference. *Statistical Science* **4**, 188–234 (with discussion).

225. Kass, R.E. and Raftery, A.E. (1995) Bayes factors. *Journal of the American Statistical Association* **90**, 773–795.

226. Kempthorne, O. and Folks, L. (1971) *Probability, Statistics, and Data Analysis.* Ames: Iowa State University Press.

227. Kiefer, J. and Wolfowitz, J. (1956) Consistency of the maximum likelihood estimator in the presence of infinitely many incidental parameters. *Annals of Mathematical Statistics* **27**, 887–906.

228. King, G. (1989) *Unifying Political Methodology. The Likelihood Theory of Statistical Inference.* Cambridge: Cambridge University Press.

229. Kullback, S. and Leibler, R.A. (1951) On information and sufficiency. *Annals of Mathematical Statistics* **22**, 79–86.

230. Lavine, M. (1991) Sensitivity in Bayesian statistics: the prior and the likelihood. *Journal of the American Statistical Association* **86**, 396–399.

231. Lavine, M., Wasserman, L., and Wolpert, R.L. (1991) Bayesian inference with specified prior marginals. *Journal of the American Statistical Association* **86**, 964–971.

232. Lawley, D.N. (1956) A general method for approximating to the distribution of the likelihood ratio criteria. *Biometrika* **43**, 295–303.

233. Lawson, A.B. (1993) A deviance residual for heterogeneous spatial Poisson processes. *Biometrics* **49**, 889–897.

234. Lehmann, E.L. (1983) *Theory of Point Estimation.* Pacific Grove: Wadsworth.

235. Lehmann, E.L. (1986) *Testing Statistical Hypotheses.* London: Chapman and Hall.

236. Letac, G. and Mora, M. (1990) Natural exponential families with cubic variances. *Annals of Statistics* **18**, 1–38.

237. Levenbach, H. (1972) Estimation of autoregressive parameters from a marginal likelihood function. *Biometrika* **59**, 61–71.

238. Liang, K.Y. (1983) On information and ancillarity in the presence of a nuisance parameter. *Biometrika* **70**, 607–612.

239. Lindley, D.V. (1957) A statistical paradox. *Biometrika* **44**, 187–192.

240. Lindley, D.V. (1958) Fiducial distributions and Bayes' theorem. *Journal of the Royal Statistical Society* **B20**, 102–107.

241. Lindley, D.V. (1965) *Introduction to Probability and Statistics from a Bayesian Viewpoint.* Part 2. *Inference.* Cambridge: Cambridge University Press.

242. Lindley, D.V. (1972) *Bayesian Statistics, A Review.* Philadelphia: SIAM.

243. Lindley, D.V. (1978) The Bayesian approach. *Scandinavian Journal of Statistics* **5**, 1–26 (with discussion).

244. Lindley, D.V. (1985) *Making Decisions.* New York: John Wiley.

245. Lindley, D.V. and Smith, A.F.M. (1972) Bayes estimates for the linear model. *Journal of the Royal Statistical Society* **B30**, 1–41 (with discussion).

246. Lindsay, B.G., Clogg, C.C., and Grego, J. (1991) Semiparametric estimation in the Rasch model and related exponential response models, including a simple latent class model for item analysis. *Journal of the American Statistical Association* **86**, 96–107.

247. Lindsey, J.K. (1973) *Inferences from Sociological Survey Data: A Unified Approach.* Amsterdam: Elsevier.

248. Lindsey, J.K. (1974a) Comparison of probability distributions. *Journal of the Royal Statistical Society* **B36**, 38–47.

249. Lindsey, J.K. (1974b) Construction and comparison of statistical models. *Journal of the Royal Statistical Society* **B36**, 418–425.

250. Lindsey, J.K. (1975) Likelihood analyses and tests for binary data. *Journal of the Royal Statistical Society* **C24** 1–16.

251. Lindsey, J.K. (1993) *Models for Repeated Measurements.* Oxford: Oxford University Press.

252. Lindsey, J.K. (1995a) Fitting parametric counting processes by using log linear models. *Journal of the Royal Statistical Society* **C44**, 201–212.

253. Lindsey, J.K. (1995b) *Modelling Frequency and Count Data.* Oxford: Oxford University Press.

254. Lindsey, J.K. (1995c) *Introductory Statistics: The Modelling Approach.* Oxford: Oxford University Press.

255. Lindsey, J.K. and Mersch, G. (1992) Fitting and comparing probability distributions with log linear models. *Computational Statistics and Data Analysis* **13**, 373–384.

256. Little, R.J.A. (1995) Modeling the drop-out mechanism in repeated-measures studies. *Journal of the American Statistical Association* **90**, 1112–1121.

257. Little, R.J.A. and Rubin, D.B. (1987) *Statistical Analysis with Missing Data.* New York: John Wiley.

258. Liu, Q. and Pierce, D.A. (1993) Heterogeneity in Mantel-Haenszel-type models. *Biometrika* **80**, 543–556.

259. Liu, Q. and Pierce, D.A. (1994) A note on Gauss-Hermite quadrature. *Biometrika* **81**, 624–629.

260. Lugannani, R. and Rice, S. (1980) Saddle point approximations for the distribution of the sum of independent random variables. *Advances in Applied Probability* **12**, 475–490.

261. McNeil, D.R. (1966) Consistent statistics for estimating and testing hypotheses from grouped samples. *Biometrika* **53**, 545–557.

262. McPherson, G. (1989) The scientists' view of statistics – a neglected area. *Journal of the Royal Statistical Society* **A152**, 221–240 (with discussion).

263. Meeker, W.Q. and Escobar, L.A. (1995) Teaching approximate confidence regions based on maximum likelihood estimation. *American Statistician* **49**, 48–53.

264. Metropolis, N. and Ulam, S. (1949) The Monte Carlo method. *Journal of the American Statistical Association* **44**, 335–341.

265. Morgan, B.J.T. (1992) *Analysis of Quantal Response Data.* London: Chapman and Hall.

266. Morris, C.N. (1982) Natural exponential families with quadratic variance functions. *Annals of Statistics* **10**, 65–80.

267. Morris, C.N. (1983a) Natural exponential families with quadratic variance functions: statistical theory. *Annals of Statistics* **11**, 515–529.

268. Morris, C.N. (1983b) Parametric empirical Bayes inference: theory and applications. *Journal of the American Statistical Association* **78**, 47–65 (with discussion).

269. Nelder, J.A. and Wedderburn, R.W.M. (1972) Generalized linear models. *Journal of the Royal Statistical Society* **A135**, 370–383.

270. Neyman, J. and Pearson, E.S. (1933) On the problem of the most efficient test of statistical hypotheses. *Philosophical Transactions of the Royal Society, London* **A231**, 289–337.

271. Neyman, J. and Scott, E.L. (1948) Consistent estimates based on partially consistent observations. *Econometrica* **16**, 1–32.

272. Novick, M.R. (1969) Multiparameter Bayesian indifference procedures. *Journal of the Royal Statistical Society* **31**, 29–51.

273. Novick, M.R. and Hall, W.J. (1965) A Bayesian indifference procedure. *Journal of the American Statistical Association* **60**, 1104–1117.

274. O'Brien, P.C. and Fleming, T.R. (1979) A multiple testing procedure for clinical trials. *Biometrics* **35**, 549–556.

275. Oehlert, G.W. (1992) A note on the delta method. *American Statistician* **46**, 27–29.

276. Orchard, T. and Woodbury, M.A. (1972) A missing information principle: theory and applications. *Proceedings of the Sixth Berkeley Symposium*, Volume I, pp. 697–715.

277. Patterson, H.D. and Thompson, R. (1971) Recovery of interblock information when block sizes are unequal. *Biometrika* **58**, 545–554.

278. Pearson, K. (1900) On the criterion that a given system of deviations from the probable in the case of a correlated system of variables is such that it can reasonably be supposed to have arisen from random sampling. *Philosophical Magazine* **5**, 157–175.

279. Pierce, D.A. (1973) Some difficulties in a frequency theory of inference. *Annals of Statistics* **1**, 241–250.

280. Pierce, D.A. (1982) The asymptotic effect of substituting estimators for parameters in certain types of statistics. *Annals of Statistics* **10**, 475–478.

281. Pierce, D.A. and Peters, D. (1992) Practical use of higher order asymptotics for multiparameter exponential families. *Journal of the Royal Statistical Society* **B54**, 701–737 (with discussion).

282. Pierce, D.A. and Sands, B.R. (1975) Extra-Bernoulli variation in binary data. *Technical Report 46.* Department of Statistics, Oregon State University.

283. Pierce, D.A. and Schafer, D.W. (1986) Residuals in generalized linear models. *Journal of the American Statistical Association* **81**, 977–986.

284. Polson, N. and Wasserman, L. (1990) Prior distributions for the bivariate binomial. *Biometrika* **77**, 901–904.

285. Pratt, J.W. (1965) Bayesian interpretation of standard inference statements. *Journal of the Royal Statistical Society* **B27**, 169–203 (with discussion).

286. Pratt, J.W., Raiffa, H., and Schlaifer, R. (1964) The foundations of decision under uncertainty: an elementary exposition. *Journal of the American Statistical Association* **59**, 353–375.

287. Pregibon, D. (1981) Logistic regression diagnostics. *Annals of Statistics* **9**, 705–724.

288. Press, W.H., Teukolsky, S.A., Vetterling, W.T., and Flannery, B.P. (1992) *Numerical Recipes in C. The Art of Scientific Computing.* Cambridge: Cambridge University Press.

289. Rao, C.R. (1945) Information and the accuracy attainable in the estimation of statistical parameters. *Bulletin of the Calcutta Mathematical Society* **37**, 81–91.

290. Rao, C.R. (1948) Large sample tests of statistical hypotheses concerning several parameters with applications to problems of estimation. *Proceedings of the Cambridge Philosophical Society* **44**, 50–57.

291. Rasch, G. (1960) *Probabilistic Models for some Intelligence and Attainment Tests.* Copenhagen: Danish Institute for Educational Research.

292. Reid, N. (1988) Saddlepoint methods and statistical inference. *Statistical Science* **3**, 213–238 (with discussion).

293. Rémon, M. (1984) On a concept of partial sufficiency: L-sufficiency. *International Statistical Review* **52**, 127–135.

294. Ross, G.J.S. (1990) *Nonlinear Estimation.* New York: Springer Verlag.

295. Ross, S.M. (1989) *Introduction to Probability Models.* San Diego: Academic Press.

296. Rubin, D.B. (1976) Inference and missing data. *Biometrika* **63**, 581–592 (with discussion).

297. Rubin, D.B. (1978) Bayesian inference for causal effects: the role of randomization. *Annals of Statistics* **6**, 34–58.

298. Rubin, D.B. (1984) Bayesianly justifiable and relevant frequency calculations for the applied statistician. *Annals of Statistics* **12**, 1151–1172.

299. Rudas, T., Clogg, C.C., and Lindsay, B.G. (1994) A new index of fit based on mixture methods for the analysis of contingency tables. *Journal of the Royal Statistical Society* **B56**, 623–639.

300. Sackrowitz, H. and Samuel-Cahn, E. (1994) Rationality and unbiasedness in hypothesis testing. *Journal of the American Statistical Association* **89**, 967–971.

301. Samaniego, F.J. and Reneau, D.M. (1994) Toward a reconciliation of the Bayesian and frequentist approaches to point estimation. *Journal of the American Statistical Association* **89**, 947–957.

302. Samorodnitsky, G. and Taqqu, M.S. (1994) *Stable Non-Gaussian Processes.* London: Chapman and Hall.

303. Sandved, E. (1967) A principle for conditioning on an ancillary statistic. *Skandinavisk Aktuarietidskrift* **50**, 39–47.

304. Savage, L.J. (1954) *The Foundations of Statistics.* New York: Dover.

305. Savage, L.J. (1971) Elicitation of personal probabilities and expectations. *Journal of the American Statistical Association* **66**, 783–801.

306. Savage, L.J. (1976) On rereading R.A. Fisher. *Annals of Statistics* **4**, 441–500 (with discussion).

307. Saville, D.J. (1990) Multiple comparison procedures: the practical solution. *American Statistician* **44**, 174–180.

308. Schwarz, G. (1978) Estimating the dimension of a model. *Annals of Statistics* **6**, 461–464.

309. Schwarz, N. (1995) What respondents learn from questionnaires: the survey interview and the logic of conversation. *International Statistical Review* **63**, 153–177 (with discussion).

310. Searle, S.R. (1971) *Linear Models.* New York: John Wiley.

311. Seidenfeld, T. (1979) *Philosophical Problems of Statistical Inference. Learning from R.A. Fisher.* Dordrecht: Reidel.

312. Seidenfeld, T. and Wasserman, L.A. (1993) Dilation for sets of probabilities. *Annals of Statistics* **21**, 1139–1154.

313. Seshadri, V. (1988) Exponential models, Brownian motion, and independence. *Canadian Journal of Statistics* **16**, 209–221.

314. Severini, T.A. (1991) On the relationship between Bayesian and non-Bayesian interval estimates. *Biometrika* **53**, 611–618.

315. Shafer, G. (1982) Lindley's paradox. *Journal of the American Statistical Association* **77**, 325–351 (with discussion).

316. Sheppard, W.F. (1898) On the calculation of the most probable values of frequency constants for data arranged according to equidistant divisions of a scale. *Proceedings of the London Mathematical Society* **29**, 353–380.

317. Silvey, S.D. (1959) The Lagrangian multiplier test. *Annals of Mathematical Statistics* **30**, 389–407.

318. Simpson, C.H. (1951) The interpretation of interaction in contingency tables. *Journal of the Royal Statistical Society* **B13**, 233–241.

319. Slate, E.H. (1994) Parametrizations for natural exponential families with quadratic variance functions. *Journal of the American Statistical Association* **89**, 1471–1482.

320. Sorić, B. (1989) Statistical 'discoveries' and effect-size estimation. *Journal of the American Statistical Association* **84**, 608–610.

321. Spiegelhalter, D.J., Freedman, L.S., and Parmar, M.K.B. (1994) Bayesian approaches to randomized trials. *Journal of the Royal Statistical Society* **A157**, 357–416 (with discussion).

322. Sprott, D.A. (1973) Normal likelihoods and their relation to large sample theory of estimation. *Biometrika* **60**, 457–465.

323. Sprott, D.A. (1975a) Application of maximum likelihood methods for finite samples. *Sankhyā* **B37**, 259–270.

324. Sprott, D.A. (1975b) Marginal and conditional sufficiency. *Biometrika* **62**, 599–605.

325. Sprott, D.A. (1978) Robustness and non-parametric procedures are not the only or the safe alternatives to normality. *Canadian Journal of Psychology* **32**, 180–185.

326. Sprott, D.A. (1980) Maximum likelihood in small samples: estimation in the presence of nuisance parameters. *Biometrika* **67**, 515–523.

327. Sprott, D.A. (1982) Robustness and maximum likelihood estimation. *Communications in Statistics* **A11**, 2513–2529.

328. Sprott, D.A. (1990) Inferential estimation, likelihood, and linear pivotals. *Canadian Journal of Statistics* **18**, 1–15.

329. Sprott, D.A. and Viveros-Aguilera, R. (1984) The interpretation of maximum-likelihood estimation. *Canadian Journal of Statistics* **12**, 27–38.

330. Stein, C. (1962) A remark on the likelihood principle. *Journal of the Royal Statistical Society* **A125**, 565–568.

331. Sterling, T.D. (1959) Publication decisions and their possible effects on inferences drawn from tests of significance – or vice versa. *Journal of the American Statistical Association* **54**, 30–34.

332. Sterling, T.D., Rosenbaum, W.L., and Weinkam, J.J. (1995) Publication decisions revisited: the effect of outcome of statistical tests on the decision to publish and vice versa. *American Statistician* **49**, 108–112.

333. Stone, M. (1976) Strong inconsistency from uniform priors. *Journal of the American Statistical Association* **71**, 114–125 (with discussion).

334. Student (1908) The probable error of the mean. *Biometrika* **6**, 1–25.

335. Tibshirani, R. and Wasserman, L. (1994) Some aspects of the reparametrization of statistical models. *Canadian Journal of Statistics* **22**, 163–173.

336. Tierney, L. (1994) Markov chains for exploring posterior distributions. *Annals of Statistics* **22**, 1701–1762 (with discussion).

337. Tierney, L. and Kadane, J.B. (1986) Accurate approximations for posterior moments and marginal densities. *Journal of the American Statistical Association* **81**, 82–86.

338. Tierney, L., Kass, R.E., and Kadane, J.B. (1989) Fully exponential Laplace approximations to expectations and variances of nonpositive functions. *Journal of the American Statistical Association* **84**, 710–716.

339. Tijms, H.C. (1994) *Stochastic Models. An Algorithmic Approach.* New York: John Wiley.

340. Tsou, T.S. and Royall, R.M. (1995) Robust likelihood. *Journal of the American Statistical Association* **90**, 316–320.

341. Tukey, J.W. (1962) The future of data analysis. *Annals of Mathematical Statistics* **33**, 1–67.

342. Tweedie, M.C.K. (1947) Functions of a statistical variate with given means, with special reference to Laplacian distributions. *Proceedings of the Cambridge Philosophical Society* **43**, 41–49.

343. Viveros, R. and Sprott, D.A. (1987) Allowance for skewness in maximum-likelihood estimation with application to the location–scale model. *Canadian Journal of Statistics* **15**, 349–361.

344. Wald, A. (1939) Contributions to the theory of statistical estimation and testing hypotheses. *Annals of Mathematical Statistics* **10**, 299–326.

345. Wald, A. (1943) Tests of statistical hypotheses concerning several parameters when the number of observations is large. *Transactions of the American Mathematical Society* **54**, 426–482.

346. Wald, A. (1945) Sequential tests of statistical hypotheses. *Annals of Mathematical Statistics* **16**, 117–196.

347. Wald, A. (1949) Statistical decision functions. *Annals of Mathematical Statistics* **20**, 165–205.

348. Wasserman, L.A. (1989) A robust Bayesian interpretation of likelihood regions. *Annals of Statistics* **17**, 1387–1393.

349. Wasserman, L.A. and Kadane, J.B. (1992) Computing bounds on expectations. *Journal of the American Statistical Association* **87**, 516–522.

350. Wedderburn, R.W.M. (1974) Quasi-likelihood functions, generalized linear models, and the Gauss-Newton method. *Biometrika* **61**, 439–447.

351. Welch, B.L. and Peers, H.W. (1963) On formulae for confidence points based on integrals of weighted likelihoods. *Journal of the Royal Statistical Society* **B25**, 318–329.

352. West, M., Harrison, P.J., and Migon, H.S. (1985) Dynamic generalized linear models and Bayesian forecasting. *Journal of the American Statistical Association* **80**, 73–97.

353. Wetherill, G.B. (1961) Bayesian sequential analysis. *Biometrika* **48**, 281–292.

354. Whitehead, J. (1980) Fitting Cox's regression model to survival data using GLIM. *Journal of the Royal Statistical Society* **C29**, 268–275.

355. Whittle, P. (1992) *Probability via Expectation.* New York: Springer Verlag.

356. Wilks, S.S. (1938) The large-sample distribution of the likelihood ratio for testing composite hypotheses. *Annals of Mathematical Statistics* **9**, 60–62.

357. Williams, D. (1991) *Probability with Martingales.* Cambridge: Cambridge University Press.

358. Williams, D. A. (1987) Generalized linear model diagnostics using the deviance and single case deletions. *Journal of the Royal Statistical Society* **C36**, 181–191.

359. Willmot, G.E. (1990) On the construction of a parameter orthogonal to the mean. *Biometrika* **77**, 424–428.

360. Yates, F. (1934) Contingency tables involving small numbers and the χ^2 test. *Journal of the Royal Statistical Society* (Supplement) **1**, 217–235.

361. Yates, F. (1984) Tests of significance for 2×2 contingency tables. *Journal of the Royal Statistical Society* **A147**, 426–463.

362. Zeger, S.L. and Liang, K.Y. (1986) Longitudinal data analysis for discrete and continuous outcomes. *Biometrics* **42**, 121–130.

363. Zhu, Y. and Reid, N. (1994) Information, ancillarity, and sufficiency in the presence of nuisance parameters. *Canadian Journal of Statistics* **22**, 111–123.

Index